Advanced Textbooks in Contro

For further volumes:
www.springer.com/series/4045

Karel J. Keesman

System Identification

An Introduction

Karel J. Keesman
Systems and Control Group
Wageningen University
Bornse Weilanden 9
6708 WG, Wageningen
Netherlands
karel.keesman@wur.nl

ISSN 1439-2232
ISBN 978-0-85729-521-7 e-ISBN 978-0-85729-522-4
DOI 10.1007/978-0-85729-522-4
Springer London Dordrecht Heidelberg New York

British Library Cataloguing in Publication Data
A catalogue record for this book is available from the British Library

Library of Congress Control Number: 2011929048

Mathematics Subject Classification: 93E12, 93E24, 93E10, 93E11

Cover design: eStudio Calamar S.L.

Printed on acid-free paper

Springer is part of Springer Science+Business Media (www.springer.com)

To Wil, Esther, Carlijn, and Rick

. . .

Series Editors' Foreword

The topics of control engineering and signal processing continue to flourish and develop. In common with general scientific investigation, new ideas, concepts and interpretations emerge quite spontaneously, and these are then discussed, used, discarded or subsumed into the prevailing subject paradigm. Sometimes these innovative concepts coalesce into a new sub-discipline within the broad subject tapestry of control and signal processing. This preliminary battle between old and new usually takes place at conferences, through the Internet and in the journals of the discipline. After a little more maturity has been acquired by the new concepts, then archival publication as a scientific or engineering monograph may occur.

A new concept in control and signal processing is known to have arrived when sufficient material has evolved for the topic to be taught as a specialised tutorial workshop or as a course to undergraduate, graduate or industrial engineers. *Advanced Textbooks in Control and Signal Processing* are designed as a vehicle for the systematic presentation of course material for both popular and innovative topics in the discipline. It is hoped that prospective authors will welcome the opportunity to publish a structured and systematic presentation of some of the newer emerging control and signal processing technologies in the textbook series.

An aim of *Advanced Textbooks in Control and Signal Processing* is to create a library that covers all the main subjects to be found in the control and signal processing fields. It is a growing but select series of high-quality books that now covers some fundamental topics and many more advanced topics in these areas. In trying to achieve a balanced library of course books, the Editors have long wished to have a text on system identification in the series. Although we often tend to think of system identification as a still-maturing subject, it is quite surprising to realise that the first International Federation of Automatic Control symposium on system identification was held as long ago as 1967 and that some of the classic textbooks on this topic were published during the 1970s and 1980s. Consequently, the existing literature and diversity of theory and applications areas is now quite extensive and provide a significant challenge to any prospective system identification course textbook author. The Series Editors were therefore pleased to discover that Associate Professor Karel Keesman of Wageningen University in the Netherlands, was proposing to take on this task and produce such a course textbook for the series entitled *System Identification: An Introduction*. We are now very pleased to welcome this finished textbook to the library of *Advanced Textbooks in Control and Signal Processing*.

Although a wide literature exists for systems identification, there is a traditional classification of techniques into non-parametric and parametric methods, and Professor Keesman reflects this with Part I of his book focussed on the non-parametric methods, and Parts II and III emphasizing the parametric methods. Since every identification practitioner wishes to know if the estimated model is a good model for the process, a novel feature of the textbook is Part IV that systematically presents a number of validation techniques for answering that very question.

As befits a course textbook, the material develops in increasing technical depth as the reader progresses through the text, but there are starred sections to identify material that is more advanced technically or presents more recent technical developments in the field. The presentational style is discursive with the integrated use of examples to illustrate technical and practical issues as they arise along the way. As part of this approach many different system examples have been used ranging from mechanical systems to more complex biological systems. Each chapter has a Problems section, and some solutions are available in the book. To support the mathematical content (system identification involves elements of systems theory, matrices, statistics, transform methods (for example, Laplace and Fourier transforms), Bode diagrams, and shift operators), there are five accessible, short, focussed mathematical appendices at the end of the book to aid the reader if necessary. This has the advantage of making the textbook fully self-contained for most readers.

In terms of processes, Professor Keesman's approach takes a broad view, and the textbook should be readily appreciated by readers from either the engineering or the scientific disciplines. Final-year undergraduate and graduate readers will find the book provides a stimulating tutorial-style entry to the field of system identification. For the established engineer or scientist, the mathematical level of the text and the supporting mathematical appendices will allow a speedy and insightful appreciation of the techniques of the field. This is an excellent addition to the *Advanced Textbooks in Control and Signal Processing* series.

Industrial Control Centre
Glasgow, Scotland, UK

M.J. Grimble
M.A. Johnson

Preface

St. Augustine of Hippo in De Civitate Dei writes
'Si [· · ·] fallor, sum' ('If I am mistaken, I am')
(book XI, 26)

'I can therefore gladly admit that falsificationists
like myself much prefer an attempt to solve
an interesting problem by a bold conjecture,
even (and especially) if it soon turns out
to be false, to any recital of a sequence
of irrelevant truisms. We prefer this
because we believe that in this way
we learn from our mistakes; and
that in finding that our
conjecture was false, we
shall have learnt much
about the truth, and
shall have got
nearer to the
truth.'

POPPER, K. 1962
Conjectures and Refutations,
New York: Basic Books, p. 231

Learning from mistakes, that according to Karl Popper brings us closer to the truth and, if prediction errors are interpreted as "mistakes", it is the basic principle underlying the majority of system identification methods. System identification aims at the construction of mathematical models from prior knowledge of the system under study and noisy time series data. Essentially, system identification is an art of modeling, where appropriate choices have to be made concerning the level of approximation given the final modeling objective and given noisy data. The scientific methods described in this book and obtained from statistics and system theory may help to solve the system identification problem in a systematic way. In general, system identification consists of three basic steps: experiment design and data acquisition, model structure selection and parameter estimation, and model validation. In the past, many methods have been developed to support each of these steps. Initially, these methods were developed for each specific case. In the seventies, a more systematic approach to system identification arose with the start of the IFAC Symposia on Identification and System Parameter Estimation and the appearance of the books of Box and Jenkins on Time Series Analysis (1970), of Schweppe on Uncertain Dynamic Systems (1973) and Eykhoff's book on System Identification (1974). Since then some ten books and many, most technical, papers have appeared on identification. Especially, the books of Norton entitled 'An Introduction to Identification' (1986) and Ljung's 'System Identification—Theory for the User' (1987, 1999) became widely used introductory text books for students, at several levels. However, still the problem of system identification has not been completely solved. Consequently, nowadays new ideas and methods to solve the system identification problem or parts of it are introduced.

This book is designed to help students and practitioners to understand the system identification process, to read the identification literature and to make appropriate choices in this process. As such the identified mathematical model will help to gain insight into processes, to effectively design experiments, to make better predictions or to improve the performance of a control system.

In this book the starting point for identification is the prior system knowledge, preferably in terms of a set of algebraic or differential equations. This prior knowledge can often be found in text books or articles related to the process phenomena under study. In particular, one may think of constitutive laws from physics, chemistry, biology and economics together with conservation laws, like material and population balances. Hence, the focus of this book is basically on 'semi-physical' or 'grey-box' modeling approaches, although data-based modeling approaches using transfer function descriptions of the system are treated at an introductory level, as well. However, the reader will not find any data-based methods related to fuzzy models, neural nets, support vector machines and the like, as these require detailed specialist knowledge and as such can be seen as special nonlinear regression structures.

The methods described in this book are not treated at a thoroughly advanced mathematical level, and thus no attention will be paid to asymptotic theory; the book is essentially problem oriented using finite input–output data. As such, the contents of the book range from classical (frequency domain) to modern (time domain) identification, from static to dynamic, from linear to nonlinear and from time-invariant to time-varying systems. Hence, for reading this book, an elementary knowledge of matrix algebra and statistics suffices. For more technical identification books, which focus on, for instance, asymptotic theory, nonlinear regression, time series analysis, frequency domain techniques, subspace identification, H_∞-approaches, infinite-dimensional systems and the increasing popularity of Bayesian estimation methods, we refer to the literature, as indicated at the end of each chapter. In this book these subjects are covered at an elementary level and basically illustrated by simple examples, so that every reader is able to redo the estimation or identification step. Some examples are more complex, but these have been introduced to demonstrate the practical applicability of the methods. All the more complex examples have been derived from 'real-world' physical/chemical applications with, in most cases, a biological component. Moreover, in all these applications heat, mass (in particular, water vapor, carbon, nitrogen and oxygen concentration) or momentum transfer processes play a key role. A list of all examples can be found in the subject index. Some of the sections and subsections have been marked with an asterisk (*) in the title, indicating useful background material related to special topics. This material, presented at a more advanced level, can be easily skipped without losing sight of the main stream of system identification methods for practical use.

The book is structured as follows. First, some introduction into system theory, and in particular on model representations and model properties, is given. Then, in Part I the focus is on data-based identification, also known as the non-parametric methods. These methods are especially useful when the prior system knowledge is very limited and only good data sets are available. Essentially, the basic assumptions are that the dynamic system is linear and time-invariant, properties that are further explained in the Introduction. Part II focuses on time-invariant system identification methods, assuming constant parameters. We start with classical linear regression related to static, time-invariant systems and end this part with the identification of nonlinear dynamic systems. In Part III, the emphasis is on time-varying system identification methods, which basically rely on recursive estimation techniques. Again, the approach is from linear to nonlinear and from static to dynamic. In Part IV, model validation techniques are discussed using both the prior knowledge and the noisy time series data. Finally, the book contains appendices with background material on matrix algebra, statistics, integral transforms, Bode diagrams, shift operator calculus and the derivation of the recursive least-squares method. In addition to this, Appendix G contains hourly measurements of the dissolved oxygen (DO) concentration in g/m^3, the saturated dissolved oxygen concentration (C_S) in g/m^3 and the radiation (I) in W/m^2, from the lake 'De Poel en 't Zwet', situated in the western part of the Netherlands, for the period 21–30 April 1983.

Solutions to the first problems of each chapter are presented in a password-protected online solutions manual, for the convenience of both the student and the tutor. Each solution, as a supplement to the many examples, is extensively described to give further insight into the problems that may appear in the identification of uncertain static or dynamic systems. For those who are starting at a very elementary level, it is recommended to study the many examples given in this book for a thorough grounding in the subject.

Finally, I would like to end this preface with a suggestion to the reader. Try to read the book as a road map for anybody who wants to wander through the diverse system identification landscape. No cycle path, let alone bush tracks, only the main roads are indicated with some nice, almost picturesque stops, which are the many simple examples that brighten up the material. Enjoy your trip!

Wageningen University
Wageningen, The Netherlands

Karel J. Keesman

Acknowledgements

Here I would like to acknowledge the contribution of some people who, maybe without knowing, implicitly or explicitly stimulated me in writing this book. It was in the early 1980s when Peter C. Young, at a workshop on real-time river flow forecasting, got my full attention to what he then called a data-based modeling approach. His "let the data speak" has been a starting point for writing this text. However, as many others, I always felt that our a priori knowledge of the system's behavior should not be overlooked. This is especially true when we only have access to small data sets. Identification of models from small data sets has been the subject of my Ph.D. work, that was (partly) supervised by Gerrit van Straten, Arun Bagchi, John Rijnsdorp and Huib Kwakernaak and that started in the early summer of 1985. From this period, when the bounded-error or set-membership approach became mature, I still remember the inspiring meetings at symposia with, in alphabetic order, Gustavo Belforte, John Norton, Helene Piet-Lahanier, Luc Pronzato and Eric Walter. Also, the contact with Jan van Schuppen on the connection between system theory and system identification for applications on systems with a biological component, in particular related to structural or theoretical identifiability and rational systems, should be mentioned, although not much of it was directly processed into a publication. In addition to this, I would like to mention the on-going discussions with Hans Stigter on identifiability and optimal input design (OID) and with Hans Zwart on estimation problems related to infinite-dimensional systems. As this last subject is far too advanced for this introductory text, it will not be covered by this book, although some reference is made to the identification of large scale models. These discussions helped me to make the final decisions with respect to the material that should be included. Our approach to solve OID problems, although very relevant for the identification of dynamic systems, is based on Pontryagin's minimum principle and uses singular optimal control theory. Because of the completely different angle of attack, I considered this to be out of the scope of the book. The regular visits to Tony Jakeman's group at the Australian National University and with a focus on identification of uncertain, time-varying environmental systems again allowed me to bring the theory into practice.

With respect to the correction of the script at first I would like to mention the students of the System Identification course at the Wageningen University. In addition to this, and more in particular, I would like to thank Rachel van Ooteghem for her calculations on the heating system example and Jimmy Omony, Dirk Vries, Hans Stigter, John Norton and Mike Johnson for their detailed comments and suggestions on the text. At last, I would like to mention Oliver Jackson and Charlotte Cross (Springer, UK) who guided me through all the practical issues related to the final publication of this book.

Contents[1]

[1]Sections marked with an asterisk (*) contain material at a more advanced level and, if desired, may be omitted by the reader, without loss of continuity of the main text.

Notations

Variables and functions

a_k	k-th coefficient in polynomial $A(q)$
b	bias
b_k	k-th coefficient in polynomial $B(q)$
c_k	k-th coefficient in polynomial $C(q)$
d_k	k-th coefficient in polynomial $D(q)$
$d_{\mathscr{M}}$	dimension of parameter vector
$e(t)$	white noise error
$f(\cdot)$	system function
f_k	k-th coefficient in polynomial $F(q)$
$g(t)$	impulse response function
h	amplitude relay output
$h(\cdot)$	output function
$h_{ij}(\cdot)$	derivative ith output w.r.t. jth parameter
j	complex number, $j = \sqrt{-1}$
l	time lag or lead
m	center of set
n	system dimension
n_a	order of polynomial $A(q)$
n_b	order of polynomial $B(q)$
n_c	order of polynomial $C(q)$
n_d	order of polynomial $D(q)$
n_f	order of polynomial $F(q)$
n_k	number of time delays
p	dimension parameter vector
p_0	switching probability
$p(\xi)$	probability density function (pdf) of ξ
r_{uu}	autocorrelation function of u
r_{uy}	cross-correlation function between u and y
r_{vv}	autocorrelation function of noise v
r_{vy}	cross-correlation function between v and y

r_{yy}	autocorrelation function of y
$r_{\varepsilon\varepsilon}$	autocorrelation function of ε
$r_{u\varepsilon}$	cross-correlation function between u and ε
$r_{\widehat{y}\varepsilon}$	cross-correlation function between $\widehat{y}$ and ε
s	Laplace variable
$s^{(i)}$	search direction at the ith iteration
t	time index/variable
u	eigenvector
$u(t)$	control input
$v(t)$	output disturbance/noise
$w(t)$	disturbance input
$x(t)$	system state
$y(t)$	system output
z	complex number, $z = e^{j\omega}$
F_x	gradient system function w.r.t. state x
F_u	gradient system function w.r.t. input u
$H_s(t)$	Heaviside step function
H_u	gradient output function w.r.t. input u
H_x	gradient output function w.r.t. state x
$J(\vartheta)$	scalar objective function
$J_W(\vartheta)$	weighted least-squares objective function
K	static gain
$L(\cdot)$	real-valued expansion coefficient
N	number of data points
$N(\alpha)$	describing function
T	specific time instant
T_s	sampling interval
α	constant
$\alpha^{(i)}$	step size at the ith iteration
β	constant
$\beta(t,k)$	tuning parameter function
$\delta(t)$	Dirac distribution
$\varepsilon(t)$	(estimated) prediction error
ϕ	phase of transfer function $\phi = \arg(G(\cdot))$
$\gamma(t)$	gain function
$\lambda(t)$	forgetting factor
λ	eigenvalue
ρ	correlation coefficient
σ_ε	standard deviation of ε
σ_i	ith singular value
τ	time delay
$\psi(t,\vartheta)$	gradient of the prediction
ω	frequency
ξ	random variable
$\xi(t,\vartheta)$	noise-free model output

Vectors and matrices

(a_{ij})	matrix A
e	noise vector, $e := [e(1), \ldots, e(N)]^T$
y	output vector, $y := [y(1), \ldots, y(N)]^T$
A	system matrix
B	input matrix
C	observation matrix
D	feed-through matrix
D	weighting matrix (TLS)
E	observation noise matrix (TLS)
$E(t)$	white noise vector (subspace)
F	Jacobi matrix with elements f_{ij}
G	disturbance input matrix
H	Hankel matrix
H	Jacobi matrix with elements h_{ij}
I	identity matrix
K	(Kalman) filter gain matrix
L	lower triangular matrix
0	null matrix
$P(\Phi)$	(orthogonal) projection matrix
P	covariance matrix recursive estimate
P_∞	steady state covariance matrix
R	covariance matrix measurement noise
$R^{(i)}$	approximation of J'' at ith iteration
Q	covariance matrix system noise
S	matrix with singular values
S_x	state sensitivity matrix
S_y	output sensitivity matrix
T	weighting matrix (TLS)
U	matrix with left-hand singular vector
$U(t)$	system input vector (subspace)
V	matrix with right-hand singular vector
W	positive-definite weighting matrix
X	sensitivity matrix
Y	observation matrix (TLS)
$Y(t)$	system output vector (subspace)
Z	instrumental variables matrix
Z	matrix containing errors in Φ (TLS)
$Z(t)$	input error vector (subspace)
δ	small positive scalar
ε	vector with residuals
ϕ	regressor vector
ϑ	parameter vector, $\vartheta := [\vartheta(1), \ldots, \vartheta(p)]^T$
χ	regressor vector extended with its derivatives
Φ	regressor matrix

Γ observability matrix
Π disturbance input matrix parameter model
Ω controllability matrix
Ξ system matrix parameter model

Polynomials and transfer functions

$f_k(q)$ kth basis function
$A(q)$ denominator polynomial related to y
$B(q)$ numerator polynomial related to u
$C(q)$ numerator polynomial related to e
$D(q)$ denominator polynomial related to e
$F(q)$ denominator polynomial related to y
$G(\cdot)$ rational transfer function in ω, q, s or z related to u
$H(q)$ rational transfer function related to e
$H(q)$ pulse-transfer operator of LTI system (Appendix E)
$L(q)$ stable linear filter
$P(q)$ rational transfer function of plant P
$Q(q)$ rational transfer function of controller Q
$U(s)$ Laplace transform of u
$W_l(q)$ l-steps ahead prediction weighting filter
$Y(s)$ Laplace transform of y
$U(z)$ z-transform of u
$Y(z)$ z-transform of y
$U_N(\omega)$ Fourier transform of $u(t)$, $t = 1, \ldots, N$
$Y_N(\omega)$ Fourier transform of $y(t)$, $t = 1, \ldots, N$

Sets and operators

$\mathrm{adj}(A)$ adjoint of matrix A
b_i ith parameter interval
$\mathrm{diag}(\phi)$ forms diagonal matrix from vector ϕ
$\mathrm{diag}(A)$ diagonal of matrix A
$\det(A)$ determinant of matrix A
q forward-shift operator
q^{-1} backward-shift operator
$\mathrm{ran}(A)$ range of matrix A
$\mathrm{Tr}(A)$ trace of matrix A
δ delta-operator
π differential operator $\frac{\mathrm{d}}{\mathrm{d}t}$
σ set of singular values

$\mathscr{B}$ orthotopic outer-bounding set
$\mathscr{E}$ ellipsoidal bounding set
$\mathscr{F}$ Fourier transform
$\mathscr{L}$ Laplace transform
$\mathscr{Z}$ z-transform

$\mathbb{N}$	set of natural numbers
$\mathbb{Q}$	set of rational numbers
$\mathbb{R}$	set of real numbers
$\mathbb{R}^n$	n-dimensional column vector of real numbers
$\mathbb{R}^{n \times n}$	$n \times n$-dimensional matrix of real numbers
$\mathbb{Z}$	set of integers
Cov	covariance
$E(\cdot)$	expectation operator
$\text{Lm}(\cdot)$	log magnitude
Var	variance
Vec	Vec operator stacking column vectors
Ω_e	error set
Ω_y	measurement uncertainty set
$\Omega_{\hat{y}}$	image set
Ω_ϑ	feasible parameter set

Special characters

$\widehat{}$	estimate
$'$	first derivative
$''$	second derivative
$+$	Moore–Penrose pseudo-inverse
$*$	transformed variable or reference variable
(i)	ith iteration
T	transpose
$\lvert \cdot \rvert$	absolute value (or modulus or magnitude) of a complex number
$\lvert A \rvert$	determinant of matrix A
$\Vert \cdot \Vert_1$	1-norm
$\Vert \cdot \Vert_2$	2-norm
$\Vert \cdot \Vert_\infty$	∞-norm
$\Vert \cdot \Vert_F$	Frobenius-norm
$\Vert \cdot \Vert_{2,Q}^2$	weighted Euclidean squared norm
$\forall$	for all
$\langle \cdot, \cdot \rangle$	inner product of matrices (Sect. 6.1.7)
$\angle(\cdot)$	phase shift

Acronyms

4SID	Subspace State-Space System IDentification
AIC	Akaike's Information Criterion
AR	Auto-Regressive
ARIMA	Auto-Regressive Integrated Moving Average
ARMA	Auto-Regressive Moving Average
ARMAX	Auto-Regressive Moving Average eXogenous
ARX	Auto-Regressive eXogenous
BJ	Box–Jenkins
CLS	Constrained Least-Squares

DGPS	Differential Global Positioning System
DO	Dissolved Oxygen
EKF	Extended Kalman Filter
EnKF	Ensemble Kalman Filter
ETFE	Empirical Transfer Function Estimate
FFT	Fast Fourier Transform
FIM	Fisher Information Matrix
FIR	Finite Impulse Response
FOPDT	First-Order Plus Dead Time
FPE	Final Prediction Error
FPS	Feasible Parameter Set
GLS	Generalized Least-Squares
IIR	Infinite Impulse Response
IV	Instrumental Variable
KF	Kalman Filter
LPV	Linear Parameter-Varying
LTI	Linear Time-Invariant
MA	Moving Average
ML	Maximum Likelihood
MUS	Measurement Uncertainty Set
NLS	Nonlinear Least-Squares
OE	Output-Error
OLS	Ordinary Least-Squares
RBS	Random Binary Signal
RLS	Recursive Least-Squares
RPE	Recursive Prediction Error
RRSQRT	Reduced Rank SQuare RooT
SVD	Singular Value Decomposition
tLS	truncated Least-Squares
TLS	Total Least-Squares
UKF	Unscented Kalman Filter
ZOH	Zero-Order Hold

Chapter 1
Introduction

The main topic of this textbook is how to obtain an appropriate mathematical model of a dynamic system on the basis of *observed* time series and *prior knowledge* of the system. Therefore first some background of dynamic systems and the modeling of these systems is presented.

1.1 System Theory

1.1.1 Terminology

Many definitions of a system are available, ranging from loose descriptions to strict mathematical formulations. In what follows, a *system* is considered to be an object in which different variables interact at all kinds of time and space scales and that produces observable signals. Systems of this type also called *open systems*. A graphical representation of a general open system, suitable for the system identification problem, is represented in Fig. 1.1. The system variables may be scalars or vectors (see Appendix A for details on vector and matrix operations), continuous or discrete functions of time. The sensor box, which will be considered as a static element, is added to emphasize the need of monitoring the systems to produce observable signals. In what follows, the sensor is considered to be a part of the dynamic system. In Fig. 1.1 the following system variables can be distinguished.

Input u: the input u is an exogenous, measurable signal. This signal can be manipulated directly by the user.

Disturbance w: the disturbance w is an exogenous, possibly measurable signal, which cannot be manipulated. It originates from the environment and directly effects the behavior of the system. If the disturbance is not measurable, it is considered as possibly structured uncertainty in the input u or in the relationship between u and x, and indicated as system noise.

State x: the system state x summarizes all the effects of the past inputs u and disturbances w to the system. Generally the evolution of the states is described by

K.J. Keesman, *System Identification*,
Advanced Textbooks in Control and Signal Processing,
DOI 10.1007/978-0-85729-522-4_1,

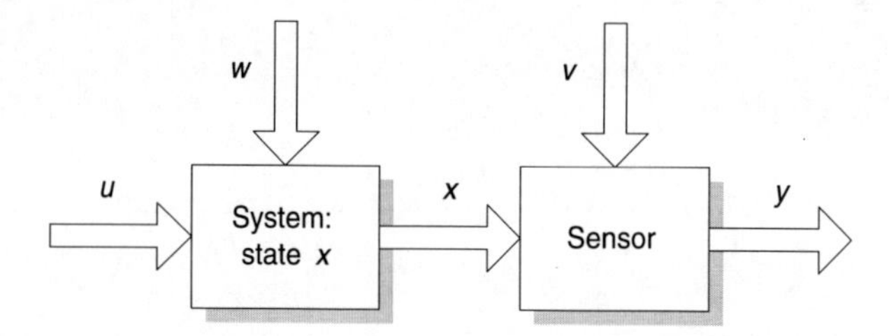

Fig. 1.1 General system representation

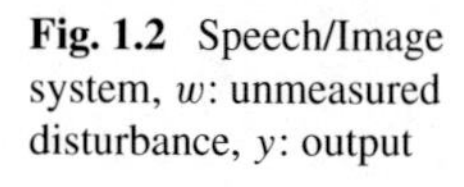
Fig. 1.2 Speech/Image system, w: unmeasured disturbance, y: output

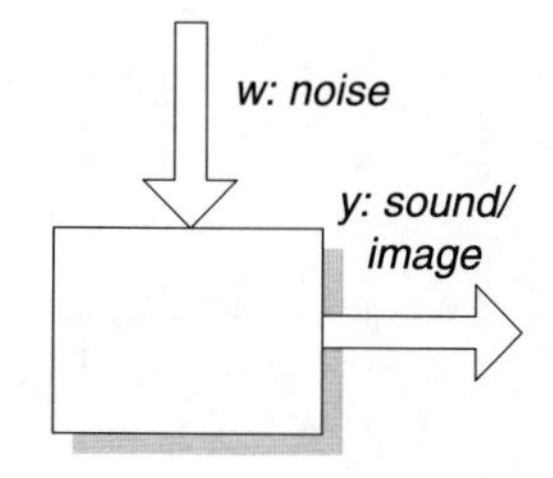

differential or difference equations. Hence, the dynamic behavior of the system is affected by variations of the exogenous signals u and w and laws describing the internal mechanism of the system. In what follows, static systems, which do not show a dynamic behavior, are considered as special cases of dynamic systems and are simply described by algebraic relationships between u, w, and x.

Disturbance v: as w, the output disturbance v is an exogenous signal, which cannot be manipulated. It represents the uncertainty (noise) introduced by the sensor, and is generally indicated as sensor noise.

Output y: the output y is the output of the sensors. It represents all the observable signals that are of interest to the user. In general, y is modeled as a function of the other signals. Since the sensor dynamics are ignored, the static relationship between y and x, v is expressed in terms of algebraic equations.

Let us illustrate the system concept by a number of "real-world" examples.

Example 1.1 Signal processing: In many speech or image processing applications there is only an output signal: time series of sound vibrations or a collection of images. The aim is to find a compact description of this signal, which after transmission or storage can be used to reconstruct the original signal. The problem here is the presence of noise (unmeasurable disturbances) in the output signal. The system can be depicted as in Fig. 1.2.

Example 1.2 Bioreactor: In the process industry bioreactors are commonly modeled for design and operation. A fed-batch reactor system is one specific type of bioreactor with no outflow. In the initial stage the reactor is filled with a small amount of nutrient substrate and biomass. After that the fed-batch reactor is progressively filled with the influent substrate. In this stage the influent flow rate is the input to

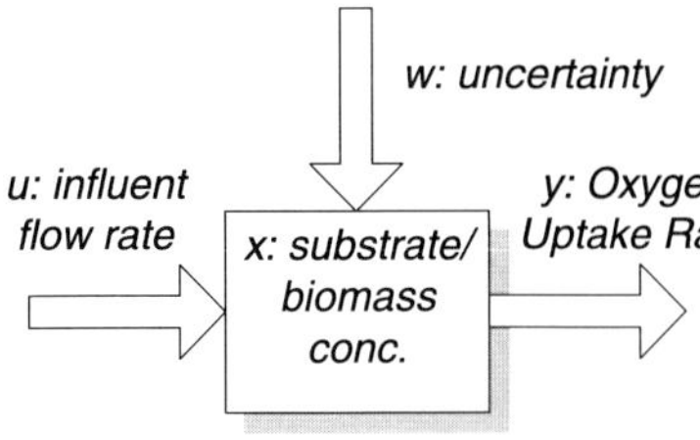

Fig. 1.3 Fed-batch reactor system, u: controlled input, w: unmeasured disturbances, y: output

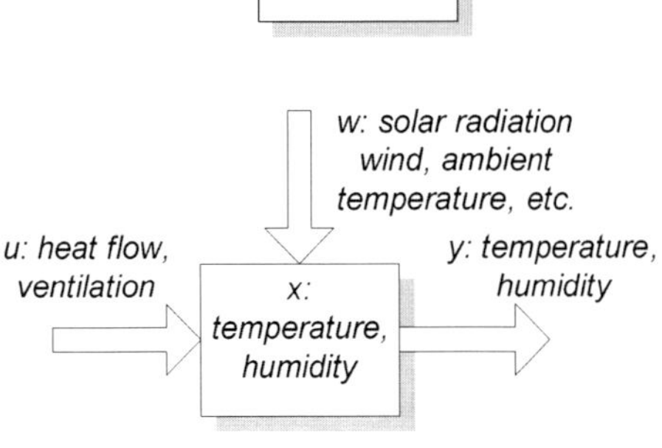

Fig. 1.4 Greenhouse climate system, u: input, w: (un)measured disturbances, y: output

the system, and substrate and biomass concentrations are the system states. Since both substrate and biomass are difficult to measure directly, dissolved oxygen is commonly used to reconstruct the Oxygen Uptake Rate (OUR), which can be considered as the output of the system. The signal w represents the uncertainties in the influent flow rate and influent substrate concentrations, and also substantial modeling errors due to the limited knowledge of the biochemical process, see Fig. 1.3.

Example 1.3 Greenhouse climate: Greenhouse climate control is one of the challenging problems at the interface of agriculture and control theory. It is common practice to restrict the modeling of the greenhouse climate to temperature and humidity. A typical feature of these type of systems is the major effect of the disturbances, like wind, ambient temperature, solar radiation, etc., on the system states. Heating and ventilation are the only manipulated variables that directly affect the climate. Under constant window aperture conditions, the system can be depicted as in Fig. 1.4.

1.1.2 Basic Problems

Basically four problem areas in system theory can be distinguished: *modeling, analysis, estimation, control*. Between these areas several interrelationships can be noted. From a system identification point of view, especially modeling and estimation are important, as these are directly related to the system identification problem. The following gives more details of this classification.

Modeling: A critical step in the application of system theory to a real process is to find a mathematical model which adequately describes the physical situation.

Several choices have to made. First, the system boundaries and the system variables have to be specified. Then relations between these variables have to be specified on the basis of prior knowledge, and assumptions about the uncertainties in the model have to be made. This alltogether defines the *model structure*.

Still, the model may contain some unknown or incompletely known coefficients, the *model parameters*, in the following denoted by ϑ, which define an additional set of system variables. Much more can be said about the modeling step. However, as yet, it suffices to say that in what follows it is explicitly assumed that a model structure, albeit not the most appropriate one, is given.

Analysis: Usually, the first step after having obtained a model structure, with corresponding parameter values, is to analyze the system output behavior by simulation. In addition to this, the stability of the system and the different time scales governed by the system dynamics are important issues to be investigated. Since most often not all the system parameters are known, a sensitivity analysis using statistical (see Appendix B for details on statistics) or unknown-but-bounded information about the parameters can be very helpful to detect crucial system properties. A central question in system identification, and the key issue of *identifiability analysis*, is: "can the unknown model parameters ϑ be uniquely, albeit locally, identified?" Other issues, important for the design of estimation schemes, are the observability aspects of a system.

Estimation: A next step, after having obtained an appropriate (un)stable, identifiable, and observable model structure, is concerned with the estimation of the unknown variables from a given data set of input–output variables. Basically, we distinguish between state estimation and parameter estimation or identification.

In state estimation problems one tries to estimate the states x from the outputs y under the assumption that the model is perfect and thus the parameters are exactly known. Similarly, parameter identification focuses on the problem of estimating the model parameters ϑ from u and y. In the early 1960s, when modern system concepts were introduced, it has also been recognized that the state and parameter estimation problems show a clear resemblance. Therefore, parameter identification problems have also been treated as state estimation problems. If in state estimation problems the condition of a perfect model is not fulfilled, one simultaneously tries to identify the unknown model parameters; this is known as the adaptive estimation problem. In addition to the state and parameter estimation problems, in some applications there is also a need for estimating or recovering the system disturbance w. Moreover, for further analysis of the uncertainty in the estimates, there is a need to infer the statistical properties of the disturbances v, w from the data. However, in this book the focus is on parameter estimation, where parameters can be time-dependent variables and thus can be considered as unobserved states.

Still, the term state or parameter estimation is not always specific enough. For example, when time is considered as the independent variable, we can categorize the state estimation problem as:

1. *Filtering*: estimation of $x(T)$ from $y(t)$, $0 \leq t \leq T$.
2. *Smoothing*: estimation of $x(\tau)$, $0 \leq \tau \leq T$, from $y(t)$, $0 \leq t \leq T$.
3. *Prediction*: estimation of $x(T+\tau)$ from $y(t)$, $0 \leq t \leq T$, $\tau > 0$.

Recall that in these specific problems the state x can be easily substituted by the (time-varying) model parameter ϑ. Details will be discussed in subsequent chapters.

Control: The control problem focuses on the calculation (determination) of the input u such that the controlled system shows the desired behavior. Basically, one defines two types of control strategies, open-loop and closed-loop controls.

The main difference between open- and closed-loop controls is that, in contrast to closed-loop control, open-loop control does not use the actual observations of the output for the calculation of the control input. In open-loop control the control input trajectory is precomputed, for instance, as a result of an optimization problem or model inversion. Consequently, a very accurate mathematical model is needed. In the situations where uncertainty is definitely present, however, closed-loop control is preferred, since it usually results in a better performance. In a number of closed-loop control schemes, state estimation is included. When the system is not completely specified, that is, it contains a number of unknown parameters, most often an adaptive control scheme is applied. Hence, those schemes require the incorporation of a parameter estimation procedure.

Clearly, in the design procedure of these types of model-based controllers, the previously stated problems of modeling, analysis, and estimation all play a role. Moreover, in modern control theory, which also treats the robustness aspect explicitly, not only a mathematical model of the system but also a model including uncertainty descriptions is a prerequisite. Hence, analysis of the uncertainties should not be forgotten.

1.2 Mathematical Models

Mathematical models can take very different forms depending on the system under study, which may range from social, economic, or environmental to mechanical or electrical systems. Typically, the internal mechanisms of social, economic, or environmental systems are not very well known or understood, and often only small data sets are available, while the prior knowledge of mechanical and electrical systems is at a high level, and experiments can be easily done. Apart from this, the model form also strongly depends on the final objective of the modeling procedure. For instance, a model for process design or operation should contain much more detail than a model used for long-term prediction.

Generally, models are developed to:

- Obtain or enlarge insight in different phenomena, for example, recovering physical laws or economic relationships.
- Analyze process behavior using simulation tools for, for example, process training of operators or weather forecasts.
- Control processes, for example, process control of a chemical plant or control of a robot.
- Estimate state variables that cannot be easily measured in real time on the basis of available measurements for, for instance, online process information.

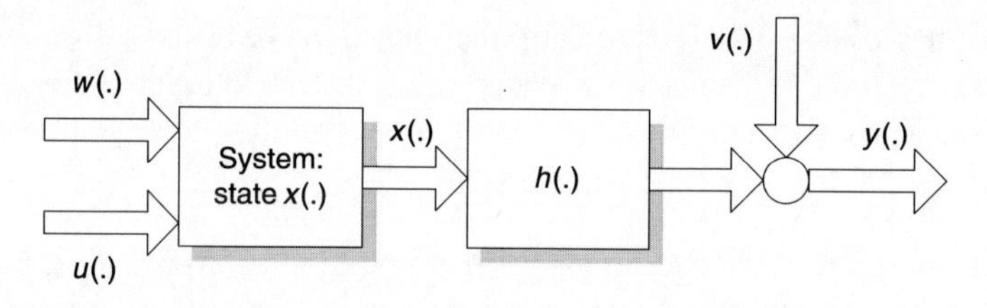

Fig. 1.5 Basic structure of mathematical model

1.2.1 Model Properties

In this textbook, the following basic model structure, based on first (physical, chemical, or biological) principles, is adopted (see also Fig. 1.5):

Discrete-time:

$$\begin{aligned} x(t+1) &= f\big(t, x(t), u(t), w(t); \vartheta\big), \quad x(0) = x_0 \\ y(t) &= h\big(t, x(t), u(t); \vartheta\big) + v(t), \quad t \in \mathbb{Z}^+ \end{aligned} \tag{1.1}$$

Continuous-time:

$$\begin{aligned} \frac{\mathrm{d}x(t)}{\mathrm{d}t} &= f\big(t, x(t), u(t), w(t); \vartheta\big), \quad x(0) = x_0 \\ y(t) &= h\big(t, x(t), u(t); \vartheta\big) + v(t), \quad t \in \mathbb{R} \end{aligned} \tag{1.2}$$

where the variables and vector functions have appropriate dimensions.

In Fig. 1.5, $v(\cdot)$ is an additive sensor noise term, which basically represents the errors originating from the measurement process. Modeling errors as a result of model simplifications (the real system is too complicated) and input disturbances are represented by $w(\cdot)$. In the following, it is often assumed that $v(\cdot)$, and also $w(\cdot)$, is a *white noise* signal. Here it suffices to give a very general description of a white noise signal as a signal that has no time structure. In other words, the value at one instant of time is not related to any past or future value of this signal. A formal description will be given later, and since white signals in continuous time are not physically realizable, the focus will then be on discrete-time white signals.

Typically (1.1)–(1.2) present a general description of a *finite-dimensional* system, represented by a set of ordinary difference/differential equations with additive sensor noise. Hence, so-called *infinite-dimensional* systems, described by partial differential equations (for an introductory text, see [CZ95]), will not be explicitly treated in this text. One way to deal with these systems is by discretization of the space or time variables, which will ultimately lead to a set of ordinary differential or difference equations.

The continuous-time representation will only be used for demonstration. For identification, usually the discrete-time form will be implemented due to the availability of sampled data and the ultimate transformation of a mathematical model into a simulation code. In addition to these classifications, we also distinguish between *linear* and *nonlinear*, *time-invariant* and *time-varying*, *static* and *dynamic* systems. Let us further define these classification terms.

Linearity: Let, under zero initial conditions, $u_1(t)$ and $u_2(t)$ be inputs to a system with corresponding outputs $y_1(t)$ and $y_2(t)$. Then, this system is called linear if its response to $\alpha u_1(t) + \beta u_2(t)$, with α and β constants, is $\alpha y_1(t) + \beta y_2(t)$. In other words, for linear systems, the properties of superposition or additivity and scaling hold. Since $f(\cdot)$ and $h(\cdot)$ in (1.1) and (1.2) represent general functions, linearity will not hold, and thus the basic model structure represents a nonlinear system.

Time-invariance: Let $u_1(t)$ be an input to a system with corresponding output $y_1(t)$. Then, a system is called time-invariant if the response to $u_1(t+\tau)$, with τ a time shift, is $y_1(t+\tau)$. In other words, the system equations do not vary in time. The notation $f(t,\cdot)$ and $h(t,\cdot)$ indicates that both functions are explicit functions of the time variable t and thus represent time-varying systems.

Causality: Let $u_1(t) = u_2(t)\ \forall t < t_1$, that is, two signals with equivalent historic behavior. Then, a system is called causal if $y_1(t_1) = y_2(t_1)$ and is called strictly causal if this equality holds for $u_1(t) = u_2(t)\ \forall t \leq t_1$. In other words, the output of a strictly causal system depends on current and past inputs. Hence, as the output of a causal system, it does not depend on future values of the input. In fact, this property holds for all physical systems. Smoothers, for instance, do not have this causality property.

Dynamics: If a system output at any time instant depends on its history, and not just on the present input, it is called a dynamic system. In other words, a dynamic system has a memory and is usually described in terms of a difference or differential equation. A static system, on the other hand, has no memory and is usually described by algebraic equations.

For what follows, this classification suffices.

1.2.2 Structural Model Representations

Notice that the system represented by (1.1) or (1.2) is very general and covers all the special cases mentioned in the previous section. Let us be more specific and illustrate the mathematical modeling process by application to a simple system, a storage tank with level controller.

Example 1.4 Storage tank: Consider the following storage tank (see Fig. 1.6).

Let us start with specifying our prior knowledge of the internal system mechanisms. The following mass balance can be defined in terms of the continuous-time state variable, the volume of the liquid in the storage tank (V), inflows $u(t)$, and outflows $y(t)$:

$$\frac{\mathrm{d}V(t)}{\mathrm{d}t} = u(t) - y(t)$$

and, in addition to this and as a result of a proportional level controller (L.C.),

$$y(t) = KV(t)$$

Fig. 1.6 Graphical scheme of storage tank

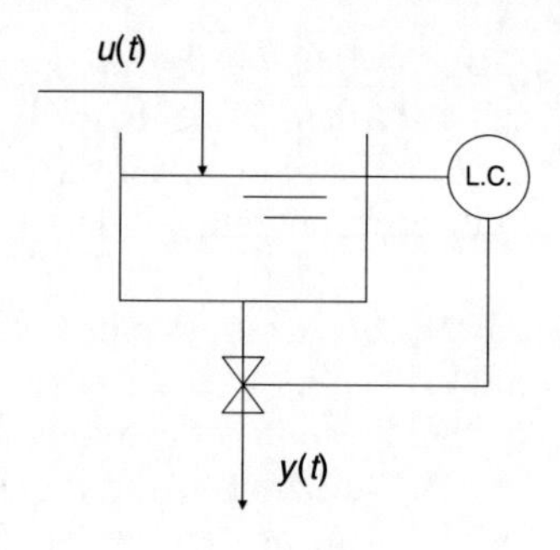

with K a real constant. Hence, the so-called *state-space* model representation of the system with $x(t) = V(t)$ is given by

$$\frac{\mathrm{d}x(t)}{\mathrm{d}t} = -Kx(t) + u(t)$$
$$y(t) = Kx(t)$$

which is a particular noise-free (deterministic) form of (1.2). Consequently, in this case where $w(t) = v(t) = 0$,

$$f\big(t, x(t), u(t), w(t); \vartheta\big) = -Kx(t) + u(t)$$
$$h\big(t, x(t), u(t); \vartheta\big) = Kx(t)$$

with system parameter $\vartheta = K$.

The specific system properties will be analyzed in the next example, in which an alternative representation is introduced.

Example 1.5 Storage tank: The so-called *differential equation* model representation between u and y after eliminating x is given by

$$\frac{1}{K}\frac{\mathrm{d}y(t)}{\mathrm{d}t} + y(t) = u(t)$$

which can be presented more explicitly after assuming that $y(0) = 0$ and $u(t) = 0$, $t < 0$. After first solving the homogenous equation, that is, with $u(t) = 0\ \forall t$, and then applying the principle of variation of constants, we arrive at the following result:

$$y(t) = y(0)\mathrm{e}^{-Kt} + \int_0^t K\mathrm{e}^{-K(t-\tau)}u(\tau)\,\mathrm{d}\tau$$

with τ the variable of integration. Implementing the initial condition, that is, $y(0) = 0$, leads to the input–output relationship

$$y(t) = \int_0^t K\mathrm{e}^{-K(t-\tau)}u(\tau)\,\mathrm{d}\tau$$

which has the following properties:

1. linear, because integration is a linear operation
2. time-invariant, because

$$y(t+l) = \int_0^{t+l} K\mathrm{e}^{-K(t+l-\tau)} u(\tau)\,\mathrm{d}\tau$$

$$=_{[v:=\tau-l]} \int_{-l}^{t} K\mathrm{e}^{-K(t-v)} u(v+l)\,\mathrm{d}v$$

$$=_{[u(t)=0 \text{ for } t<l]} \int_0^{t} K\mathrm{e}^{-K(t-v)} u(v+l)\,\mathrm{d}v$$

3. causal, because the output does not depend on future input values.

From this continuous-time example it is important to note that two specific model representations became visible, the state-space and differential model representation. A general state-space model of a linear, time-invariant (LTI) dynamic systems is

$$\begin{aligned} \frac{\mathrm{d}x(t)}{\mathrm{d}t} &= Ax(t) + Bu(t) \\ y(t) &= Cx(t) + Du(t) \end{aligned} \tag{1.3}$$

where the matrices A, B, C, and D have appropriate dimensions.[1] Consequently, in the storage tank example: $A = -K$, $B = 1$, $C = K$, and $D = 0$. Alternatively, a general differential equation model is represented by

$$a_n \frac{\mathrm{d}^n y(t)}{\mathrm{d}t^n} + \cdots + a_1 \frac{\mathrm{d}y(t)}{\mathrm{d}t} + y(t) = b_0 u(t) + b_1 \frac{\mathrm{d}u(t)}{\mathrm{d}t} + \cdots + b_m \frac{\mathrm{d}^m u(t)}{\mathrm{d}t^m} \tag{1.4}$$

Hence, in Example 1.5 we obtain: $a_n = a_{n-1} = \cdots = a_2 = 0$, $a_1 = 1/K$ and $b_0 = 1, b_1 = b_2 = \cdots = b_m = 0$. In addition to these two representations, other representations will follow in subsequent sections and chapters.

Example 1.6 Moving average filter: A discrete-time example is provided by the three-point moving average filter with input u and output y:

$$y(t) = \frac{1}{3}\big[u(t) + u(t-1) + u(t-2)\big], \quad t \in \mathbb{Z}^+$$

which is a *difference equation* model representation. It can be easily verified that this is another example of a linear, time-invariant system. A discrete-time state-space representation is obtained by defining, for example, $x_1(t) = u(t-1)$ and

[1]The analytical solution of (1.3), for $x(0) = x_0$ and $u(t) = 0$ for $t < 0$, is given by $y(t) = C[\mathrm{e}^{At}x_0 + \int_0^t \mathrm{e}^{A(t-\tau)} Bu(\tau)\,\mathrm{d}\tau] + Du(t)$ (see, for instance, [GGS01]). Commonly, this expression is evaluated when simulating an LTI system.

$x_2(t)=u(t-2)$, so that

$$\begin{aligned} x_1(t+1) &= u(t) \\ x_2(t+1) &= u(t-1) = x_1(t) \\ y(t) &= \frac{1}{3}\big[u(t)+x_1(t)+x_2(t)\big], \quad t \in \mathbb{Z}^+ \end{aligned}$$

or in matrix form:

$$\begin{pmatrix} x_1(t+1) \\ x_2(t+1) \end{pmatrix} = \begin{pmatrix} 0 & 0 \\ 1 & 0 \end{pmatrix} \begin{pmatrix} x_1(t) \\ x_2(t) \end{pmatrix} + \begin{pmatrix} 1 \\ 0 \end{pmatrix} u(t)$$

$$y(t) = \begin{pmatrix} \frac{1}{3} & \frac{1}{3} \end{pmatrix} \begin{pmatrix} x_1(t) \\ x_2(t) \end{pmatrix} + \frac{1}{3}u(t), \quad t \in \mathbb{Z}^+$$

so that $A=\left(\begin{smallmatrix} 0 & 0 \\ 1 & 0 \end{smallmatrix}\right)$, $B=\left(\begin{smallmatrix} 1 \\ 0 \end{smallmatrix}\right)$, $C=\left(\begin{smallmatrix} \frac{1}{3} & \frac{1}{3} \end{smallmatrix}\right)$, and $D=\frac{1}{3}$.

It can be easily verified from this example that the state-space representation is not unique. To see this, define, for example, $x_1(t)=u(t-2)$ and $x_2(t)=u(t-1)$. Hence, the identification of state-space models needs extra care. On the other hand, the transformation from state-space to differential/difference equation models is unique.

The input–output relationships in the previous examples with single input and single output (SISO) can be represented in the following general form:

$$y(t) = \int_{-\infty}^{t} g(t-\tau)u(\tau)\,\mathrm{d}\tau, \quad t \in \mathbb{R} \tag{1.5}$$

and

$$y(t) = \sum_{k=-\infty}^{t} g(t-k)u(k), \quad t \in \mathbb{Z}^+ \tag{1.6}$$

which is also indicated as the *impulse response* model representation. The function $g(t)$ is called the continuous or discrete impulse response of a system; a name which will become clear in the next chapter when dealing with impulse response methods. In (1.5)–(1.6), the output $y(\cdot)$ is presented in terms of the convolution integral or sum, respectively, of $g(\cdot)$ and $u(\cdot)$. Therefore these models are also called *convolution models*.

1.3 System Identification Procedure

In the previous section, mathematical models with their properties and different ways of representation have been introduced. Excluding the theoretical studies on *exact* modeling of a system, a mathematical model is always an approximation of

Fig. 1.7 The system identification loop (after [Lju87])

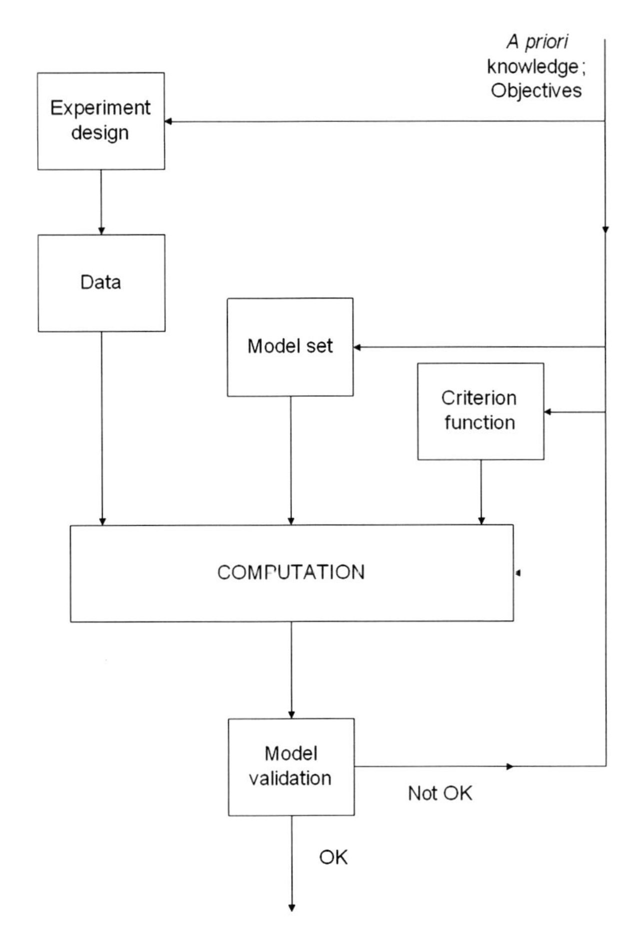

the real system. In practice, the system complexity, the limited prior knowledge of the system, and the incomplete availability of observed data prevent an exact mathematical description of the system. However, even if there is full knowledge of the system and sufficient data available, an exact description is most often not desirable, because the model would become too complex to be used in an application. Consequently, *system identification* is considered as *approximate* modeling for a specific application on the basis of observed data and prior system knowledge.

In what follows, the identification procedure, with the aim to arrive at an appropriate mathematical model of the system, is described in some detail (see Fig. 1.7). As mentioned before, prior knowledge, objectives, and data are the main components in the system identification procedure, where prior knowledge has a key role. It should be realized that these entities are not independent. Most often, data is collected on the basis of prior system knowledge and modeling objectives, leading to an appropriate experiment design. At the same time observed data may also lead to an adjustment of the prior knowledge or even to the objectives.

Figure 1.7 shows that the choice of a model set is completely determined by our prior knowledge of the system. This choice of a set of candidate models is without doubt the most important and most difficult step in a system identification procedure. For instance, in some simulator applications a very detailed model is required. A natural choice is then to base the model on physical laws and additional relationships with corresponding physical parameters, which leads to a so-called *white-box* model structure. If, however, some of these parameters are uncertain or not well known and, for instance, realistic predictions have to be obtained, the parameters can be estimated from the data. Model sets with these adjustable parameters comprise so-called *grey-box* models. In other cases, as, for instance, in control applications, it usually suffices to use linear models which do not necessarily refer to the underlying physical laws and relationships of the process. These models are generally called *black-box* models. In addition to a choice of the structure, we also have to choose the model representation, for instance, state-space, impulse response or differential equation model representation, and the model parameterization which deals with the choice of the adjustable parameters.

In order to measure the fit between model output and observed data, a criterion function has to be specified, and the identification method, which numerically solves the parameter estimation problem, has to be chosen. After that, a model validation step considers the question whether the model is good enough for its intended use. If then the model is considered as appropriate, the model can be used, otherwise the procedure must be repeated, which is most often the case in practice. However, it is important to conclude that, due to the large number of significant choices to be made by the user, the system identification procedure includes a loop in order to obtain a validated model (see Fig. 1.7).

1.4 Historical Notes and References

The literature on the system identification problem is extensive. Many congress papers on this subject can be found in, for instance, the Proceedings of the IFAC Symposia on Identification and System Parameter Estimation, which since 1994 is called System Identification and abbreviated as SYSID. The first IFAC Symposium on Identification started in 1967, which more or less indicates the time that system identification became a mature research area.

In addition to this, many books have appeared, for instance, [BJ70, Sch73, Eyk74, KR76, GP77, Sor80, You84, Nor86],[2] [Lju87, Lju99b, SS87, Joh93, LG94, WP97, CG00, PS01, Gui03, Kat05, VV07, GP08]. The basic material of this chapter is based on these books, especially [Sch73, Nor86], and [Lju87, Lju99b]. The system theoretic concepts introduced here at an elementary level can be found in many books on mathematical systems theory, for example, [PW98, HP05], and [HRvS07].

[2] An Introduction to Identification by J.P. Norton. Paperback: 320 pages; Publisher: Dover Publications (23 April 2009).

1.5 Problems

Problem 1.1 Consider again the storage tank example (1.4), but now with a slightly modified effect of the input on the output, such that $bu(t)$ flows into the system. On the basis of this a priori knowledge of the tank system, different types of representation will be investigated.

(a) Give some physical conditions under which $b \neq 1$.
(b) Give the differential equation describing this system in terms of the relationship between $u(t)$ and $y(t)$ and provide the solution, under zero initial conditions, for a unit input, such that $u(t) = 1$ for all t.
(c) Represent the system in terms of an impulse response or convolution model and give the continuous-time impulse response.
(d) Represent the system in state-space form. Are there other state-space forms that lead to the same input–output behavior? If there is, give an example. If not, motivate your answer.
(e) Is this system linear, time-invariant, and causal? Justify your answer.

Problem 1.2 Consider the moving average filter example (1.6), but now as a four-point moving average filter with input u and output y.

(a) Give the difference equation describing this system in terms of the relationship between $u(t)$ and $y(t)$ and numerically evaluate the behavior of the filter for a unit step input, such that $u(t) = 1$ for all t.
(b) Represent the system in terms of an impulse response or convolution model and give the discrete-time pulse response, i.e., for $u(t) = 1$ for $t = 0$ and $u(t) = 0$ for $t \neq 0$.
(c) Represent the system in state-space form. Give an alternative state-space form and check the corresponding input–output behaviors.
(d) Is this system linear, time-invariant, and causal? Motivate your answer.

Part I
Data-based Identification

The basic model representation for the analyzes, in this part of the book, is given by the convolution integral,

$$y(t) = \int_{-\infty}^{t} g(t-\tau)u(\tau)\,\mathrm{d}\tau, \quad t \in \mathbb{R}$$

or its discrete-time counterpart, the convolution sum,

$$y(t) = \sum_{k=-\infty}^{t} g(t-k)u(k), \quad t \in \mathbb{Z}^{+}$$

This model representation is particularly suited for SISO LTI dynamic systems and formed the basis of classical data-based or nonparametric identification methods. The adjectives "data-based" and "nonparametric" express the very limited prior knowledge used in the identification procedure; the prior knowledge is limited to assumptions with respect to linearity and time-invariance of the system under consideration, as we will see in the next chapters.

In Chap. 2 the focus is on methods that directly utilize specific responses of the system, in particular the impulse, step, and sine-wave response. The first two signals directly provide estimates of $g(t)$, while the sine-wave response forms the basis for the methods described in the following chapter.

Chapter 3 describes methods which directly provide estimates of $g(t)$ in the frequency domain. These frequency domain descriptions are particularly suited for controller design.

In many applications noise is clearly present. Under those circumstances, the reliability of the estimates can be significantly reduced. Therefore, in Chap. 4 methods that are less sensitive to noise, and thus very useful under practical circumstances, are presented.

Chapter 2
System Response Methods

2.1 Impulse Response

2.1.1 Impulse Response Model Representation

In order to motivate the general applicability of the convolution model to LTI systems, first the *unit impulse function* has to be introduced. The unit impulse function or Dirac (δ) function at time zero is defined heuristically as

$$\delta(t) := 0 \quad \text{for all } t \neq 0, \qquad \int_{-\infty}^{\infty} \delta(t)\,\mathrm{d}t = 1 \tag{2.1}$$

and can be viewed as a rectangular, unit-area pulse with infinitesimally small width. Let the unit impulse function $\delta(t)$ be input to an LTI system and denote the impulse response by $g(t)$. Then, due to the time-invariant behavior of the system, a time-shifted impulse $\delta(t-\tau)$ will result in an output signal $g(t-\tau)$. Moreover, because of the linearity, the impulse $\delta(t-\tau)u(\tau)$ will result in the output $g(t-\tau)u(\tau)$, and after integrating both the input and output impulses over the time interval $[-\infty, \infty]$, that is,

$$\int_{-\infty}^{\infty} \delta(t-\tau)u(\tau)\,\mathrm{d}\tau = u(t)$$

due to the properties of the impulse function, and

$$\int_{-\infty}^{\infty} g(t-\tau)u(\tau)\,\mathrm{d}\tau = y(t)$$

we obtain a relationship between the input $u(t)$ and output $y(t)$. Since only causal systems (see Sect. 1.2.2) are treated, the upper bound of the last convolution integral is set equal to t. In the case where $u(t) = 0$ for $t < 0$ and zero initial condition response, as a result of zero initial conditions or a stable system for which the initial condition response has died to zero by $t = 0$, the lower bound can be set to zero.

K.J. Keesman, *System Identification*,
Advanced Textbooks in Control and Signal Processing,
DOI 10.1007/978-0-85729-522-4_2,

Hence, in the derivation of the practically applicable convolution model

$$y(t) = \int_0^t g(t-\tau)u(\tau)\,\mathrm{d}\tau \tag{2.2}$$

only assumptions have been made with respect to the linearity and time-invariance of the system. Thus the convolution model, fully characterized by the impulse response function $g(t)$, is able to describe the input–output relationship of the large class of LTI systems. Consequently, if $g(t)$ is known, then for a given input signal $u(t)$, the corresponding output signal can be easily computed. This feature explains the interest in impulse response model representations, especially if there is limited prior knowledge about the system behavior.

2.1.2 Transfer Function Model Representation

In the analysis of linear systems the Laplace transformation (see Appendix C for details) forms one of the basic tools. Recall that the Laplace transform is defined as

$$\mathcal{L}[f(t)] \equiv F(s) := \int_0^\infty f(t)\mathrm{e}^{-st}\,\mathrm{d}t \tag{2.3}$$

Laplace transformation of the convolution model (2.2) gives

$$Y(s) = G(s)U(s) \tag{2.4}$$

which defines an algebraic relationship between transformed output signal $Y(s)$ and transformed input signal $U(s)$. The function $G(s)$ is the Laplace transformed impulse response function, that is, $G(s) \equiv \mathcal{L}[g(t)]$, and is called the transfer function. Consequently, representation (2.4) is called the *transfer function* model representation, which will be treated in more detail in the chapter on frequency response methods. The various model representations with their connections, in terms of transformations and back-transformations, are shown in Fig. 2.1, where the impulse response model has a central place.

Let us further illustrate the application of the transfer function model representation to Example 1.4 and indicate the different connections with the other representations.

Example 2.1 Storage tank: Recall that the input–output relationship of the storage tank, after solving a first-order linear differential equation, was given by

$$y(t) = \int_0^t K\mathrm{e}^{-K(t-\tau)}u(\tau)\,\mathrm{d}\tau$$

Consequently, comparison with the convolution model (2.2) reveals that the impulse response function $g(t)$ is equal to $K\mathrm{e}^{-Kt}$, and thus the transfer function is

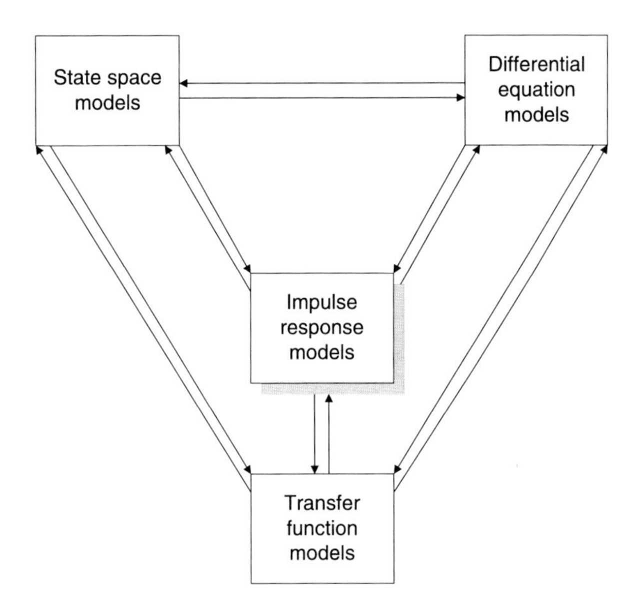

Fig. 2.1 Various model representations for LTI systems

given by $G(s) = \mathcal{L}[K\mathrm{e}^{-Kt}] = \frac{K}{s+K}$, so that

$$Y(s) = G(s)U(s) = \frac{K}{s+K}U(s)$$

An alternative way to find the transfer function and the impulse response function of the storage tank in Example 1.4 is via Laplace transformation of the differential equation

$$\frac{1}{K}\frac{\mathrm{d}y(t)}{\mathrm{d}t} + y(t) = u(t)$$

as given in Example 1.5. For zero initial conditions, $y(0) = 0$, and after applying the rules of Laplace transformation (see Appendix C for details on the Laplace transform), we find that

$$\frac{1}{K}sY(s) + Y(s) = U(s)$$

Hence, the transfer function, $G(s)$, of this SISO system is found from

$$G(s) = \frac{Y(s)}{U(s)} = \frac{K}{s+K}$$

which, as we have seen before, is the Laplace transform of $g(t)$. Thus $g(t)$ can be directly found by inverse Laplace transformation of $G(s)$. In the same way, $g(t)$ and $G(s)$ can be found from the state-space model.[1] Assuming the zero initial conditions

[1] The transfer function of a general LTI state-space model (1.3) with $x(0) = 0$, possibly obtained after a state correction when $x(0) = x_0 \neq 0$, is given by $G(s) = C[sI - A]^{-1}B + D$ (see, for

on $y(t)$ and $u(t)$ and on all their derivatives and after introducing the differential operator $\pi := \frac{\mathrm{d}}{\mathrm{d}t}$, we can also write the input–output relationship as

$$y(t) = G(\pi)u(t)$$

with $G(\pi) = \frac{K}{\pi+K}$, which shows a clear resemblance with the transfer function $G(s)$.

So far, no real data have been involved; the impulse response function and transfer function have been evaluated on the basis of prior knowledge only. However, in a system identification procedure, this could be the first step in the selection of a proper sampling scheme if there is also some knowledge about the parameter values.

2.1.3 Direct Impulse Response Identification

In what follows, it is indicated how to obtain an estimate of the impulse response function from real data. Since data acquisition is typically performed in discrete time, in the remainder of this chapter and the next chapters, the focus will be on discrete-time representations. In particular, for $u(t) = 0$, $t < 0$, and zero initial condition response, the convolution sum is given by

$$y(t) = \sum_{k=0}^{t} g(t-k)u(k) = \sum_{k=0}^{t} g(k)u(t-k), \quad t \in \mathbb{Z}^+ \tag{2.5}$$

where $g(0)$ is usually equal to zero, because no real system responds instantly to an input. Hence, if we are able to generate a unit pulse, the coefficients of $g(t)$ can be directly found from the measured output. Let, for instance, the pulse input be specified as

$$u(t) = \begin{cases} \alpha, & t = 0 \\ 0, & t \neq 0 \end{cases} \tag{2.6}$$

where α is chosen in accordance with the physical limitations on the input signal. The corresponding output will be

$$y(t) = \alpha g(t) + v(t) \tag{2.7}$$

instance, [GGS01] and, for infinite-dimensional systems, [Zwa04]). For the state correction, introduce $\Delta x(t) := x(t) - \tilde{x}(t)$, where $\tilde{x}(t)$ obeys $\frac{\mathrm{d}\tilde{x}(t)}{\mathrm{d}t} = A\tilde{x}(t)$, $\tilde{x}(0) = x_0$, and thus $\Delta x(0) = 0$, while x and $\tilde{x}$ share the same dynamics. Hence, for this specific example with $x(0) = 0$, $A = -K$, $B = 1$, $C = K$, and $D = 0$, we obtain $G(s) = \frac{K}{s+K}$.

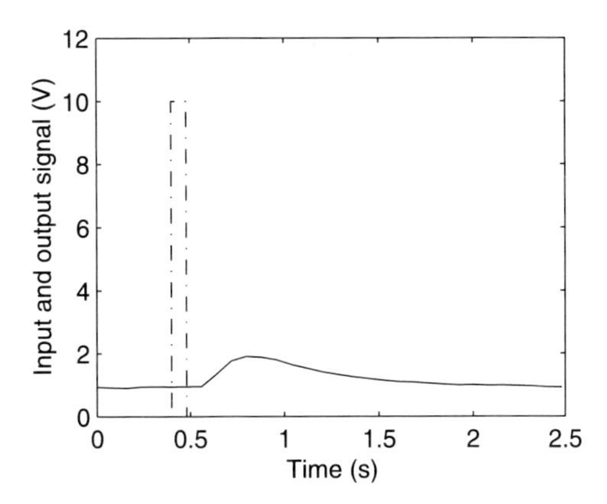

Fig. 2.2 Heating system: pulse input (*dash-dotted line*) at $t = 0.4$ s and measured output (*solid line*)

where $v(t)$ represents the measurement noise of the output signal. Consequently, an estimate of the impulse function, or better the unit-pulse response, is

$$\widehat{g}(t) = \frac{y(t)}{\alpha} \tag{2.8}$$

and the estimation errors are $v(t)/\alpha$. The main advantage of the method is its simplicity, but there are some severe restrictions. Commonly, the estimated unit-pulse response describes the sampled behavior of the continuous-time system. Thus the unit-pulse response may miss significant fast dynamics when the sampling interval is chosen too large, or it may miss the slow dynamics when the duration of the experiment is too small. If dead time (pure delay) is present, it can only be determined within one sampling period. However, its main weakness is that α is limited in practice, which usually prevents a significant reduction of the measurement noise in the estimates, since the estimation errors are inversely proportional with the value of α.

Example 2.2 Heating system: The following pulse response has been measured at a simple lab-scale heating system (see Fig. 2.2). The input of the system is the voltage applied to the heating element. The output is measured with a thermistor. Hence, the output is also in volts. The maximum allowable magnitude of the input is 10 V, and the sampling interval is 0.08 s. To avoid unwanted effects of the initial condition of the system, the pulse input has been applied at $t = 0.4$ s.

The smooth initial curvature in the impulse response indicates that the system is approximately second-order with dead time. Notice from Fig. 2.2 that the dead time is approximately 0.2 s, that is, two to three sampling intervals. After removing the steady-state value, the impulse response coefficients can be directly computed from (2.8).

Consequently, for the identification of LTI systems described by convolution models, the following algorithm can be used.

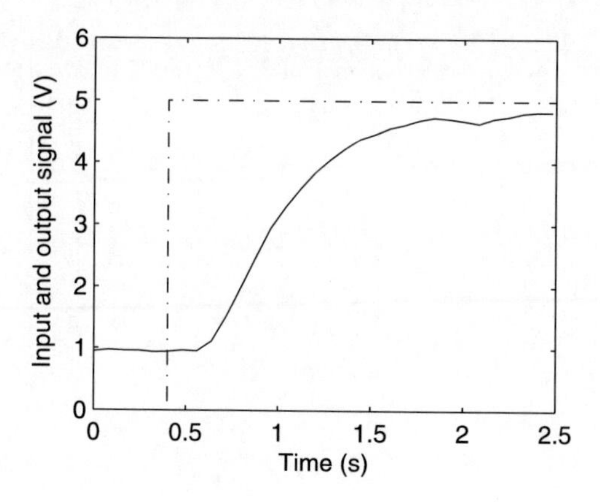

Fig. 2.3 Heating system: step input (*dash-dotted line*) starting at $t = 0.4$ s and measured output (*solid line*)

Algorithm 2.1 Identification of $g(t)$ from a pulse input

1. Generate a pulse with maximum allowable magnitude, α
2. Apply this pulse to the system
3. Use (2.8) to determine estimates of the components of the impulse response $g(t)$

2.2 Step Response

2.2.1 *Direct Step Response Identification*

A step can be considered as an indefinite succession of contiguous, equal, short, rectangular pulses. Hence, in a similar way as the pulse input, the step input is specified as

$$u(t) = \begin{cases} 0, & t < 0 \\ \alpha, & t \geq 0 \end{cases} \tag{2.9}$$

Example 2.3 Heating system: The effect of applying a step input to the lab-scale heating system can be seen in Fig. 2.3.

Analysis of the step response reveals again that the system is approximately second-order with a dead time of about 0.2 s. For a further analysis of the system, which can be easier obtained from the step response (Fig. 2.2) than from the pulse response (Fig. 2.3), we neglect the second-order dynamics in the graph of the step response. Hence, the dominant time constant, thus neglecting the smooth initial curvature in the step response, can be found by extrapolating the initial slope to the steady-state value. The time intercept is the time constant and is approximately equal to 0.4 s. The static gain is found by dividing the difference between the steady-state values of the output by the difference between the steady-state values of the input, i.e., $(4.8 - 1.0)/(5 - 0) = 0.76$ V/V. Recall that this information about dead time, dominant time constant, and static gain is sufficient for the tuning of PID controllers using the famous Ziegler–Nichols tuning rules (see, for instance, [GGS01]). However, in the design of some predictive controllers for linear systems, as the Dynamic Matrix Controller (DMC), all the step response coefficients are used.

2.2.2 *Impulse Response Identification Using Step Responses*

Applying the step input of (2.9) to an LTI system described by (2.5) gives

$$y(t) = \alpha \sum_{k=0}^{t} g(k) + v(t) \tag{2.10}$$

Since $y(t-1) = \alpha \sum_{k=0}^{t-1} g(k) + v(t-1)$, estimates of $g(t)$ can be found by taking differences in the step response

$$\widehat{g}(t) = \frac{y(t) - y(t-1)}{\alpha} \tag{2.11}$$

with corresponding error equal to $[v(t) - v(t-1)]/\alpha$. Since differentiation amounts to filtering with a gain proportional to the frequency, differentiation of a noisy step response will generally lead to unacceptable estimates of the impulse response coefficients. Hence, the suggestion is to make α as large as possible.

Summarizing, if for the identification of an LTI system, an impulse input cannot be applied, a step input can be chosen using the following algorithm.

Algorithm 2.2 Identification of $g(t)$ from a step input

1. Generate a step with maximum allowable magnitude, α
2. Apply this step to the system
3. From the step response the dead time, dominant time constant, and static gain can be graphically determined
4. Use (2.11) to determine estimates of the components of the impulse response $g(t)$

However, as stated before, if the goal is to obtain some basic response characteristics, such as dead time, dominant time constant, and static gain, analysis of step responses suffices.

Example 2.4 Heating system: The reconstruction of the impulse response from the previously presented step response (Fig. 2.3), using Algorithm 2.2, shows the following result (see Fig. 2.4). For comparison, the measured impulse response (—) is plotted in Fig. 2.4 as well.

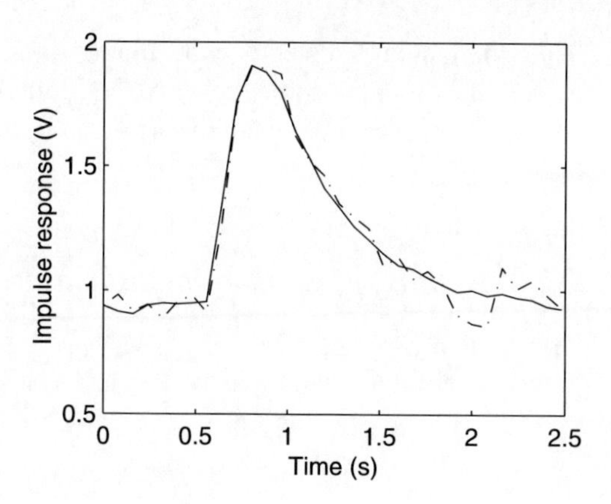

Fig. 2.4 Measured (*solid line*) and reconstructed (*dash-dotted line*) impulse response

2.3 Sine-wave Response

2.3.1 Frequency Transfer Function

Another elementary signal that can identify LTI systems is the sine-wave, which is specified as

$$u(t) = \alpha \sin \omega t \tag{2.12}$$

Before analyzing the output, we must first introduce the frequency transfer function or frequency function, $G(j\omega)$ with j the complex number. This frequency function is the Fourier transform (see Appendix C) of $g(t)$, which can be found by simply substituting $j\omega$ for s in the transfer function $G(s)$. For *sampled systems*, instead of the Laplace or Fourier transform, the discrete Fourier transform (DFT) of $g(t)$ has to be used, that is,

$$G(\mathrm{e}^{j\omega}) = \sum_{t=-\infty}^{\infty} g(t)\mathrm{e}^{-j\omega t} \tag{2.13}$$

The DFT can be interpreted as a discrete version of the Fourier transform.

2.3.2 Sine-wave Response Identification

Recall that $\sin \omega t = \mathrm{Im}(\mathrm{e}^{j\omega t})$. Since $G(\mathrm{e}^{j\omega})$ is a complex number, it can be written as $|G(\mathrm{e}^{j\omega})|\mathrm{e}^{j\phi}$, where $|G(\cdot)|$ indicates the magnitude and $\phi = \arg(G(\cdot))$. Hence, using (2.5) with $k = -\infty, \ldots, \infty$, the sine-wave input gives an output

$$y(t) = \alpha \sum_{k=-\infty}^{\infty} g(k)\,\mathrm{Im}\big(\mathrm{e}^{j\omega(t-k)}\big) = \alpha\,\mathrm{Im} \sum_{k=-\infty}^{\infty} g(k)\mathrm{e}^{-j\omega(t-k)}$$

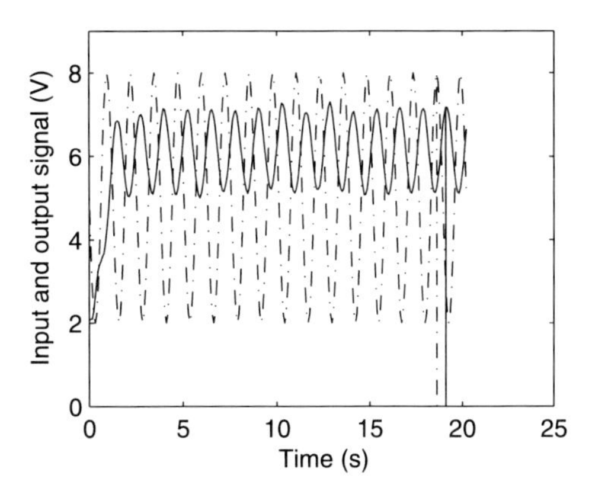

Fig. 2.5 Heating system: sine-wave input (*dash-dotted line*) and output (*solid line*)

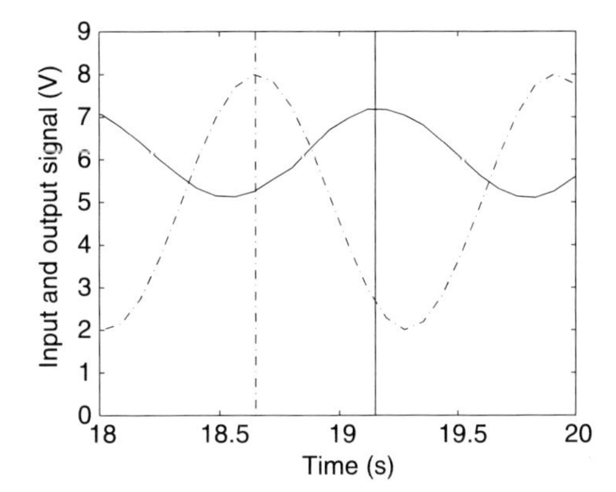

Fig. 2.6 Heating system: snapshot of sine-wave input (*dash-dotted line*) and output (*solid line*)

$$
\begin{aligned}
&= \alpha \operatorname{Im}\left\{ \mathrm{e}^{j\omega t} \sum_{k=-\infty}^{\infty} g(k)\mathrm{e}^{-j\omega k} \right\} = \alpha \operatorname{Im}\left\{ \mathrm{e}^{j\omega t} G\left(\mathrm{e}^{j\omega}\right)\right\} \\
&= \alpha \left| G\left(\mathrm{e}^{j\omega}\right)\right| \sin(\omega t + \phi)
\end{aligned}
\tag{2.14}
$$

Consequently, the output is a sine-wave of the same frequency of $u(t)$, but multiplied in magnitude by $|G(\mathrm{e}^{j\omega})|$ and shifted in phase by ϕ. Notice that the result implies that the input is an everlasting sine-wave, which can never be true in practice. Therefore, if it is assumed that $u(t) = 0,\ t < 0$, an initial transient must be accepted in the response. In general, a convenient way to deal with this is neglecting the first part of the response, which is also demonstrated by the following example.

Example 2.5 Heating system: The effect of a sine-wave input signal with a frequency of 5 rad/s on the system output is presented in the following figure (see Fig. 2.5).

The magnitude and phase shift of the frequency function at 5 rad/s is determined at the end of the signal (see Fig. 2.6 for the details). The gain $|G(\mathrm{e}^{j\omega})|$ for $\omega =$ 5 rad/s, is 0.256 V/V, and the phase shift $\phi = -\omega \Delta t = -5 \times 0.50 = -2.50$ rad.

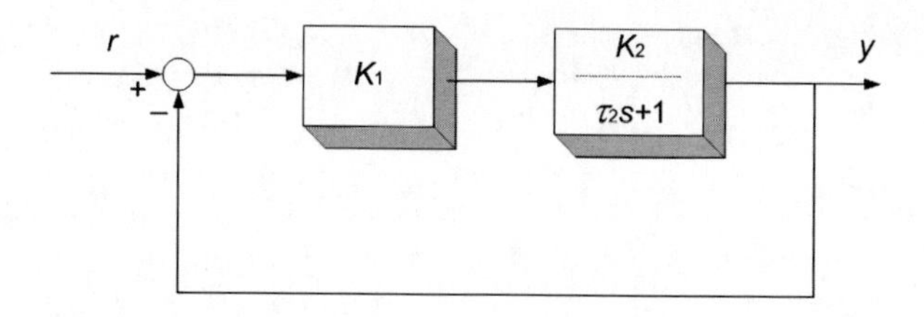

Fig. 2.7 Schematic presentation of closed-loop system under P-control

Since from the signals individual points were taken, this result is very sensitive to noise in both signals, especially at extreme values.

2.4 Historical Notes and References

The methods in this chapter have already a long history with applications on especially electrical and mechanical systems. A general overview of the class of non-parametric identification methods have been given by [Rak80, Wel81]. In particular, impulse response identification has attained a lot of attention in the past and also in recent years [FBT96, SC97, YST97, GCH98, TOS98, SL03, DDk05]. The step response is important in many industrial applications and especially in relation with PID controller tuning. Step response identification methods have been covered by [MR97, WC97]. Sine-wave response identification in the time domain has not received too much attention. Its relevance is much higher in the frequency domain, as we will see in the next chapter.

The more experienced readers, with a background in systems and control theory, may miss the behavioral model representation of Willems [Wil86a, Wil86b, Wil87] in Fig. 2.1. This model representation (see also [PW98]) is out of the scope of this book, as it is too advanced for this introductory text. Nevertheless, the behavioral approach is of interest for further research and application in the system identification field, see, for instance, [JVCR98, JR04].

2.5 Problems

Problem 2.1 In practice we often have to deal with feedback control systems. For instance, in process industry it frequently occurs that a process is controlled by feedback. A schematic example of a first-order system under simple proportional feedback is presented in Fig. 2.7.

On the basis of a priori knowledge of the systems' behavior (see Fig. 2.7), different types of representation will be investigated.

(a) Give the transfer function from (reference) input r to output y.
(b) Give the (set of) differential equation(s) of this system on the basis of the transfer functions presented in the figure.

(c) Derive from the overall transfer function the impulse response of this system, analytically using the inverse Laplace transform (MATLAB: *ilaplace*). Explain/interpret your result.
(d) Represent the system in terms of its convolution or impulse response model.
(e) Plot the unit step response for $K_1 = 1$, $K_2 = 2$, and $\tau_2 = 0.5$ hours. Explain your result.
(f) Represent this system in state-space form.

Problem 2.2 Consider the storage tank example (Example 1.4) with $K = 0.8$.

(a) Define the system (*sys1*) in state-space form using the MATLAB command *ss*.
(b) Define the system (*sys2*) in transfer function form using the MATLAB command *tf*.
(c) Check both representations with the commands *ss2tf* and *tf2ss*.
(d) For this system, determine the impulse response $g(t)$ using the MATLAB command *impulse*.
(e) Determine the step response (y) as well, using the MATLAB command *step*.
(f) Differentiate the step response using the command *diff*. Note: perform a scaling of the differentiated response (yd) by multiplying it with $g(1)/yd(1)$ and add a zero (why?). Plot both impulse responses.
(g) Generate a step input using *zeros* and *ones*. Use the command *lsim* to calculate the corresponding output. Plot the result and explain the result.

Problem 2.3 Let us evaluate the sine-wave response in some more detail. Consider, for this purpose, the system with transfer function

$$G(s) = \frac{2}{10s + 1}$$

(a) Define the system in MATLAB
(b) Generate and plot a sine-wave signal with a user-defined frequency.
(c) Determine the sine-wave response using *lsim* and plot it together with the input in one figure. Interpret the result.

Problem 2.4 Investigate the effects of a nonideal input in an impulse response test by plotting the response of the system with impulse response

$$g(t) = \exp(-t) - \exp(-5t)$$

to a rectangular pulse input of unit area and duration (i) 0.1, (ii) 0.2, and (iii) 0.5. Compare each response with $g(t)$ (after [Nor86]).

Chapter 3
Frequency Response Methods

3.1 Empirical Transfer-function Identification

3.1.1 Sine Wave Testing

From Sect. 2.3.2 the following algorithm for the identification of the frequency function can be deduced.

Algorithm 3.1 Identification of $G(e^{j\omega})$ from sine waves

1. Generate for a specific frequency a sine-wave with maximum allowable magnitude.
2. Apply this sine wave to the system.
3. Record the resulting sine-wave response.
4. Determine magnitude and phase shift of $G(e^{j\omega})$ for the specific frequency from the two signals.
5. Repeat this for a number of interesting frequencies $\omega \in \{\omega_1, \omega_2, \dots, \omega_N\}$.

As mentioned in the previous chapter, the complex-valued function $G(e^{j\omega})$, $-\pi \leq \omega \leq \pi$, is called the frequency transfer function, or in short the frequency function , of a discrete-time LTI system. The frequency function is used in many frequency domain methods for controller design. Consequently, there has always been much interest in the direct identification of the frequency function from the data. The previously described procedure for the identification of the frequency function using a single frequency sine-wave at a time, also called sine-wave testing, is one of the simplest methods. However, this procedure may be time-consuming. As we will see in what follows, the frequency function can also be reconstructed on the basis of multifrequency inputs. Therefore, we first have to introduce the Discrete Fourier transform of signals.

K.J. Keesman, *System Identification*,
Advanced Textbooks in Control and Signal Processing,
DOI 10.1007/978-0-85729-522-4_3,

3.1.2 Discrete Fourier Transform of Signals

The Discrete Fourier Transform (DFT) of the signal $y(t)$, sampled at $t = 1, 2, \ldots, N$, is given by

$$Y_N(\omega) = \frac{1}{\sqrt{N}} \sum_{t=1}^{N} y(t) \mathrm{e}^{-j\omega t} \tag{3.1}$$

where $\omega = 2\pi k/N,\ k = 1, 2, \ldots, N$. Notice that for a specific k, N/k is the period associated with the specific frequency ω_k. Similarly, the DFT of $u(t)$ can be found. The absolute square value of $Y(\omega_k)$, $|Y(2\pi k/N)|^2$, is a measure of the energy contribution of this frequency to the energy of the signal. The plot of values of $|Y(\omega)|^2$ as a function of ω is called the periodogram of the signal $y(t)$.

Example 3.1 Sine-wave signal: Consider the signal

$$y(t) = A \cos \omega_0 t$$

where $A \in \mathbb{R}$ and $\omega_0 = 2\pi/N_0$ for some integer $N_0 > 1$. Let N be a multiple of N_0 such that $N = mN_0$, and let us consider the time instants $t = 1, 2, \ldots, N$. Since

$$\cos \omega_0 t = \frac{1}{2}\left[\mathrm{e}^{j\omega_0 t} + \mathrm{e}^{-j\omega_0 t}\right]$$

after substitution of the expression into (3.1) we find

$$Y_N(\omega) = \frac{1}{\sqrt{N}} \sum_{t=1}^{N} \frac{A}{2}\left[\mathrm{e}^{j(\omega_0-\omega)t} + \mathrm{e}^{-j(\omega_0+\omega)t}\right]$$

This expression can be simplified using the following relationship:

$$\frac{1}{N} \sum_{k=1}^{N} \mathrm{e}^{j2\pi(nk/N)} = \begin{cases} 1, & n = 0 \\ 0, & 1 \le n < N \end{cases}$$

so that

$$\left|Y_N(\omega)\right|^2 = \begin{cases} N\,\frac{A^2}{4} & \text{if } \omega = \pm\omega_0 = \frac{2\pi}{N_0} = \frac{2\pi m}{N} \\ 0 & \text{if } \omega = \frac{2\pi k}{N}, k \neq m \end{cases}$$

Hence, the periodogram has two spikes, at frequencies $\omega = -\omega_0$ and $\omega = \omega_0$, on the interval $[-\pi, \pi]$. Figure 3.1 presents the periodogram of the signal $y(t) = \cos(\omega_0 t)$ with $\omega_0 = 2$, $t = 1, \ldots, N$ and $N = 629$.

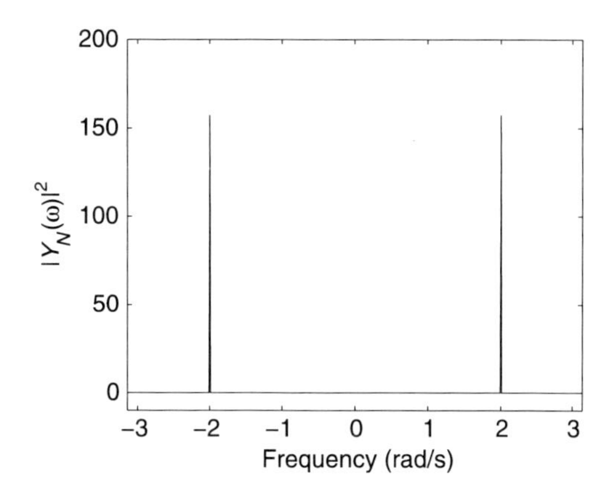

Fig. 3.1 Periodogram of the signal $y(t) = \cos(2t)$

3.1.3 Empirical Transfer-function Estimate

Recall from (2.4) that $Y(s) = G(s)U(s)$, so that after substitution of $s = j\omega$ we obtain the relationship

$$Y(j\omega) = G(j\omega)U(j\omega) \tag{3.2}$$

which can also be derived after Fourier transforming the convolution model (2.2). This type of algebraic relationship also holds for sampled systems. Hence, for a given input $u(t)$ and an output signal $y(t)$, $t = 1, 2, \ldots, N$, and after taking the DFT of both $u(t)$ and $y(t)$, the following estimate of the transfer function can be found:

$$\widehat{G}\left(e^{j\omega}\right) = \frac{Y_N(\omega)}{U_N(\omega)} \tag{3.3}$$

This estimate is indicated as the Empirical Transfer-Function Estimate (ETFE), and the expression also holds for the case where the input is not a single sine-wave. In fact, both U_N and Y_N are series expansions of the input and output signals in terms of sines and cosines. Thus, roughly speaking, for each of the frequencies contained in $u(t)$ and $y(t)$, the relationship of (2.14) holds, which allows the reconstruction of both the magnitude and phase shift of the frequency function for a number of frequencies. In order to avoid the effect of the initial conditions, in practice one often removes the first numbers of the input and output data vectors. The DFT of these modified data vectors again provides vectors that after component-wise division give the estimates of $G(e^{j\omega})$ for $\omega = \frac{2\pi}{N}, \ldots, \pi$ rad/s, as in (3.3). Let us demonstrate the application of the MATLAB function *etfe* by the following example.

Example 3.2 ETFE: Let a binary input signal $u(t)$ with expanding pulses, to facilitate the estimation of the static gain via visual inspection, produce the following output data; see Table 3.1.

From a first visual inspection of Fig. 3.2 we notice that the system is approximately first-order with unit time delay, since the output follows the input after one

Table 3.1 Input–output data

$u(t)$	0	1	0	0	1	1	0	0	0	1	1	1	0	0	0	0
$y(t) \cdot 10^2$	4.50	0	87.53	11.56	5.50	89.30	97.76	8.47	5.01	0	87.65	101.09	103.97	15.88	0	0

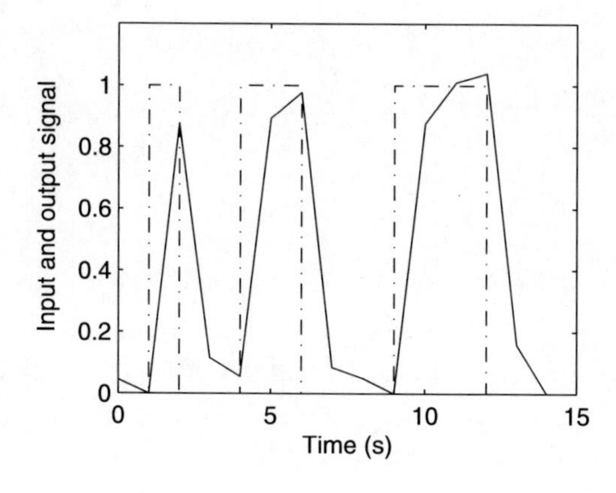

Fig. 3.2 Graphical presentation of input (*dash-dotted line*) and output (*solid line*) signals

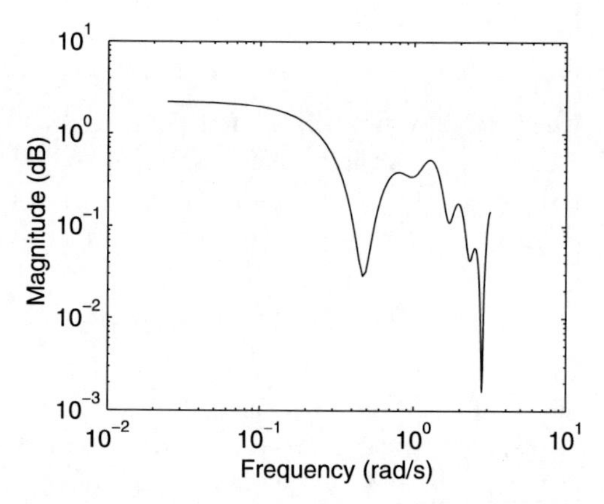

Fig. 3.3 Periodogram of output signal

sampling interval. Furthermore, the dominant time constant is approximately 0.5 s, and the static gain is close to one.

The periodogram of the output signal, using the MATLAB function *etfe* which evaluates the output vector at 128 equally spaced frequencies between 0 (excluded) and π, is presented in Fig. 3.3.

The Bode plot (see Appendix D), presenting magnitudes (log-scale) and phase shifts (linear scale) as a function of the frequency (log-scale), is a useful tool for graphical evaluation of the frequency function. Again *etfe* is used but now for the estimation of the empirical transfer function (3.3). The results are plotted in a Bode plot (see Figs. 3.4 and 3.5).

It can be easily verified from the magnitude plot that the static gain is approximately equal to 1 and that at high frequencies no useful information about the system dynamics can be obtained due to the significant presence of high-frequency noise

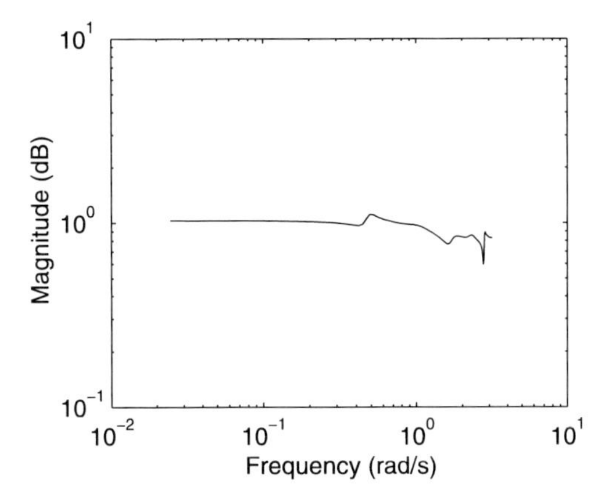

Fig. 3.4 Magnitude plot of empirical transfer function estimate

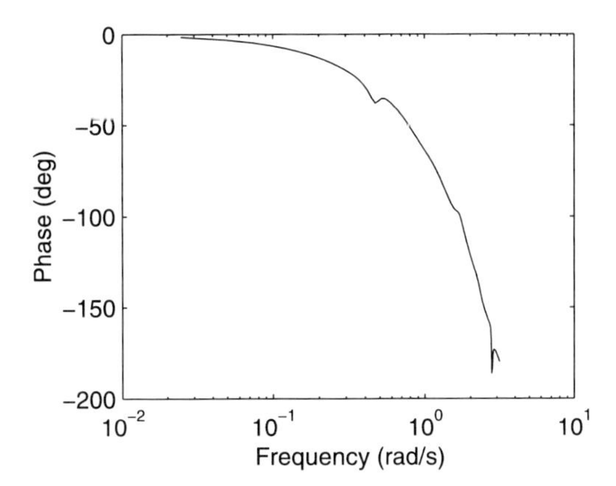

Fig. 3.5 Phase plot of empirical transfer function estimate

components in the output data. Again this result is very sensitive to noise, especially at those frequencies which coincide with the dominant noise frequencies.

Recall that $\widehat{G}(\mathrm{e}^{j\omega})$ for $-\pi \leq \omega \leq \pi$ is an estimate of the discrete Fourier transform of the impulse function. Hence, an estimate of the impulse response can be recovered in theory from $\widehat{G}(\mathrm{e}^{j\omega})$, by an inverse Fourier transformation. In practice, however, the Bode plot is analyzed in terms of some well-defined, elementary frequency responses, such as first- or second-order and pure time delays, see Appendix D for details.

Hence, we can deduce the following algorithm for the identification of $G(\mathrm{e}^{j\omega})$ from input–output data.

Algorithm 3.2 Identification of $G(\mathrm{e}^{j\omega})$ from input–output data

1. Generate an arbitrary input signal $u(t)$, $t = 1, 2, \ldots, N$.
2. Apply this input signal to the system, assuming a zero-order hold (ZOH) on the inputs.
3. Record the input $u(t)$ and corresponding output signal $y(t)$.

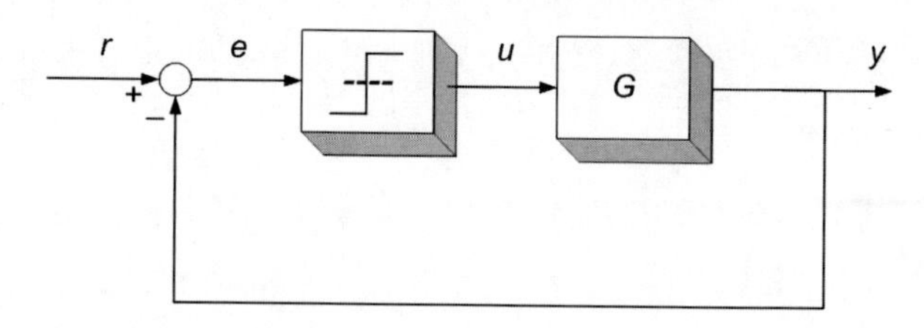

Fig. 3.6 Conventional relay feedback system

4. Take the DFT of both $u(t)$ and $y(t)$, resulting in $U_N(\omega)$ and $Y_N(\omega)$, respectively.
5. Divide component-wise $Y_N(\omega)$ by $U_N(\omega)$ for $\omega = \frac{2\pi}{N}, \ldots, \pi$ rad/s to obtain an estimate of $G(e^{j\omega})$.
6. Optionally, use elementary frequency responses to get an estimate of the transfer function $G(s)$ (see Appendix D).

Notice that for a given (disturbance) input signal, we can directly start from step 3.

3.1.4 Critical Point Identification

For the automatic tuning of PID controllers for simple systems, however, it often suffices to have an estimate of the critical point on the Nyquist curve. As opposed to the Bode plot, the Nyquist plot is a single graph in polar coordinates in which the gain and phase of a frequency response are plotted. This plot shows the phase as the angle and the magnitude as the distance from the origin, and thus it combines the two types of Bode plot on a single graph with frequency as a parameter along the curve. Hence, the critical point consists of a critical frequency and a critical gain.

Nowadays, the relay identification experiment for the estimating the critical point is one of the most popular methods in process control. The key idea behind this identification experiment is that many industrial processes exhibit stable limit cycle oscillations for a relay feedback system. A conventional relay feedback system for a process with transfer function $G(s)$ is presented in Fig. 3.6.

For the estimation of the critical point, most often the so-called describing function method is applied. In the describing function method the relay is replaced by an "equivalent" LTI system, which will be derived in the following. Let in the self-oscillation mode of the overall feedback system the system oscillate with the period T_{osc}. For the derivation of the describing function, a sinusoidal relay input $e(t)$ is considered. Let this input be given by

$$e(t) = \alpha \sin \omega t \tag{3.4}$$

Consequently, the relay output $u(t)$ is a square wave with frequency ω and an amplitude h, which is equal to the relay output level. Using a Fourier series expansion in terms of sines and cosines, $u(t)$ can be written as

$$u(t) = \frac{4h}{\pi} \sum_{n=0}^{\infty} \frac{\sin(2n+1)\omega t}{2n+1} \tag{3.5}$$

The describing function of the relay, denoted by $N(\alpha)$, is simply the complex ratio of the fundamental component of $u(t)$, for $n = 0$, to the sinusoidal relay input, that is,

$$N(\alpha) = \frac{4h}{\pi\alpha} \tag{3.6}$$

Hence, the describing function ignores the harmonics beyond the fundamental component

$$\omega_{\text{osc}} = \frac{2\pi}{T_{\text{osc}}} \tag{3.7}$$

Let $G(s)$ denote the transfer function of the process, as in Fig. 3.6. Then, for $r = 0$ and with the system in the self-oscillating mode, we have

$$e = -y \tag{3.8}$$

$$u = N(\alpha)e \tag{3.9}$$

$$y = G(j\omega_{\text{osc}})u \tag{3.10}$$

and thus,

$$G(j\omega_{\text{osc}}) = -\frac{1}{N(\alpha)} \tag{3.11}$$

The critical point of a linear system is found from the intersection of the Nyquist curve of $G(j\omega)$ and $-\frac{1}{N(\alpha)}$ in the complex plane. Hence, the critical point is given by $(\omega_{\text{osc}}, \frac{4h}{\pi\alpha})$. The critical point can be identified using the following algorithm.

Algorithm 3.3 Identification of the critical point using a relay experiment

1. Implement a relay feedback loop with amplitude of the relay element h around the process (see Fig. 3.6).
2. Start the feedback system and wait until it is in its self-oscillation mode.
3. Measure the output signal $y(t)$.
4. Derive from $y(t)$ the oscillating frequency ω_{osc} and the amplitude α.
5. Evaluate $(\frac{4h}{\pi\alpha})$ to obtain the critical point $(\omega_{\text{osc}}, \frac{4h}{\pi\alpha})$.

The fundamental assumption of the describing function method, also known as the filtering hypothesis, is that the amplitudes of the third, fifth, and higher harmonics are much smaller than that of the fundamental component. In addition to this, the conventional relay method is not directly applicable to certain classes of processes, as those with a large dead time or nonminimum phase (NMP) processes. Moreover, it is not able to extract other points of the process frequency response.

Let us demonstrate the critical point identification method by an FOPDT (first-order plus dead time) example.

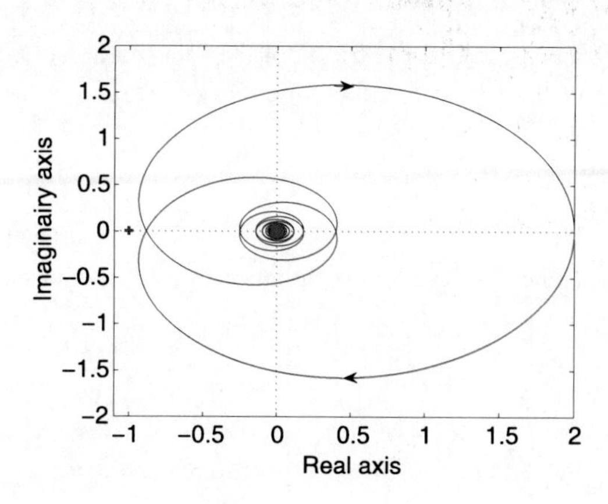

Fig. 3.7 Nyquist plot of FOPDT process with $\tau = 1$

Table 3.2 Critical point identification

τ	Real process		Relay experiment	
	K_c	ω_c	K_c	ω_c
0.5	1.903	3.673	1.640	3.740
1	1.131	2.029	1.012	2.114
5	0.566	0.531	0.551	0.641
10	0.520	0.286	0.637	0.293

Example 3.3 FOPDT process: Let the transfer function of an FOPDT process be given by

$$G(s) = \frac{2}{s+1} \mathrm{e}^{-\tau s} \tag{3.12}$$

For $\tau = 1$, the corresponding Nyquist plot is presented in Fig. 3.7.

The critical points (ω_c, K_c), related to the "real" process and found from the relay experiment, are presented in Table 3.2. Hence, reasonable estimates are found from a relay experiment.

3.2 Discrete-time Transfer Function

3.2.1 *z-Transform*

Recall that $G(\mathrm{e}^{j\omega})$ is the discrete Fourier transform (DFT) of $g(t)$. In other words, the complex number $G(\mathrm{e}^{j\omega})$ is the transfer function of a sampled system evaluated at the point $z = \mathrm{e}^{j\omega}$. As can be seen in the previous chapter, this number gives full information as to what will happen under stationary conditions, when the input is a sine-wave of frequency ω. In general, the transfer function for discrete-time systems is defined as

$$G(z) := \sum_{k=0}^{\infty} g(k)z^{-k} \tag{3.13}$$

which is the z-transform (see Appendix C) of the impulse response $g(t)$. Similarly, the z-transform of the sampled data vectors $u(t)$ and $y(t)$ is defined as $U(z) := \sum_{k=0}^{\infty} u(k)z^{-k}$ and $Y(z) := \sum_{k=0}^{\infty} y(k)z^{-k}$, respectively. Substitution of the convolution sum (2.5) in $Y(z)$ leads to

$$\begin{aligned} Y(z) &= \sum_{k=0}^{\infty} \sum_{l=0}^{k} g(k-l)u(l)z^{-k} \\ &= \sum_{l=0}^{\infty} \left(\sum_{k=l}^{\infty} g(k-l)z^{-(k-l)} \right) u(l)z^{-l} \\ &= G(z)U(z) \end{aligned} \tag{3.14}$$

Clearly, (3.14) is the discrete-time counterpart of (2.4).

3.2.2 Impulse Response Identification Using Input–output Data

Writing (3.14) as

$$Y(z) = y(0) + y(1)z^{-1} + y(2)z^{-2} + \cdots \tag{3.15}$$

and

$$\begin{aligned} Y(z) = g(0)u(0) &+ \left[g(1)u(0) + g(0)u(1)\right]z^{-1} \\ &+ \left[g(2)u(0) + g(1)u(1) + g(0)u(2)\right]z^{-2} + \cdots \end{aligned} \tag{3.16}$$

and collecting all corresponding terms, we directly find that

$$y(0) = g(0)u(0), \qquad y(1) = g(1)u(0) + g(0)u(1), \ldots$$

so that $g(0), g(1), g(2), \ldots$ can be solved successively from the input–output data.

The algorithm for the direct estimation of $g(t)$ from input–output data is given by the following:

Algorithm 3.4 Identification of $g(t)$ from input–output data

1. Generate an arbitrary input signal $u(t)$, $t = 1, 2, \ldots, N$.
2. Measure the input $u(t)$ and corresponding output signal $y(t)$.
3. Solve successively $g(0), g(1), g(2), \ldots$ from $y(0) = g(0)u(0)$, $y(1) = g(1)u(0) + g(0)u(1), \ldots$ using (3.16).

Recall that we can also start directly from step 3 if the input–output data is given.

Example 3.4 Impulse response identification: In contrast to the preceding procedure, we are also able to reconstruct the unit-pulse response $g(t)$ directly from any observed input–output data set using the expression for the convolution sum (2.5). Recall from (2.5) that

$$y(t) = \sum_{k=0}^{t} g(k)u(t-k), \quad t \in \mathbb{Z}^+ \tag{3.17}$$

Let furthermore both the inputs $u(0), u(1), \ldots, u(N)$ and corresponding outputs $y(0), y(1), \ldots, y(N)$ be recorded. Substituting the input values into the convolution sum (3.17) and assuming that the input is zero before time zero gives

$$y(t) = g(0)u(t) + g(1)u(t-1) + g(2)u(t-2) + \cdots$$

so that

$$\begin{aligned}
&\text{for } t = 0: \quad y(0) = g(0)u(0) \\
&\text{for } t = 1: \quad y(1) = g(0)u(1) + g(1)u(0) \\
&\text{for } t = 2: \quad y(2) = g(0)u(2) + g(1)u(1) + g(2)u(0) \\
&\quad\vdots
\end{aligned}$$

Consequently, in matrix form we obtain

$$\begin{bmatrix} y(0) \\ y(1) \\ y(2) \\ \cdot \\ \cdot \\ y(N) \end{bmatrix} = \begin{bmatrix} u(0) & 0 & \cdot & \cdot & \cdot & 0 \\ u(1) & u(0) & 0 & \cdot & \cdot & 0 \\ u(2) & u(1) & u(0) & 0 & \cdot & 0 \\ \cdot & \cdot & \cdot & & & \cdot \\ \cdot & \cdot & \cdot & & & 0 \\ u(N) & u(N-1) & \cdot & u(2) & u(1) & u(0) \end{bmatrix} \begin{bmatrix} g(0) \\ g(1) \\ g(2) \\ \cdot \\ \cdot \\ g(N) \end{bmatrix}$$

from which the elements $g(0), g(1), \ldots, g(N)$ can be solved successfully if the matrix with the inputs is invertible. Usually, for asymptotically stable systems and if N has been chosen large enough, it suffices to determine only the first $s < N$ elements of the unit-pulse response, as $g(t)$ for $t > s$ is close to zero. Hence, if $s \ll N$, this choice may decrease the dimensions of vectors and matrix significantly. Notice here that by setting $u(t)$ equal to a unit pulse, that is, $u(0) = 1$ and $u(t) = 0$ for $t \neq 0$, we directly find the unit-pulse response coefficients $g(t)$. However, in this special case and also in a more general case, the presence of noise may spoil the idea, since output noise directly affects the unit-pulse response coefficients.

At this point it should be noted that z^{-1} in discrete-time cases can be interpreted as a compressed notation for e^{-sT_s}, where T_s is the sampling interval. Recall that e^{-sT_s} is the Laplace transform of a unit time delay, and thus z^{-1} can be interpreted as the delay operator. For simplicity and from an operational point of view, under the assumption of zero initial conditions on $y(t)$ and $u(t)$, in what follows the *forward shift operator* q (see Appendix E) with

$$qu(t) = u(t+1)$$

and the *backward shift operator* q^{-1}: $q^{-1}u(t) = u(t-1)$ will be used instead of the complex variable z. Consequently, similar to the introduction of the differential operator in system descriptions (2.1), the convolution sum (2.5) can be written as

$$y(t) = G(q)u(t) \tag{3.18}$$

where $G(q) = \sum_{k=0}^{\infty} g(k)q^{-k}$, an infinite polynomial in q^{-1}, which in the sequel will be called the transfer function of a discrete-time LTI system.

3.2.3 *Discrete-time Delta Operator*

Given the interpretation of z and the introduction of the forward shift operator q in the previous subsection, it is a small step to approximate a derivative in terms of q. Thus,

$$\begin{aligned}\frac{\mathrm{d}y}{\mathrm{d}t} &\approx \frac{y(t+1)-y(t)}{T}\\ &= \frac{(q-1)}{T}y(t)\\ &= \delta y(t)\end{aligned}$$

where $\delta := \frac{(q-1)}{T}$ is the so-called delta operator, also indicated as the δ-operator. The δ-operator allows a unified treatment of continuous-time and discrete-time systems, since, as $T \to 0$, a discrete-time system in the δ-operator form smoothly converges to a system in continuous-time. Let us illustrate this by Example 1.4.

Example 3.5 Storage tank: Recall that the system in input–output form is given by

$$\frac{1}{K}\frac{\mathrm{d}y(t)}{\mathrm{d}t} + y(t) = u(t)$$

A discrete-time approximation in q, using the Euler backward method, leads to

$$\begin{aligned}&\frac{1}{K}\frac{(q-1)}{T}y(t) + y(t) = u(t)\\ \Longrightarrow\quad &(q-1+KT)y(t) = KTu(t)\\ \Longrightarrow\quad &y(t) = \frac{KT}{q-1+KT}u(t)\end{aligned}$$

With $\delta := \frac{(q-1)}{T}$, we obtain

$$\frac{1}{K}\delta y(t) + y(t) = u(t)$$

$$\Longrightarrow \quad y(t) = \frac{K}{\delta + K} u(t)$$

which shows a transfer function in δ with a similar structure as the transfer function of the continuous-time transfer function $G(s)$.

The main advantage of using the δ-operator formulation is that it shows better numerical properties and causes fewer conditioning problems than the conventional shift-operator form. Consequently, it may be worthwhile to investigate alternative discrete-time operators, for instance, based on the Euler forward method or central difference approximations.

3.3 Historical Notes and References

In addition to step response identification methods, sine-wave testing on SISO LTI systems [BMS+04, Har91, Fre80], as presented in Sect. 3.1.1, is also very popular in industry. In the process industry, and in particular for the auto-tuning of PID controllers, the identification of the critical point of the frequency response using relay feedback [ÅH84, ÅH88, JM05] is very popular. To handle processes with large time delays, noisy data, underdamping, or NMP behavior, several modifications of the conventional relay feedback system, as prefiltered relay, preload relay, and relay with hysteresis, have been suggested [TLH+06, MCS08, LG09]. For an overview of frequency response methods and Fourier techniques for system identification, see [Rak80]. Nowadays, frequency response methods are still frequently applied in, for instance, chemical and hydraulic engineering studies.

As an alternative to the z-transform of discrete-time models, the δ-operator form has been introduced by Middleton and Goodwin, see [MG86, MG90].

3.4 Problems

Problem 3.1 Given experimental data $\{u(0), y(0), u(1), y(1), \ldots, u(N)y(N)\}$. Show that for an LTI system, the following holds:

$$\begin{bmatrix} y(N) \\ \vdots \\ y(1) \\ y(0) \end{bmatrix} = \begin{bmatrix} u(N) & \cdots & u(1) & u(0) \\ \vdots & & u(0) & 0 \\ u(1) & \ddots & \ddots & \vdots \\ u(0) & 0 & \cdots & 0 \end{bmatrix} \begin{bmatrix} g(0) \\ g(1) \\ \vdots \\ g(N) \end{bmatrix}$$

with $g(0), g(1), \ldots, g(N)$ the impulse response coefficients.

Problem 3.2 For an LTI system, the following holds:

$$y(t) = \sum_{k=0}^{t} g(k)u(t-k), \quad t \in \mathbb{Z}^+$$

with impulse response coefficients $g(0), g(1), \ldots, g(N)$. Given this input–output relationship, show that in matrix form the following holds:

$$y = gU$$

with $g = [g(0), g(1), \ldots, g(N)]$ and y, U of appropriate dimensions.

Problem 3.3 Show that for an LTI system, the following holds:

$$y(t) = \sum_{k=0}^{t} g(k)u(t-k) = \sum_{k=0}^{t} g(t-k)u(k), \quad t \in \mathbb{Z}^+$$

Chapter 4
Correlation Methods

4.1 Correlation Functions

4.1.1 Autocorrelation Function

From the previous chapters the conclusion can be drawn that the system response and the frequency response methods are all more or less simple to use. However, the main disadvantage is that the results are sensitive to noise as raw input–output data sets are used. Therefore, in the past, so-called correlation methods have been developed to overcome this noise sensitivity.

In order to arrive at these correlation methods, let us first introduce the *autocorrelation function* $r_{uu}(\tau, t)$ of a signal $u(t)$ (see also Appendix B),

$$r_{uu}(\tau, t) = E\big[u(t)u(t+\tau)\big] \tag{4.1}$$

where τ is the lag time. The notation $E[\cdot]$ stands for the expectation operator, or in other words, it signifies the mean value of the particular function. In what follows, this expectation will always be interpreted as the time average, and with abuse of notation,

$$r_{uu}(\tau) = \lim_{T\to\infty} \frac{1}{2T} \int_{-T}^{T} u(t)u(t+\tau)\,\mathrm{d}t \tag{4.2}$$

Notice that this function is now only a function of lag τ and not of t. Hence, it includes some time-invariance or stationarity property. The integral is taken over the interval $[-T, T]$ with $T \to \infty$, because at this stage transient responses will be excluded. The discrete-time counterpart, applicable to sampled data, is given by

$$r_{uu}(l) = \lim_{N\to\infty} \frac{1}{2N+1} \sum_{i=-N}^{N} u(i)u(i+l) \tag{4.3}$$

Notice that for a finite sampled sequence $u(t)$ with N elements, the sample autocorrelation function $r_{uu}(l)$ can be calculated as $r_{uu}(l) = E[u(i)u(i+l)^T]$, where

K.J. Keesman, *System Identification*,
Advanced Textbooks in Control and Signal Processing,
DOI 10.1007/978-0-85729-522-4_4,

$u(i)$ is the subsequence from $-N$ to $N-l$, and $u(i+l)$ is the subsequence from $-N+l$ to N. In order to obtain a reliable estimate of the autocorrelation function values, the lag l is mostly chosen to be smaller than $N/4$. It can be easily verified that the autocorrelation function includes both negative and positive lags and that it is an even (i.e., symmetric around 0) function. Notice hereto that

$$\begin{aligned} r_{uu}(-l) &= \lim_{N\to\infty} \frac{1}{2N+1} \sum_{i=-N}^{N} u(i)u(i-l) \\ &=_{[j:=i-l]} \lim_{N\to\infty} \frac{1}{2N+1} \sum_{j=-N-l}^{N-l} u(j+l)u(j) \\ &=_{[l\ll N]} \lim_{N\to\infty} \frac{1}{2N+1} \sum_{j=-N}^{N} u(j)u(j+l) \end{aligned} \tag{4.4}$$

which, for $l \ll N$, is equivalent to (4.3). Furthermore, $|r_{uu}(l)| \le r_{uu}(0)\ \forall l \in \mathbb{Z}$, and if $u(t)$ has a periodic component, then $r_{uu}(l)$ has a periodic component as well, as is demonstrated in the following example.

Example 4.1 Sine-wave signal: Consider the sine-wave signal $u(t) = \sin(\omega t)$ with $t \in \mathbb{R}$. Then, after substituting this specific function in (4.2) and applying the goniometric rules,

$$\begin{aligned} \sin\alpha \sin\beta &= \frac{1}{2}\cos(\alpha-\beta) - \frac{1}{2}\cos(\alpha+\beta) \\ \sin(\alpha+\beta) &= \sin\alpha\cos\beta + \sin\beta\cos\alpha \end{aligned}$$

we arrive at the following result:

$$\begin{aligned} r_{uu}(\tau) &= \lim_{T\to\infty} \frac{\cos\omega\tau * (T - \frac{1}{2}\sin 2\omega T)}{2T} \\ &= \frac{1}{2}\cos\omega\tau \end{aligned}$$

which is again a sine function with frequency ω. Let us generate the sampled signal $u(t) = \sin(\frac{2\pi}{32}t)$ on the finite interval $t = 0, 1, \ldots, 128$. Using the MATLAB function *xcorr*, the corresponding sample autocorrelation function is calculated. See Fig. 4.1 with original sine-wave signal and normalized autocorrelation function for a graphical representation of the result. The attenuation of the autocorrelation function with increasing lag is caused by the fact that a finite signal is considered. A finite signal on the interval $[0, N]$ can be considered as the multiplication of the infinite signal and a block function with amplitude 1 and with its basis on $[0, N]$. Since a block function has a triangular autocorrelation function and the superposition principle holds, the amplitude of every autocorrelation function of a finite sequence will decrease with increasing lag.

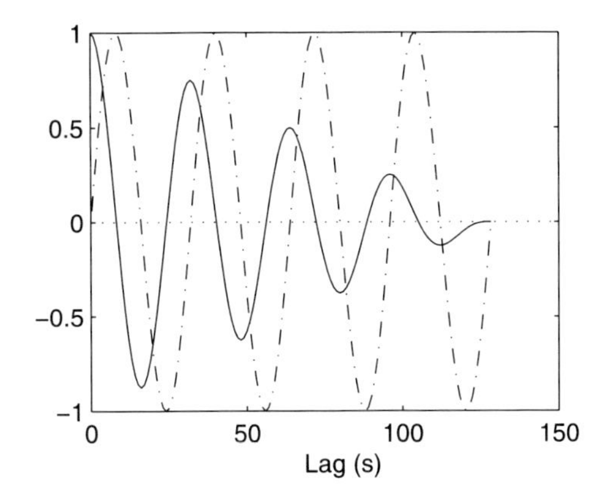

Fig. 4.1 Sine-wave signal (*dash-dotted line*) with its autocorrelation function (*solid line*)

4.1.2 White Noise Sequence

A signal that needs further attention is the so-called *white noise sequence*. White noise is one of the most significant signals when identifying LTI systems. A sequence with zero mean, finite variance, and serially uncorrelated terms is called a white noise sequence. In other words: a white noise sequence has no time structure. However, a continuous-time white noise signal does not exist in any physical sense, as it would require an infinite amount of power to generate it. Therefore, only discrete-time white noise signals are considered. A formal definition of discrete-time white noise $w(t)$ is given by

$$E\big[w(t)\big] = 0 \tag{4.5}$$

$$E\big[w(t)w^T(t+l)\big] = \begin{cases} Q, & l = 0 \\ 0, & l \neq 0 \end{cases} \tag{4.6}$$

In the next example a computer-generated white noise sequence will be further examined.

Example 4.2 White noise: A uniformly distributed white noise sequence, generated with the MATLAB function *rand*, is presented in Fig. 4.2.

The associated normalized autocorrelation function, that is, $r_{uu}(l)/r_{uu}(0)$ for $l = 0, 1, 2, \ldots$ is also presented (see Fig. 4.3), indicating that only at zero lag the autocorrelation is significant. The dotted lines indicated the 99% confidence limits, as calculated by the MATLAB function *xcorr*.

4.1.3 Cross-correlation Function

In addition to the autocorrelation function, the cross-correlation function $r_{uy}(\tau, t)$ between two different signals $u(t)$ and $y(t)$ is introduced and is defined as

$$r_{uy}(\tau, t) := E\big[u(t)y(t+\tau)\big] \tag{4.7}$$

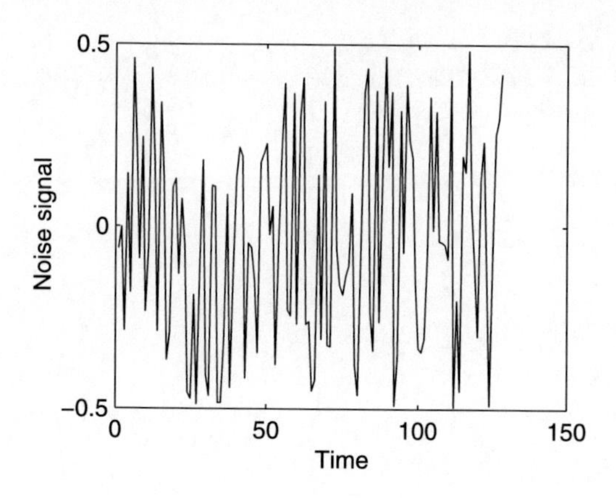

Fig. 4.2 Generated white noise sequence

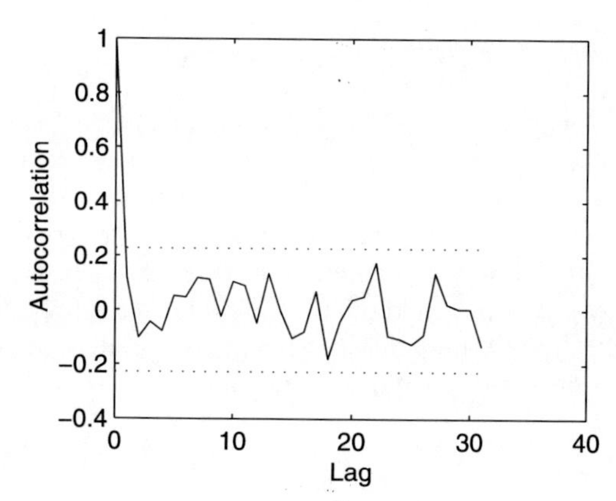

Fig. 4.3 Sample autocorrelation function (*solid line*) and corresponding 99% confidence limits (*dotted lines*) of the white noise sequence

Similarly,

$$r_{uy}(\tau) = \lim_{T\to\infty} \frac{1}{2T} \int_{-T}^{T} u(t)y(t+\tau)\,\mathrm{d}t \tag{4.8}$$

and in discrete-time,

$$r_{uy}(l) = \lim_{N\to\infty} \frac{1}{2N+1} \sum_{i=-N}^{N} u(i)y(i+l) \tag{4.9}$$

In practice, with sampled data and thus N finite, we call $r_{uy}(l)$ the sample cross-correlation function. Although the cross-correlation function also exists for negative lags, it is not an even function. Notice that for negative lags, the correlation between inputs at time instant i and outputs at $i+l$, with $l<0$, is calculated. These correlations are seldom of interest, because in causal systems the output does not depend on future inputs. Hence, for practical interpretation, only the function values for positive lags are of interest.

It is also important to note here that both the auto- and cross-correlation functions are important in the data-based identification of LTI systems, because they are closely related to the unit-pulse response of the system as will be seen in the next section.

4.2 Wiener–Hopf Relationship

4.2.1 Wiener–Hopf Equation

Recall that the output $y(t) = \sum_{k=0}^{\infty} g(k)u(t-k)$, as a result of the input $u(t)$ which started an indefinitely long time ago, at time instant $i+l$ is given by

$$y(i+l) = \sum_{k=0}^{\infty} g(k)u(i+l-k) \tag{4.10}$$

Consequently, the cross-correlation between the sequences $\{u\}$ and $\{y\}$ is

$$\begin{aligned} r_{uy}(l) &= \lim_{N\to\infty} \frac{1}{2N+1} \sum_{i=-N}^{N} u(i) \sum_{k=0}^{\infty} g(k)u(i+l-k) \\ &= \sum_{k=0}^{\infty} g(k) \lim_{N\to\infty} \frac{1}{2N+1} \sum_{i=-N}^{N} u(i)u(i+l-k) \\ &= \sum_{k=0}^{\infty} g(k) r_{uu}(l-k) \end{aligned} \tag{4.11}$$

This relationship is called the *Wiener–Hopf equation*. Notice here the similarity with the convolution sum (2.5), where $r_{uy}(\cdot)$ is substituted by $y(\cdot)$ and $r_{uu}(\cdot)$ by $u(\cdot)$.

4.2.2 Impulse Response Identification Using Wiener–Hopf Equation

In the following example an alternative method for the reconstruction of the unit-pulse response $g(t)$ from an observed input–output data set by using auto- and cross-correlation estimates is presented.

Example 4.3 Impulse response identification: For asymptotically stable systems, it suffices to determine only the first s elements of $g(t)$, so that

$$r_{uy}(l) = \sum_{k=0}^{s} g(k) r_{uu}(l-k)$$

Let both the inputs $u(0),\ldots,u(N)$ and corresponding outputs $y(0),\ldots,y(N)$ be recorded. After removal of the initial conditions effect, the following sequences remain: $u(M),u(M+1),\ldots,u(N)$ and $y(M),\,y(M+1),\ldots,\,y(N)$. The correlation functions can then be calculated as

$$r_{uu}(l)\simeq\frac{1}{N-M+1-l}\sum_{i=M}^{N-l}u(i)u(i+l)$$

and

$$r_{uy}(l)\simeq\frac{1}{N-M+1-l}\sum_{i=M}^{N-l}u(i)y(i+l)$$

Substituting the correlation function values into the Wiener–Hopf equation gives, for $l=0,1,\ldots,s$,

$$\begin{aligned}
r_{uy}(0)&=g(0)r_{uu}(0)+g(1)r_{uu}(-1)+g(2)r_{uu}(-2)+\cdots+g(s)r_{uu}(-s)\\
r_{uy}(1)&=g(0)r_{uu}(1)+g(1)r_{uu}(0)+g(2)r_{uu}(-1)+\cdots+g(s)r_{uu}(1-s)\\
r_{uy}(2)&=g(0)r_{uu}(2)+g(1)r_{uu}(1)+g(2)r_{uu}(0)+\cdots+g(s)r_{uu}(2-s)\\
&\;\vdots\\
r_{uy}(s)&=g(0)r_{uu}(s)+g(1)r_{uu}(s-1)+g(2)r_{uu}(s-2)+\cdots+g(s)r_{uu}(0)
\end{aligned}$$

Rewriting this result in matrix form

$$\begin{bmatrix} r_{uy}(0)\\ r_{uy}(1)\\ r_{uy}(2)\\ \vdots\\ r_{uy}(s)\end{bmatrix}=\begin{bmatrix} r_{uu}(0) & r_{uu}(-1) & r_{uu}(-2) & . & r_{uu}(-s)\\ r_{uu}(1) & r_{uu}(0) & r_{uu}(-1) & . & r_{uu}(1-s)\\ r_{uu}(2) & r_{uu}(1) & r_{uu}(0) & . & r_{uu}(2-s)\\ \vdots & \vdots & \vdots & & \vdots\\ r_{uu}(s) & r_{uu}(s-1) & r_{uu}(s-2) & . & r_{uu}(0)\end{bmatrix}\begin{bmatrix} g(0)\\ g(1)\\ g(2)\\ \vdots\\ g(s)\end{bmatrix}$$

clearly suggests that the elements $g(0),g(1),\ldots,g(s)$ can be solved by matrix inversion, if the matrix is invertible, noting that $r_{uu}(-l)=r_{uu}(l)$. Notice again that by setting $u(t)$ equal to a unit pulse, that is, $r_{uu}(0)=1$ and $r_{uu}(l)=0$ for $l\neq 0$, we directly find the unit-pulse response coefficients $g(t)$.

This example reveals another property of the Wiener–Hopf equation, that is, if we are able to find signals for which $r_{uu}(l-k)=0$ for $l\neq k$, the computation of the impulse response coefficients will become much easier. From this example we can also derive the following algorithm.

Algorithm 4.1 Identification of $g(t)$ from input–output data using the Wiener–Hopf relationship

1. Generate an arbitrary input signal $u(t),\ t=1,2,\ldots,N$.

2. Measure the input $u(t)$ and corresponding output signal $y(t)$.
3. Calculate both the sample autocorrelation function

$$r_{uu}(l) \simeq \frac{1}{N-l} \sum_{i=1}^{N-l} u(i)u(i+l)$$

and sample cross-correlation function

$$r_{uy}(l) \simeq \frac{1}{N-l} \sum_{i=1}^{N-l} u(i)y(i+l)$$

4. For $l = 0, 1, \ldots, s$, form the vector $r_{uy} = [r_{uy}(0), r_{uy}(1), r_{uy}(2), \ldots, r_{uy}(s)]$ and the corresponding $(s+1) \times (s+1)$ matrix R_{uu} filled with sample autocorrelation function values, as in the previous example.
5. Find $g = [g(0), g(1), g(2), \ldots, s]^T$ from $g = R_{uu}^{-1} r_{uy}$.

Again, we may start directly at step 3 if the input–output data is given.

4.2.3 Random Binary Sequences

Recall that the condition: $r_{uu}(l-k) = 0$ for $l \neq k$, in addition to unit-pulse signals, also holds for white noise sequences. Under this condition, the Wiener–Hopf relationship reduces to $r_{uy}(l) = g(l)r_{uu}(0)$. In practice, however, using a white noise input $\{u\}$, which is known as white-noise testing, still has some restrictions. For instance, when using a Gaussian distribution (see Appendix B), very large input values may occur, which cannot be implemented due to physical restrictions. In addition to this, a signal from a genuinely random noise source is not reproducible. Therefore, amplitude constrained signals are preferred in practice as, for instance, a uniformly distributed signal. A good choice for practical applications is a binary input so long as the autocorrelation function shows the desired characteristics. Random Binary Signals (RBS), generated by

$$u(t) = u(t-1) * \text{sign}\big(w(t) - p_0\big) \tag{4.12}$$

where $w(t)$ is a computer-generated white-noise process for $t = 1, 2, \ldots, N$ (MATLAB: *rand*) and $0 \leq p_0 \leq 1$ the switching probability, have these properties.

Example 4.4 *RBS*: Let us generate an RBS for $p_0 = 0.5$ and with $N = 128$ (see Fig. 4.4).

As already indicated, the associated autocorrelation function shows the desired property (see Fig. 4.5), albeit that for lag three, the autocorrelation coefficient is equal to the lower 99% confidence limit.

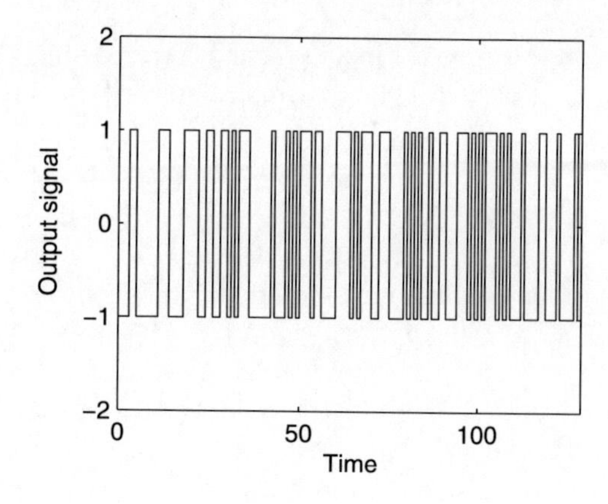

Fig. 4.4 Random Binary Signal ($p_0 = 0.5$ and $N = 128$)

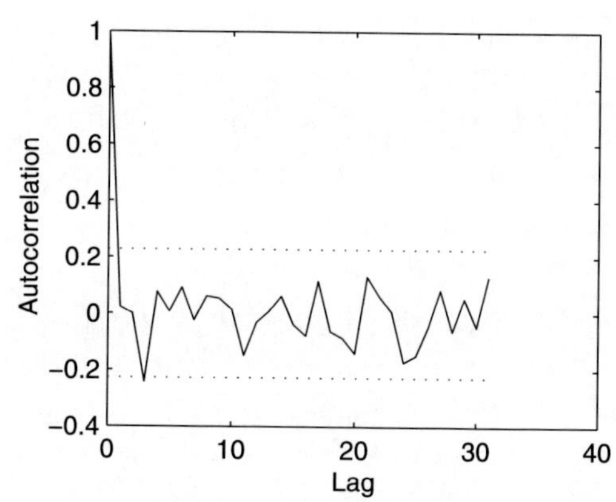

Fig. 4.5 Autocorrelation function of RBS ($p_0 = 0.5$ and $N = 128$, *solid line*) and corresponding 99% confidence limits (*dotted lines*)

4.2.4 Filter Properties of Wiener–Hopf Relationship

However, the question may arise why to go to the trouble of computing the sequences $\{r_{uy}\}$ and $\{r_{uu}\}$, even if it can be made simpler by choosing appropriate input signals, if the elements of $g(t)$ can also be determined directly from the observed data. The answer to this question is presented in what follows.

Assume that the observed output is composed of a noise-free part $\{\overline{y}\}$ and a noise part $\{v\}$, so that

$$y(t) = \overline{y}(t) + v(t) \tag{4.13}$$

Computation of the cross-correlation function gives

$$r_{uy}(l) \simeq \frac{1}{N - M + 1 - l} \sum_{i=M}^{N-l} u(i)\big[\overline{y}(i + l) + v(i + l)\big] \tag{4.14}$$

$$\simeq r_{u\overline{y}}(l) + r_{uv}(l) \tag{4.15}$$

so that as long as $\{v\}$ is unrelated to $\{u\}$ and has zero mean, the long-term average of $u(i)v(i + l)$ is very likely to be close to zero. Hence, using the Wiener–Hopf relationship filters out the effect of the noise on the estimates of the unit-pulse response, unlike the direct methods of the previous chapter.

4.3 Frequency Analysis Using Correlation Techniques

4.3.1 Cross-correlation Between Input–output Sine Waves

In Sect. 3.1.3 on frequency response methods, the empirical transfer-function estimate (ETFE) has been introduced. Noticed that, since the methods introduced in Chap. 3 are based on raw input–output data sets, both the ETFE and the estimates of $|G(e^{j\omega})|$ and $\arg(G(e^{j\omega}))$ obtained from graphic methods cannot be estimated very accurately under the presence of noise. Since for a given input $u(t) = \alpha \sin \omega t$, the output $y(t)$ of an LTI system is dominated by a sine function of known frequency ω, it is possible to correlate it out from the noise in the following way. Compute

$$I_s(N) = \frac{1}{NT}\sum_{t=0}^{NT} y(t)\sin \omega t, \qquad I_c(N) = \frac{1}{NT}\sum_{t=0}^{NT} y(t)\cos \omega t \tag{4.16}$$

that are the averages of the transformed output of N cycles of the output with sample time T. Inserting (2.14) plus an additional noise term $v(t)$ into (4.16) gives

$$\begin{aligned}
I_s(N) &= \frac{1}{NT}\sum_{t=0}^{NT} \alpha\left|G\left(e^{j\omega}\right)\right| \sin(\omega t + \phi)\sin \omega t \\
&\quad + \frac{1}{NT}\sum_{t=0}^{NT} v(t)\sin \omega t \\
&= \alpha\left|G\left(e^{j\omega}\right)\right| \frac{1}{NT}\sum_{t=0}^{NT} \frac{1}{2}\left[\cos\phi - \cos(2\omega t + \phi)\right] \\
&\quad + \frac{1}{NT}\sum_{t=0}^{NT} v(t)\sin \omega t \\
&= \frac{\alpha|G(e^{j\omega})|}{2}\cos\phi - \frac{\alpha|G(e^{j\omega})|}{2}\frac{1}{NT}\sum_{t=0}^{NT}\cos(2\omega t + \phi) \\
&\quad + \frac{1}{NT}\sum_{t=0}^{NT} v(t)\sin \omega t
\end{aligned} \tag{4.17}$$

Notice that in general the second term will diminish as N tends to infinity. The last term, containing the noise $v(t)$, will disappear if $v(t)$ does not contain a pure periodic component of the input frequency. Even for random noise, the last term tends to zero as N tends to infinity. Similarly, $I_c(N)$ can be approximated by the term $\frac{1}{2}\alpha|G(\mathrm{e}^{j\omega})|\sin\phi$.

4.3.2 *Transfer-function Estimate Using Correlation Techniques*

From the previous results it can be easily verified that both $|G(\mathrm{e}^{j\omega})|$ and ϕ can be estimated from $I_s(N)$ and $I_c(N)$, that is,

$$|\widehat{G}(\mathrm{e}^{j\omega})| = 2\sqrt{I_c^2(N) + I_s^2(N)}/\alpha \tag{4.18}$$

$$\widehat{\phi} = \arg \widehat{G}(\mathrm{e}^{j\omega}) = -\arctan\big(I_s(N)/I_c(N)\big) \tag{4.19}$$

Frequency transfer function analyzers that work on this principle of frequency analysis by correlation methods are commercially available.

Algorithm 4.2 Identification of $G(\mathrm{e}^{j\omega})$ using correlation techniques

1. Generate for a specific frequency a sine wave with maximum allowable magnitude.
2. Apply this sine wave to the system.
3. Measure the resulting sine-wave response.
4. Determine, from N cycles of the output, $I_s(N)$ and $I_c(N)$, according to (4.16).
5. Calculate magnitude and phase shift of $G(\mathrm{e}^{j\omega})$ for the specific frequency from (4.18)–(4.19).
6. Repeat this for a number of interesting frequencies $\omega \in \{\omega_1, \omega_2, \ldots, \omega_N\}$.

Application of this method to the sine-wave response of the heating system is illustrated in the following example.

Example 4.5 Heating system: Recall that, using the graphic method, it has been found that for $\omega = 5$ rad/s, $|\widehat{G}(\mathrm{e}^{j\omega})| = 0.256$ V/V and $\widehat{\phi} = -2.50$ rad. Application of (4.18) for $N = 12$, that is, when averaging occurs only over the last 12 periods, gives $|\widehat{G}(\mathrm{e}^{j\omega})| = 0.266$ V/V and $\widehat{\phi} = -2.76$ rad. According to the analysis presented in Sect. 4.2.4, these estimates are expected to be more reliable than those obtained from the graphic method.

4.4 Spectral Analysis

4.4.1 *Power Spectra*

As an alternative to the time domain approach using auto- and cross-correlation functions, frequency domain methods based on spectral analysis have been de-

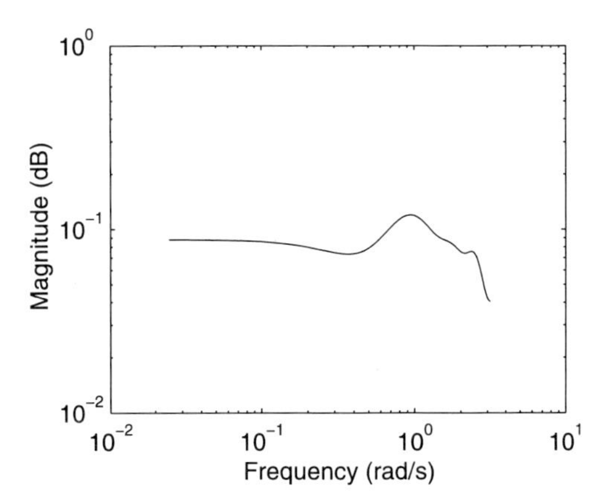

Fig. 4.6 Power spectrum white noise sequence ($N = 128$)

veloped as well. These spectral analysis methods for the determination of frequency functions of LTI systems have been initiated in the statistical literature. For given auto- and cross-correlation functions, the so-called *power (auto-)spectrum* and *cross-spectrum* are defined as

$$\Phi_{uu}(\omega) := \sum_{l=-\infty}^{\infty} r_{uu}(l)\mathrm{e}^{-j\omega l} \tag{4.20}$$

$$\Phi_{uy}(\omega) = \sum_{l=-\infty}^{\infty} r_{uy}(l)\mathrm{e}^{-j\omega l} \tag{4.21}$$

Since the autocorrelation function is always an even function, the Fourier transform of this function only contains cosine functions, and thus $\Phi_{uu}(\omega)$ is always real, while $\Phi_{uy}(\omega)$ is in general a complex-valued function of ω. Consequently, $\Phi_{uy}(\omega)$ has a real part, called the *cospectrum,* and an imaginary part, called the *quadrature spectrum*. In terms of magnitude and argument, one distinguishes between *amplitude spectrum* $|\Phi_{uy}(\omega)|$ and *phase spectrum* $\arg \Phi_{uy}(\omega)$. By definition,

$$Eu^2(t) = r_{uu}(0) = \frac{1}{2\pi}\int_{-\pi}^{\pi} \Phi_{uu}(\omega)\,\mathrm{d}\omega \tag{4.22}$$

which is a measure for the energy in the signal $u(t)$. Let us demonstrate the spectral analysis to a white noise sequence.

Example 4.6 White noise: Recall that white noise $w(t)$ has the following autocorrelation function: $r_{ww}(0) = E[w(t)w(t)] = \sigma_w^2$ and $r_{ww}(l-k) = 0$ for $l \neq k$, so that the spectrum is given by $\Phi_{ww}(\omega) = \sigma_w^2$, which is a flat spectrum. However, a white noise sequence generated in practice will always deviate from this theoretical spectrum. For instance, the RBS generated in Example 4.4 has the following spectrum (see Fig. 4.6), which especially deviates from the desired flat spectrum at high frequencies.

The RBS with $p_0 = 0.5$ and $N = 128$ shows a similar spectrum. By selecting a lower value of p_0 we are able to shape the spectrum so that this deviation from the theoretical flat spectrum especially occurs at lower frequencies.

4.4.2 Transfer-function Estimate Using Power Spectra

The relationship between $\Phi_{uy}(\omega)$ and $\Phi_{uu}(\omega)$ can be derived as follows:

$$\begin{aligned}
\Phi_{uy}(\omega) &= \sum_{l=-\infty}^{\infty} r_{uy}(l)\mathrm{e}^{-j\omega l} \\
&= \sum_{l=-\infty}^{\infty} \sum_{k=0}^{\infty} g(k) r_{uu}(l-k)\mathrm{e}^{-j\omega l} \\
&= \sum_{l=-\infty}^{\infty} \sum_{k=0}^{\infty} g(k)\mathrm{e}^{-j\omega k} r_{uu}(l-k)\mathrm{e}^{-j\omega(l-k)} \\
&= \sum_{k=0}^{\infty} g(k)\mathrm{e}^{-j\omega k} \sum_{l=-\infty}^{\infty} r_{uu}(l-k)\mathrm{e}^{-j\omega(l-k)} \\
&=_{[\lambda:=l-k]} \sum_{k=0}^{\infty} g(k)\mathrm{e}^{-j\omega k} \sum_{\lambda=-\infty}^{\infty} r_{uu}(\lambda)\mathrm{e}^{-j\omega\lambda} \\
&= G(\mathrm{e}^{j\omega})\Phi_{uu}(\omega)
\end{aligned} \tag{4.23}$$

From this it can be easily derived that for finite input–output data sets, an alternative to the ETFE is given by

$$\widehat{G}(\mathrm{e}^{j\omega}) = \frac{\Phi_{uy}(\omega)}{\Phi_{uu}(\omega)} \tag{4.24}$$

Algorithm 4.3 Identification of $G(\mathrm{e}^{j\omega})$ using spectral analysis

1. Generate for a specific frequency a sine wave with maximum allowable magnitude.
2. Apply this sine wave to the system.
3. Measure the resulting sine-wave response.
4. Determine, for $l = 0, 1, \ldots, s$, the power spectrum and cross-spectrum, according to (4.20)–(4.21).
5. Calculate $G(\mathrm{e}^{j\omega})$ for the specific frequency from (4.24).
6. Repeat this for a number of interesting frequencies $\omega \in \{\omega_1, \omega_2, \ldots, \omega_N\}$.

The application of frequency analysis by correlation methods and spectral analysis is presented in the following example.

Table 4.1 Heating system data

Frequency ω (rad/s)	Gain (V/V)	Phase shift (rad)
0.25	0.55	−0.43
0.5	0.54	−0.46
0.75	0.51	−0.65
1.0	0.52	−0.79
2.5	0.42	−1.46
5.0	0.27	−2.76
7.5	0.13	−2.71
10.0	0.07	−3.23
12.5	0.02	−3.49

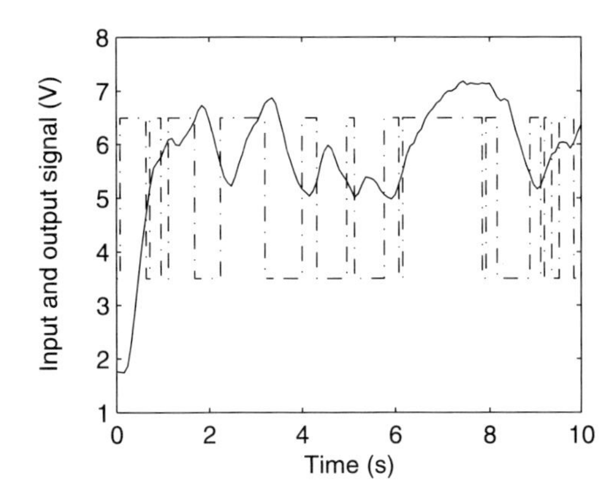

Fig. 4.7 RBS input (*dash-dotted line*) and output (*solid line*)

Example 4.7 Heating system: For the reconstruction of the frequency transfer function using sine-wave testing with correlation techniques, nine sweeps have been made. The results are presented in Table 4.1.

The spectral estimate is based on an RBS input with $p_0 = 0.2$ and $N = 1000$ with corresponding output. The input and output are presented in Fig. 4.7 for the first 10 s only.

The Bode plot of the estimated frequency transfer function as a result of spectral analysis and the individual estimates from the sine-wave testing (see Figs. 4.8 and 4.9) reveals that the estimates do not deviate too much, except for higher frequencies, where a significant difference is observed. However, it should be then realized that the effect of measurement noise is most apparent in the higher-frequency region, so that in this region the estimates are not fully reliable.

4.4.3 Bias-variance Tradeoff in Transfer-function Estimates

Given the power spectrum Φ_{vv} of the noise v, we are able to investigate the mean and variance (Appendix B) of the ETFE, as presented in Sect. 3.1.3. It has been

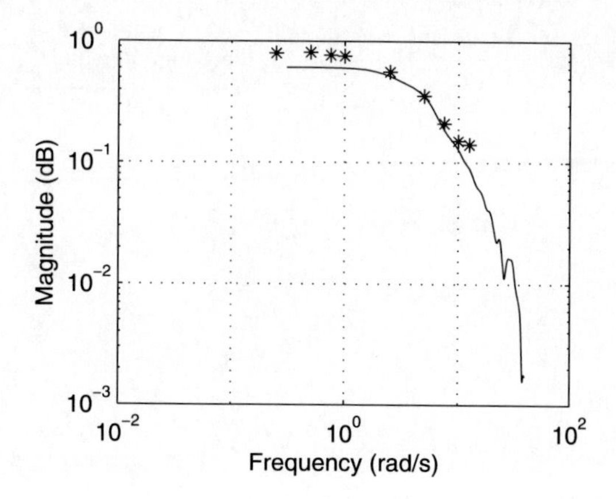

Fig. 4.8 Magnitude plot of transfer function estimates from spectral analysis (*solid line*) and from sine-wave testing (*)

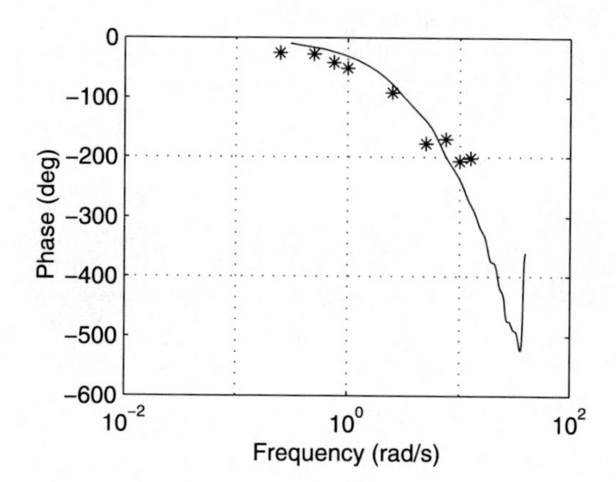

Fig. 4.9 Phase plot of transfer function estimates from spectral analysis (*solid line*) and from sine-wave testing (*)

shown in [Lju99b] that the ETFE approximately satisfies

$$E\big[\widehat{G}\big(\mathrm{e}^{j\omega}\big)\big] = G\big(\mathrm{e}^{j\omega}\big) + R_N^{(1)} \tag{4.25}$$

$$\operatorname{Var}\widehat{G}\big(\mathrm{e}^{j\omega}\big) = \frac{1}{|U(\omega)|^2}\big(\Phi_{vv}(\omega) + R_N^{(2)}\big) \tag{4.26}$$

where $R_N^{(i)} \to 0$ as $N \to \infty$ for $i = 1, 2$. Consequently, the ETFE is an asymptotically unbiased estimate of $G(\mathrm{e}^{j\omega})$, where in this statistical context bias is defined as the difference between the estimator's expected value ($E[\widehat{G}(\mathrm{e}^{j\omega})]$) and the true value of the ETFE for a specific frequency ($G(\mathrm{e}^{j\omega})$). However, the variance will not tend to zero for N large. It approaches the noise-to-input signal ratio at the specific frequency ω. A common approach to improve the variance properties of the ETFE is to apply a local averaging procedure,

$$\widehat{G_w}\big(\mathrm{e}^{j\omega}\big) = \frac{1}{\sum_k w_k(\omega)} \sum_k w_k(\omega)\widehat{G}\big(\mathrm{e}^{\omega}\big) \tag{4.27}$$

where the frequency weights $w_k(\omega)$ follow from a good trade-off between bias and variance. Typically, the weights are selected according to a frequency window. Within this context, the so-called Hamming window is very popular (for further information, see, for instance, [EO68]).

The determination of the ETFE from finite data is not straightforward; problems of aliasing, leakage, and windowing always occur. Therefore, the time domain alternative from Sect. 4.3, which does not have these problems, in general prevails for practical application.

4.5 Historical Notes and References

A survey of correlation techniques for identification can be found in [God80]. The application of correlation methods in identification studies is described in, for instance, [ES81, CC94].

For background material on spectral analysis, we refer to the books [JW68, Bri81, Mar87, Kay88, SM97]. An overview of different frequency domain techniques for time series analysis is given in [BE83].

There is an extensive literature on frequency domain identification; for details, see [PS01] and the references therein. However, we would like to mention here a few to stress the development in this field, starting in the 1970s; see [KKZ77, Kol93, PS97, RSP97, SVPG99, SGR+00, Bai03b, Bai03a, GL09b]. The frequency domain techniques have mostly been applied to mechatronic systems, see [TY90, HvdMS02, AMLL02, CHY02]. During the last decade, much emphasis has also been put on control-oriented identification methods that focus on a direct determination of the frequency function and the associated uncertainty in the estimates from open- and closed-loop data; see, for instance, [LL96, SOS00, WZG01, Wel77, WG04].

4.6 Problems

Problem 4.1 Let us evaluate the Random Binary Signal (RBS) response in some more detail. Consider the continuous-time system with transfer function:

$$G(s) = \frac{2}{10s + 1}$$

and generate a random binary input signal (MATLAB command: *idinput*) of length N, preferably $N = 2^n$ with n an integer as MATLAB uses FFT for frequency domain calculations, and a relative frequency band of $[0, 0.5]$.

Table 4.2 Normalized data chemical reactor

Time	1	2	3	4	5	6	7	8	9	10	11
$u(t)$ (m^3/s)	1	1	1	1	1	−1	−1	−1	−1	−1	−1
$y(t)$ (kg/m^3)	0	0.13	0.09	0.10	0.10	0.10	−0.17	−0.08	−0.11	−0.10	−0.10

(a) Calculate the system output, using, for example, the MATLAB command *lsim*,[1] and plot both input and output in one figure. Interpret the result.
(b) For the next analyzes, it is necessary to remove the mean from both discrete-time signals!! First, determine the frequency function $G(j\omega)$ using *etfe*.
(c) Plot the frequency function using *bodeplot*. Interpret the result.
(d) Determine the frequency function $G(j\omega)$ again but now by using *spa*.
(e) Plot the resulting frequency function using *bodeplot*. Interpret the result.

Problem 4.2

(a) Generate a white noise signal with zero mean and unity variance using the MATLAB function *rand*. Check whether this signal is serially uncorrelated (using *xcorr*) and plot the results of this analysis.
(b) Add a constant to the white noise signal and evaluate the auto-correlation function. Explain the result.

Problem 4.3 Let the following (normalized) data from an experiment investigating the effect of the feed rate on the substrate concentration in a reactor (see Table 4.2) be given:

(a) Plot both input and output data (MATLAB: *stairs*). Give a first interpretation of the result.
(b) Determine the impulse response function $g(t)$ from this data. For easy manipulation of the input data matrix you may use the MATLAB command $yreverse = y(length(x) : -1 : 1)$ and the MATLAB function *hankel*.
(c) Determine again the impulse response function, but now using the Wiener–Hopf equation and thus using cross- and autocorrelation functions (MATLAB: *xcorr*) to cope with the noise. Explain your result.

[1] There are several routes for obtaining a solution to the simulation problem of an LTI system in transfer function form. Most often, a so-called state-space realization is first determined, and subsequently the general analytical solution, as in footnote 1 in Chap. 1, is applied.

Part II
Time-invariant Systems Identification

In Part I the impulse response model had a central place in the identification of LTI systems on the basis of data only. In the modeling of the storage tank (Example 1.4) it appeared that the impulse response had the following form: $g(t) = K\mathrm{e}^{-Kt}$. In a discrete-time model representation this would result in a number, depending on the value of K, of impulse response coefficients. However, at the cost of imposing a specific model structure in the description of the system behavior in terms of differential and algebraic relationships, in the state-space model there is only one unknown model parameter, the proportional gain K. Hence, the parameter estimation procedure will become much easier for the same type of system. Moreover, starting from prior system's knowledge allows a wider area of application, since static, nonlinear, and time-varying systems can be covered as well.

In particular for the identification of nonlinear systems, as is quite common in applications with biological or chemical components, we must use other techniques than the ones introduced in Part I. These techniques will be introduced in this Part II and the next part. For applications with a biological component that show a time-varying system's behavior due to adaptation of the organisms, however, the time-varying system identification techniques in Part III are of most interest.

In this and the next part, we will always start with the postulation of a *model structure* followed by a *model parameter estimation* procedure. This approach, which is also indicated as a *parameterized identification* method, will then be applied to the identification of static and dynamic systems.

In Chap. 5 we will start with the identification of static linear systems, that is, no dynamics are involved. The output of a static system depends only on the input at the same instant and thus shows instantaneous responses. In particular, the so-called least-squares method will be introduced. As will be seen in following sections, the least-squares method for the static linear case forms the basis for solving nonlinear and dynamic estimation problems. For the analysis of the resulting estimates, properties like bias and accuracy will be treated. Special attention will be paid to errors-in-variables problems, which allow noise in both input and output variables, to maximum likelihood estimation as a unified approach to estimation, in particular well defined in the case of normal distributions, and to bounded-noise problems for cases with small data sets.

Chapter 6 focuses on the identification of dynamic systems, both linear and nonlinear. The selected model structure of linear dynamic systems, in particular the structure of the noise model, appears to be of crucial importance for specific applications and the estimation methods to be used. It will be stressed that both the linear and the nonlinear model structures in this chapter can be formulated in terms of (nonlinear) regression equations, which allows a unification of the estimation problems. In this chapter, special attention will be paid to subspace identification for the direct estimation of the entries of A, B, C, and D in a discrete-time, linear state-space model formulation, to the identification of discrete-time linear parameter-varying models of nonlinear or time-varying systems, to the use of orthogonal basis functions for efficient calculation, and to closed-loop identification in LTI control system configurations.

Chapter 5
Static Systems Identification

5.1 Linear Static Systems

5.1.1 Linear Regression

Essentially, in what follows the model used in parameterized or model-based identification methods relates an observable variable $y(t)$ to p explanatory variables, also called the *regressors*, $\phi_1(t), \dots, \phi_p(t)$. The independent variable t need not necessarily represent time; it may be any index variable. Furthermore, it is assumed that the model has one unknown parameter ϑ_i per explanatory variable, which may be known in advance, or which has been measured. Any linear relationship can thus be modeled as

$$y(t) = \phi_1(t)\vartheta_1 + \cdots + \phi_p(t)\vartheta_p + e(t) \tag{5.1}$$

The interpretation of this so-called *linear regression* model is that the variable y is explained in terms of the variables $(\phi_1, \dots, \phi_p)$ plus an unobserved error term e. Let $t = 1, \dots, N$, and define $y := [y(1), \dots, y(N)]^T$, $e = [e(1), \dots, e(N)]^T$, $\vartheta := [\vartheta_1, \dots, \vartheta_p]^T$, which are column vectors of appropriate dimensions. Let furthermore, Φ be an $N \times p$ matrix with elements $\Phi_{t_j} := \phi_j(t)$, $j = 1, \dots, p$. Then, the model (5.1) can be written in matrix notation (see Appendix A) as

$$y = \Phi\vartheta + e \tag{5.2}$$

Notice, however, that (5.2) can be equally interpreted in terms of a static system description with unknown static states ϑ and an observation matrix Φ with known elements relating the states to the observations. Let us illustrate this fact by a simple example.

Example 5.1 Constant process state: Consider the case where we have two measurements $y(1)$ and $y(2)$ of a process state x, which is assumed to be constant during the experiment. The model becomes

$$y(1) = x + e(1)$$

K.J. Keesman, *System Identification*,
Advanced Textbooks in Control and Signal Processing,
DOI 10.1007/978-0-85729-522-4_5,

$$y(2) = x + e(2)$$

so that, in matrix notation, $y = [y(1)\ y(2)]^T$, $\vartheta = x$, $\Phi = [1\ 1]^T$, $e = [e(1)\ e(2)]^T$.

The following example illustrates a parameter estimation problem, which is linear in the unknown parameters.

Example 5.2 Moving object: Let x be the position of an object moving in a straight line with constant acceleration a. Using the kinematic law, $x(t) = x_0 + v_0 t + \frac{1}{2}at^2$, we are able to predict the position at time instant t if the initial position x_0, the initial velocity v_0, and the acceleration a are known. However, if these variables are unknown or not exactly known, we can estimate these from given observations of y and t. Hence, in terms of a linear regression model, we define $\vartheta := [x_0\ v_0\ a]^T$ and $\phi(t) := [1\ t\ t^2/2]^T$, so that

$$y(t) = \phi(t)^T \vartheta + e(t)$$

which is not linear in t, but *linear in the unknown parameters*. Notice that the kinematic model, albeit explicitly dependent on time t, leads to a static relationship, as no differential or difference equation is used to describe the process. Notice also that the explanatory variable associated with x_0 is 1 for all samples, and thus it can be assumed with good reason that e has zero-mean.

It is important to note from these two examples that the terms states and parameters in these particular cases can be interchanged. The problem of static system state estimation can thus be regarded as a linear parameter estimation problem and vice versa. However, in the following sections, we mainly focus on parameter estimation problems.

5.1.2 Least-squares Estimation

A reasonable way to estimate the unknowns from given data is by demanding that the *prediction errors* or *residuals* $\varepsilon(t) := y(t) - \phi(t)^T\vartheta$ are small. Formally stated, we will choose the parameter vector ϑ such that the sum of squared prediction errors

$$J(\vartheta) := \sum_{t=1}^{N} \varepsilon^2(t) = \sum_{t=1}^{N} \left(y(t) - \phi(t)^T\vartheta\right)^2 \tag{5.3}$$

is minimal. The scalar function $J(\vartheta)$ is also known as the least-squares objective function. In matrix notation, (5.3) can be written as

$$J(\vartheta) := \varepsilon^T\varepsilon = \left(y^T - \vartheta^T\Phi^T\right)(y - \Phi\vartheta) \tag{5.4}$$

using the fact from matrix theory that $(\Phi\vartheta)^T = \vartheta^T\Phi^T$ (see Appendix A for details on matrix properties and operations). As in the scalar case, J is minimal if and only if the gradient of J with respect to ϑ is zero, in general a p-dimensional vector, and the second derivative is positive. In the following, two standard results for derivatives of vector-matrix expressions are used, that is,

$$\frac{\partial a^T \vartheta}{\partial \vartheta} = a \tag{5.5}$$

and

$$\frac{\partial \vartheta^T A \vartheta}{\partial \vartheta} = \left(A + A^T\right)\vartheta \tag{5.6}$$

which can be easily verified by writing out all the elements and taking the derivatives. Hence, since J is a scalar function, the individual terms are scalars so that, with $y, \Phi\vartheta \in \mathbb{R}^N$, $y^T \Phi \vartheta = \vartheta^T \Phi^T y$ and thus

$$\begin{aligned} J(\vartheta) &= y^T y - y^T \Phi \vartheta - \vartheta^T \Phi^T y + \vartheta^T \Phi^T \Phi \vartheta \\ &= y^T y - 2\vartheta^T \Phi^T y + \vartheta^T \Phi^T \Phi \vartheta \end{aligned} \tag{5.7}$$

Consequently, taking the derivative with respect to ϑ gives a zero for the first term, $-2\Phi^T y$ for the second term, and $(A + A^T)\vartheta$ with symmetric matrix $A := \Phi^T \Phi$ $(A^T = A)$ for the last term, so that

$$\frac{\partial J(\vartheta)}{\partial \vartheta} = -2\Phi^T y + 2\Phi^T \Phi \vartheta \tag{5.8}$$

(see Appendix A for details). The gradient of $J(\vartheta)$ is zero if and only if

$$\Phi^T \Phi \widehat{\vartheta} = \Phi^T y \tag{5.9}$$

which are called the *normal equations*. From (5.9) we can deduce the *ordinary least-squares estimate* by multiplying both sides with $(\Phi^T \Phi)^{-1}$:

$$\widehat{\vartheta} = \left(\Phi^T \Phi\right)^{-1} \Phi^T y \tag{5.10}$$

under the assumption that the $p \times p$ matrix $\Phi^T \Phi$ is invertible. It remains to show that this estimate gives a minimum of J. Let $\vartheta = \widehat{\vartheta} + \Delta\vartheta$; then substitution of this expression into (5.7) ultimately leads to

$$J\left(\widehat{\vartheta}\right) = J(\vartheta) - (\Delta\vartheta)^T \Phi^T \Phi (\Delta\vartheta) \tag{5.11}$$

Hence, if $(\Delta\vartheta)^T \Phi^T \Phi (\Delta\vartheta) > 0$, then $J(\vartheta)$ has a minimum at $\widehat{\vartheta}$.

Let us illustrate the least-squares method to the estimation of the unknown parameters in Example 5.2.

Example 5.3 Moving object: Let the following observations on the moving object, for which x_0, v_0, and a are unknown, be available (see [Nor86], p. 62, and Table 5.1) and thus $p = 3$ and $N = 6$.

Table 5.1 Moving object data

t (s)	0.0	0.2	0.4	0.6	0.8	1.0
y (m)	3	59	98	151	218	264

Given the moving object data,

$$\Phi = \begin{bmatrix} 1 & 0.0 & 0 \\ 1 & 0.2 & 0.02 \\ 1 & 0.4 & 0.08 \\ 1 & 0.6 & 0.18 \\ 1 & 0.8 & 0.32 \\ 1 & 1.0 & 0.50 \end{bmatrix}$$

an $N \times p$ matrix, and

$$\Phi^T y = \begin{bmatrix} 793 \\ 580 \\ 238 \end{bmatrix}, \qquad \Phi^T \Phi = \begin{bmatrix} 6 & 3 & 1.1 \\ 3 & 2.2 & 0.9 \\ 1.1 & 0.9 & 0.3916 \end{bmatrix}$$

$$\Longrightarrow \quad (\Phi^T \Phi)^{-1} = \begin{bmatrix} 0.821 & -2.95 & 4.46 \\ -2.95 & 18.2 & -33.5 \\ 4.46 & -33.5 & 67.0 \end{bmatrix}$$

Consequently, using (5.10),

$$\widehat{\vartheta} = [4.79\ 234\ 55.4]^T$$

In MATLAB the estimate can also be found by using the expression $th = \text{PHI}\backslash y$, where PHI and y are properly defined. The backslash ('\') defines the so-called left matrix division. The prediction errors can be calculated from $\varepsilon(t) = y(t) - \phi(t)^T \widehat{\vartheta}$ for $t = 1, \ldots, 6$, so that

$$\varepsilon = [-1.8\ 6.2\ -5.0\ -4.4\ 7.9\ -2.9]^T$$

Notice that the mean value of ε is equal to zero and the root of the mean-square error (MSE) ($\sqrt{\frac{\varepsilon^T \varepsilon}{N-p}}$) is equal to 7.3 m. Analysis of ε further shows that there is no clear evidence that unreliable measurements of y, so-called *outliers*, are present in the data.

From this example it can be seen that the dimension of $\Phi^T \Phi$ does not depend on the number of observations; it only depends on the number of parameters. Furthermore, $\Phi^T \Phi$ is symmetrical and positive definite, which implies that all the eigenvalues (λ_i), in this case 0.01, 0.67, and 7.9, are positive. Hence, $\det(\Phi^T \Phi) = \prod_{i=1}^{n} \lambda_i > 0$ with $n = 3$, and thus the matrix $\Phi^T \Phi$ is invertible (see Appendix A).

So far, all the prediction errors have been weighted equally. However, under certain circumstances, for instance, in the case of outliers or if recently measured data has to be weighted more heavily, there is a need to weight the errors individually. Then, for a positive definite matrix W, the criterion is modified to

$$J_W(\vartheta) := \varepsilon^T W \varepsilon = \left(y^T - \vartheta^T \Phi^T\right) W (y - \Phi\vartheta) \tag{5.12}$$

Following the previous derivation of the ordinary least-squares estimate, it can be easily verified that the so-called *weighted least-squares estimate* is given by

$$\widehat{\vartheta}_W = \left(\Phi^T W \Phi\right)^{-1} \Phi^T W y \tag{5.13}$$

Under the condition that $\Phi^T \Phi$ is invertible and with W positive definite, to ensure that J_W is positive, $\Phi^T W \Phi$ is also positive definite and thus invertible (see Appendix A). For a specific weighting of the individual data points, W is a diagonal matrix, which does not increase the computational complexity too much. However, in general, W is a nondiagonal matrix, as is illustrated in later sections.

Hence, for given experimental data, the (weighted) least-squares estimation algorithm can be summarized by the following.

Algorithm 5.1 (Weighted) Least-squares estimation of ϑ in linear static systems

1. Given $y(t)$ and $\phi_j(t)$ for $t = 1, \ldots, N$ and $j = 1, \ldots, p$, define the N-dimensional vector $y := [y(1), \ldots, y(N)]^T$.
2. Form the $N \times p$ matrix Φ with elements $\Phi_{t_j} := \phi_j(t)$, where ϕ_j is the jth regressor.
3. Calculate from (5.10) or (5.13), respectively, the ordinary or weighted least-squares estimate of the unknown p-dimensional parameter vector ϑ.

Example 5.4 Moving object: Analysis of the prediction errors or residuals may suggest a specific weighting. Given the residuals in Example 5.3, let us weight the first, fourth, and sixth measurements more heavily, because the values of these prediction errors are somewhat smaller than the other ones. For instance, we may choose $w_1 = 4$, $w_2 = 1$, $w_3 = 1$, $w_4 = 4$, $w_5 = 1$, $w_6 = 4$. Then,

$$\Phi^T W = \begin{bmatrix} 4 & 1 & 1 & 4 & 1 & 4 \\ 0 & 0.2 & 0.4 & 2.4 & 0.8 & 4 \\ 0 & 0.02 & 0.08 & 0.72 & 0.32 & 2 \end{bmatrix}$$

and

$$\Phi^T W y = \begin{bmatrix} 2047 \\ 1644 \\ 715.5 \end{bmatrix}, \qquad \Phi^T W \Phi = \begin{bmatrix} 15.0 & 7.8 & 3.14 \\ 7.8 & 6.28 & 2.724 \\ 3.14 & 2.724 & 1.2388 \end{bmatrix}$$

$$\Longrightarrow \quad \left(\Phi^T W \Phi\right)^{-1} = \begin{bmatrix} 0.234 & -0.721 & 0.994 \\ -0.721 & 5.672 & -10.64 \\ 0.994 & -10.64 & 21.69 \end{bmatrix}$$

leading to

$$\hat{\vartheta} = [3.72\ 231\ 59.3]^T$$

with prediction errors

$$\varepsilon = [-0.7\ 7.8\ -3.0\ -2.2\ 10.2\ -0.8]^T$$

Notice here that in particular the initial velocity and the acceleration are affected by the weighting. Clearly, the prediction errors associated with the first, fourth, and sixth measurements have been reduced significantly, because extra weights have been put on these.

Apart from an increase in the computational effort, the specific choice of the weighting factors is another problem associated with the weighted least-squares method, which will be solved in later sections. As we will see later, unlike the more or less arbitrary way of choosing weights as we did so far, a weighting that is related to the accuracy of a specific sensor or chosen to whiten prediction errors is more well founded.

5.1.3 Interpretation of Least-squares Method

In this section the properties of the ordinary least-squares estimation method, which originated from astronomical studies of Gauss in the early 19th century, are further analyzed. Let us first consider the dependence between the ordinary least-squares estimate and the number of output samples. In case the number of output measurements equals the number of unknown parameters, that is, $N = p$,

$$\hat{\vartheta} = \Phi^{-1} y \tag{5.14}$$

if Φ is invertible, which only holds if the columns of the square matrix Φ are independent. Notice that in this specific case with $N = p$, the noise in y is directly reflected in the estimates. Hence, from this point of view, in practice, N is preferably chosen much larger than p. As a rule of thumb, N is chosen at least five times larger than p. If $N > p$, there are more equations than unknowns, and the estimate is found from (5.10), where $(\Phi^T \Phi)^{-1} \Phi^T$ is called the *pseudo* or *generalized inverse* of Φ. If, however, $N < p$, then the number of unknowns exceeds the number of equations, and thus no unique solution exists.

The next property of orthogonal projection is illustrated by a very simple example.

Example 5.5 Orthogonal projection: The length or magnitude or norm of a column vector $a = [a_1, \ldots, a_p]^T \in \mathbb{R}^p$, commonly denoted as $\|a\|$ (see Appendix A), is defined as

$$\|a\| := \sqrt{a_1^2 + a_2^2 + \cdots + a_p^2} = \sqrt{a^T a}$$

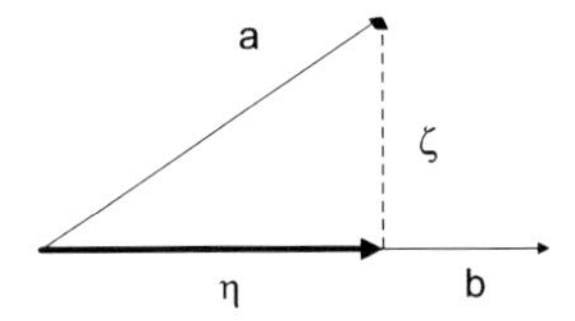

Fig. 5.1 Orthogonal projection in R^2

Let further η be the orthogonal projection of vector a on vector b in $\mathbb{R}^2$ (see Fig. 5.1) and define $\zeta := a - \eta$.

Then,

$$a^T a = (\eta + \zeta)^T (\eta + \zeta) = \eta^T \eta + \zeta^T \zeta$$

because $\eta^T \zeta = 0$ due to the orthogonality between these two vectors. Notice that this result could also have been obtained after direct application of Pythagoras' theorem. Let furthermore $\eta := \gamma b$; then

$$b^T a = \eta^T (\eta + \zeta)/\gamma =_{[\eta^T \zeta = 0]} \eta^T \eta/\gamma =_{[\eta := \gamma b]} \gamma b^T b$$

Consequently, the scalar γ is found from

$$\gamma = \frac{b^T a}{b^T b}$$

and the two-dimensional vector η is given by

$$\eta = \frac{b^T a}{b^T b} b$$

The results of this example will now be applied to the least-squares method. In case ϑ is a scalar which has to be estimated from a number of measurements collected in the vector y for a given explanatory variable whose values have been put into the vector ϕ, the least-squares estimate (5.10) is simply given by $\widehat{\vartheta} = \phi^T y/\phi^T \phi$. The similarity between this expression and the expression for γ in the example is evident. Let us further introduce the predicted model output

$$\widehat{y} = \Phi \widehat{\vartheta} \tag{5.15}$$

and define the prediction error in terms of this model output,

$$\varepsilon = y - \widehat{y} \tag{5.16}$$

Then, for the scalar case, $\widehat{y} = (\phi^T y/\phi^T \phi)\phi$, which resembles the expression for η. Recall that the expressions have been derived under different conditions: the first one has been derived by minimizing the sum of squares of the errors, and the second by orthogonal projection of the output vector onto the explanatory vector. Hence, ordinary least-squares estimation can be viewed as orthogonal projection. This result can be further stressed by looking at the sum of the products of corresponding model output and error samples for the general case,

$$\begin{aligned}\widehat{y}^T \varepsilon &= \widehat{\vartheta}^T \Phi^T \left(y - \Phi\widehat{\vartheta}\right) \\ &= \widehat{\vartheta}^T \left(\Phi^T y - \left(\Phi^T \Phi\right)\widehat{\vartheta}\right) =_{[(5.9)]} 0\end{aligned} \tag{5.17}$$

and thus,

$$\begin{aligned}\|y\|^2 &= y^T y = \widehat{y}^T \widehat{y} + 2\widehat{y}^T \varepsilon + \varepsilon^T \varepsilon \\ &= \left\|\widehat{y}\right\|^2 + \|\varepsilon\|^2\end{aligned} \tag{5.18}$$

Hence, since the inner product is zero, $\widehat{y}$ and ε are two orthogonal vectors, and ε spans the shortest distance between y and $\widehat{y}$, which is a linear combination of the explanatory variables.

From matrix theory (see Appendix A), where it is stated that a matrix P is said to be an *orthogonal projection matrix* if and only if $P^2 = P$ and $P^T = P$, a similar result is obtained. After substitution of the least-squares estimate (5.10) into (5.15) we obtain

$$\widehat{y} = \Phi\left(\Phi^T \Phi\right)^{-1} \Phi^T y = P(\Phi) y \tag{5.19}$$

Since $P(\Phi)^2 = P(\Phi)$ and $P(\Phi)^T = P(\Phi)$, it becomes immediately clear that $P(\Phi)$ is an orthogonal projection matrix (see also Appendix A for details on projection matrices). Similarly,

$$\varepsilon = y - \Phi\left(\Phi^T \Phi\right)^{-1} \Phi^T y = \left(I - P(\Phi)\right) y \tag{5.20}$$

and again the $N \times N$ matrix $I - P(\Phi)$ is an orthogonal projection matrix. Hence $\widehat{y}$, which is situated in the hyperplane spanned by the column vectors of Φ, i.e., $\phi_1, \ldots, \phi_p$, and ε are found such that the two vectors are perpendicular to each other. However, this result no longer holds for the weighted least-squares method, where

$$\widehat{y}_W = \Phi\left(\Phi^T W \Phi\right)^{-1} \Phi^T W y = P_W(\Phi) y \tag{5.21}$$

and $P_W(\Phi)^2 = P_W(\Phi)$, but $P_W(\Phi)^T \neq P_W(\Phi)$. In this case, $P_W(\Phi)$ is a general (oblique) projection matrix.

A last property of the least-squares method is found from analyzing the cross-correlation between $\widehat{y}$ and ε. Recall that for given bounded sequences of $\widehat{y}$ and ε,

$$r_{\widehat{y}\varepsilon}(l) \simeq \frac{1}{N} \sum_{k=1}^{N} \widehat{y}(k)\varepsilon(k+l) \tag{5.22}$$

so that

$$r_{\widehat{y}\varepsilon}(0) \simeq \frac{1}{N} \sum_{k=1}^{N} \widehat{y}(k)\varepsilon(k) = \frac{1}{N} \widehat{y}^T \varepsilon =_{[(5.17)]} 0 \tag{5.23}$$

In other words, the predicted model output is uncorrelated with the associated prediction error; there is no correlation between the explained part of the output and the unexplained part.

5.1.4 Bias

Since empirical data always contain some measurement uncertainty, with either stochastic or deterministic properties, each estimate of an unknown variable from given empirical data will thus contain some uncertainty. A first question is whether the resulting estimate is *unbiased*, that is, will the estimates cluster around the true value. *Bias*, denoted by b, is defined as the difference between the expected value of the estimate $\widehat{\vartheta}$ and the true value ϑ. In mathematical notation,

$$b := E\big[\widehat{\vartheta}\big] - \vartheta \tag{5.24}$$

where $E[\cdot]$ denotes the expectation operator (see Appendix B). In the following, the expectation will always be interpreted as the average. The following examples illustrate how bias can be evaluated for different estimators.

Example 5.6 Single parameter problem: Consider the model with single input and single output,

$$y(t) = \alpha u(t) + e(t)$$

in which the unknown α has to be estimated from a number of measurements at times $t = 1, \ldots, N$. At each time instant, α can be estimated from $y(t)/u(t)$. Averaging these instantaneous estimates gives

$$\widehat{\alpha} = \frac{1}{N}\sum_{t=1}^{N}\frac{y(t)}{u(t)}$$

Substituting $y(t)$ by $\alpha u(t) + e(t)$ gives

$$\begin{aligned}\widehat{\alpha} &= \frac{1}{N}\sum_{t=1}^{N}\frac{\alpha u(t) + e(t)}{u(t)}\\ &= \frac{1}{N}\sum_{t=1}^{N}\alpha + \frac{1}{N}\sum_{t=1}^{N}\frac{e(t)}{u(t)}\end{aligned}$$

so that the bias for estimator $\widehat{\alpha}$ is given by

$$\begin{aligned}b &= E\big[\widehat{\alpha}\big] - \alpha\\ &= E\left[\frac{1}{N}\sum_{t=1}^{N}\alpha + \frac{1}{N}\sum_{t=1}^{N}\frac{e(t)}{u(t)}\right] - \alpha\end{aligned}$$

$$= E\left[\frac{1}{N}\sum_{t=1}^{N}\frac{e(t)}{u(t)}\right]$$

$$= \frac{1}{N}\sum_{t=1}^{N}\frac{E[e(t)]}{u(t)}$$

Hence, $b = 0$ if $E[e(t)] = 0$, that is, $e(t)$ has zero mean. Furthermore, b becomes small if $u(t)$ is large or if N is chosen large enough.

Example 5.7 Constant process state: Consider again the case with the constant process state, which has to be estimated from a number of data. Given

$$y(t) = x + e(t)$$

a reasonable estimate is given by

$$\widehat{x} = (y_{\max} + y_{\min})/2$$

where $y_{\min}$ and $y_{\max}$ are the minimum and maximum values of the output sequence, respectively. These extreme values are associated with $e_{\min}$ and $e_{\max}$. Consequently,

$$b = E\big[(x + e_{\max} + x + e_{\min})/2\big] - x$$
$$= E\big[(e_{\max} + e_{\min})/2\big]$$

and thus $b = 0$ if $e_{\max} = -e_{\min}$. In other words, the residuals must have a symmetrical distribution with finite support, and the output must touch the boundaries during the experiment.

Since the least-squares methods plays such an important role both in estimation theory and in practice, the bias of the least-squares estimate will be of special interest. Substituting (5.10) and (5.2) into (5.24) gives

$$b = E\big[(\Phi^T\Phi)^{-1}\Phi^T y\big] - \vartheta$$
$$= E\big[(\Phi^T\Phi)^{-1}\Phi^T(\Phi\vartheta + e)\big] - \vartheta$$
$$= E\big[(\Phi^T\Phi)^{-1}\Phi^T e\big] \tag{5.25}$$

which in general is not equal to zero if Φ and e are statistically dependent. Notice here that Φ may be a matrix with stochastic or random elements (see Appendix B) due to measurement errors in the explanatory variables. In the case where Φ and e are statistically independent, $b = E[(\Phi^T\Phi)^{-1}\Phi^T]E[e]$. Hence, from this we conclude that the bias of the least-squares estimate is zero if Φ and e are statistically independent and $E[e] = 0$, a null vector in $\mathbb{R}^N$. A similar result can be obtained for the weighted least-squares estimator. Consequently, in what follows, the application of the (weighted) least-squares estimator should include these tests on possibly random regressors and residuals to guarantee unbiased estimates. Let us evaluate this for the moving object example.

Example 5.8 Moving object: If we assume that the data can be explained by the following model with zero x_0:

$$x(t) = v_0 t + \frac{1}{2} a t^2 + e(t)$$

so that $\vartheta = [v_0 \; a]^T$ and $\phi(t) = [t \; \frac{1}{2}t^2]^T$, the resulting least-squares estimate becomes

$$\widehat{\vartheta} = [252 \; 29.3]^T$$

with prediction error sequence $\varepsilon = [3.0 \; 8.1 \; -5.0 \; -5.3 \; 7.3 \; -2.3]^T$. The mean value of ε is 1 m, and thus biased estimates could have been expected if Φ and e are statistically independent.

For further analysis, let us assume that time t was measured exactly and thus Φ contains deterministic regressors. Then, given the residual vector ε as a realization of e and using the normal equations (5.9), we obtain

$$\begin{aligned} \Phi^T \varepsilon &= \Phi^T \left(y - \Phi \widehat{\theta} \right) \\ &= \Phi^T y - \Phi^T \Phi \widehat{\theta} = 0 \end{aligned}$$

Hence, for deterministic and thus exactly known Φ, least-squares estimation always leads to unbiased estimates.

In addition to the conclusion from Example 5.8 that, for deterministic regressors, the least-squares estimates are unbiased, we conclude that the prediction error vector ε is in the null space or kernel of Φ^T, that is, in abstract mathematical terms, $\varepsilon \in \ker(\Phi^T)$. Let us now focus on a third example. Unlike the previous two examples, in this example we allow noise in the explanatory variables.

Example 5.9 Single parameter problem: Consider again the least-squares estimation problem of a single parameter (i.e., ϑ is a scalar) in the linear regression model with modeling error $w(t)$,

$$y^o(t) = \alpha u^o(t) + w(t)$$

from noisy measurements of the input $u(t) = u^o(t) + z(t)$ and the output $y(t) = y^o(t) + v(t)$. In this so-called *errors-in-variables* problem, both $u^o(t)$ and $y^o(t)$ indicate the noise-free input and output, respectively. It is further assumed that the noises $z(t)$, $v(t)$, and $w(t)$ have zero mean and are mutually uncorrelated. Hence, the regression model can be written as

$$y(t) = \alpha u(t) + e(t)$$

with $e(t) = v(t) + w(t) - \alpha z(t)$. The least-squares estimate is given by

$$\widehat{\alpha} = \sum_{t=1}^{N} u(t)y(t) \Big/ \sum_{t=1}^{N} u^2(t)$$

so that the bias can be computed from

$$\begin{aligned}
b &= E\left[\sum_{t=1}^{N} u(t)\{\alpha u(t) + e(t)\} \Big/ \sum_{t=1}^{N} u^2(t)\right] - \alpha \\
&= E\left[\sum_{t=1}^{N} u(t)e(t) \Big/ \sum_{t=1}^{N} u^2(t)\right] \\
&= E\left[\sum_{t=1}^{N} \{u^o(t) + z(t)\}\{v(t) + w(t) - \alpha z(t)\} \Big/ \sum_{t=1}^{N} u^2(t)\right] \\
&= -E\left[\sum_{t=1}^{N} \{u^o(t) + z(t)\}z(t) \Big/ \sum_{t=1}^{N} \{u^o(t) + z(t)\}^2\right]\alpha
\end{aligned}$$

The last step in this derivation follows from the assumed uncorrelatedness between the noise terms and the assumed zero means. In general, the resulting bias is not equal to zero. This result could also have been seen directly by noting that $u(t) = u^o(t) + z(t)$ and $e(t) = v(t) + w(t) - \alpha z(t)$ are not statistically independent. Consequently, in these cases the least-squares estimates are biased.

Generalizing the result of this last example for the vector case leads to the following expression for the bias of the least-squares estimate:

$$\begin{aligned}
b &= E\left[(\Phi^T\Phi)^{-1}\Phi^T e\right] \\
&= -E\left[((\Phi^o + Z)^T(\Phi^o + Z))^{-1}(\Phi^o + Z)^T Z\right]\vartheta
\end{aligned} \tag{5.26}$$

where Z is an $N \times p$ matrix containing the errors in the explanatory variables. It is important to realize that bias in the estimates will directly lead to a systematic error in model predictions, which should be avoided if possible. Therefore in this section not only the expressions for bias have been evaluated, but also the conditions under which bias will occur.

5.1.5 Accuracy

In addition to bias, another important property of the estimates is the *accuracy*, also indicated as the *estimation uncertainty*. Usually, the dispersion of a random variable

y is expressed in terms of the variance (see Appendix B for details)

$$\operatorname{Var} y := E\big[\big(y - E[y]\big)^2\big] \tag{5.27}$$

denoted by σ_y^2, or its square root, the standard deviation σ_y. Generalization to the vector case gives the so-called *covariance matrix*, which is defined as

$$\operatorname{Cov} y := E\big[\big(y - E[y]\big)\big(y - E[y]\big)^T\big] \tag{5.28}$$

which also allows covariances between the different elements in vector y. It is important to note here that this definition holds for any vector with finite variance and is thus applicable to observed data sequences and estimated parameter vectors. The covariance matrix can be further evaluated as

$$\begin{aligned}\operatorname{Cov} y &= E\big[y^T y - yE[y]^T - E[y]y^T + E[y]E[y]^T\big] \\ &= E\big[yy^T\big] - E[y]E[y]^T \end{aligned} \tag{5.29}$$

If $E[y] = 0$, then $\operatorname{Cov} y = E[yy^T]$. Let us explore this special case in some more detail and let $y = [y(1)\ y(2)\ \cdots\ y(N)]^T$ be a zero-mean sequence; then

$$\begin{aligned}\operatorname{Cov} y &= E\left[\begin{bmatrix} y(1) \\ y(2) \\ \vdots \\ y(N) \end{bmatrix}\big[y(1)\ y(2) \cdots y(N)\big]\right] \\ &= \begin{bmatrix} E[y(1)y(1)] & E[y(1)y(2)] & \cdots & E[y(1)y(N)] \\ E[y(2)y(1)] & E[y(2)y(2)] & & E[y(2)y(N)] \\ \vdots & & & \vdots \\ E[y(N)y(1)] & \cdots & \cdots & E[y(N)y(N)] \end{bmatrix} \\ &= \begin{bmatrix} r_{yy}(0) & r_{yy}(1) & \cdots & r_{yy}(N-1) \\ r_{yy}(1) & r_{yy}(0) & & r_{yy}(N-2) \\ \vdots & & & \vdots \\ r_{yy}(N-1) & \cdots & \cdots & r_{yy}(0) \end{bmatrix} = R_{yy} \end{aligned} \tag{5.30}$$

Recall from Sect. 4.1.1 that $r_{yy}(l) := E[y(k)y(k+l)]$, which allows the last step in (5.30) from covariance matrix $\operatorname{Cov} y$ to equal the autocorrelation matrix R_{yy}.

Example 5.10 White noise: Consider a unit variance white noise sequence $\{e\}_{t=1}^N$, which implies that $r_{ee}(0) = 1$ and $r_{ee}(l) = 0$ for $l \neq 0$. Then, it can be easily verified that $\operatorname{Cov} e = R_{ee} = I_N$, where I_N denotes the $N \times N$ identity matrix.

In addition to the previous case of a random data sequence, in the following the covariance matrix associated with (weighted) least-squares estimates is investigated. The basic idea is that since the estimate includes a real-data vector (see, for instance,

(5.10)), which in general is corrupted with noise, the estimate will also be affected by this noise. If the data matrix Φ is statistically independent of the error e and if e has zero mean, then the covariance matrix associated with the unbiased least-squares estimate $\widehat{\vartheta}$ such that $E[\widehat{\vartheta}] = \vartheta$ is given by

$$\begin{aligned}
\operatorname{Cov}\widehat{\vartheta} &= E\big[(\widehat{\vartheta} - E[\widehat{\vartheta}])(\widehat{\vartheta} - E[\widehat{\vartheta}])^T\big] \\
&= E\big[([\Phi^T\Phi]^{-1}\Phi^T y - \vartheta)([\Phi^T\Phi]^{-1}\Phi^T y - \vartheta)^T\big] \\
&= E\big[([\Phi^T\Phi]^{-1}\Phi^T(\Phi\vartheta + e) - \vartheta)([\Phi^T\Phi]^{-1}\Phi^T(\Phi\vartheta + e) - \vartheta)^T\big] \\
&= E\big[[\Phi^T\Phi]^{-1}\Phi^T \operatorname{Cov} e\,\Phi[\Phi^T\Phi]^{-1}\big]
\end{aligned} \tag{5.31}$$

If $\{e\}_{t=1}^N$ is a white noise sequence with constant variance σ^2, then $\operatorname{Cov} e = \sigma^2 I_N$. Consequently, (5.31) reduces to

$$\operatorname{Cov}\widehat{\vartheta} = \sigma^2 E\big[[\Phi^T\Phi]^{-1}\big] \tag{5.32}$$

which in the case of deterministic Φ even further simplifies to $\operatorname{Cov}\widehat{\vartheta} = \sigma^2[\Phi^T\Phi]^{-1}$. However, the expressions cannot be directly used, since in practice σ^2 is unknown. Noting that $\sigma_\varepsilon^2 = E[\varepsilon^2(t)] - (E[\varepsilon(t)])^2 =_{[E\varepsilon(t)=0]} E[\varepsilon^2(t)]$, an unbiased estimate of σ^2 can be obtained from the prediction error sequence and is given by

$$\widehat{\sigma_\varepsilon^2} = \frac{1}{N-p}\sum_{t=1}^{N}\varepsilon^2(t) \tag{5.33}$$

Hence, in practice, σ^2 in (5.32) is replaced by (5.33).

Example 5.11 Moving object: Recall that the prediction errors are

$$\varepsilon = [-1.8\ 6.2\ -5.0\ -4.4\ 7.9\ -2.9]^T$$

and we have

$$(\Phi^T\Phi)^{-1} = \begin{bmatrix} 0.821 & -2.95 & 4.46 \\ -2.95 & 18.2 & -33.5 \\ 4.46 & -33.5 & 67.0 \end{bmatrix}$$

Then, with $N = 6$ and $p = 3$ an estimate of the prediction error variance is 52.62 m^2. Hence, the covariance matrix of the estimates is given by

$$\operatorname{Cov}\widehat{\vartheta} = \begin{bmatrix} 43.20 & -155.0 & 234.9 \\ -155.0 & 956.1 & -1762 \\ 234.9 & -1762 & 3524 \end{bmatrix}$$

where the diagonal elements are the variances of the corresponding estimates. Hence, by taking the square root of the diagonal elements the standard deviations are obtained, that is, 6.57 m, 30.92 m/s, and 59.36 m/s^2. Analysis of this result reveals that only v_0 can be accurately estimated; the other deviations are approximately equal to the estimated values indicating low accuracy.

Example 5.12 Constant process state: Consider again the case where at sampling instant t we have two measurements from two different sensors $y(1)$ and $y(2)$ of a process state x, which is considered to be constant during the experiment. Recall that the model becomes

$$y(1) = x + e(1)$$
$$y(2) = x + e(2)$$

Notice that this is in fact a *multioutput* case. If then at $t+1$ another two measurements $y(3)$ and $y(4)$ become available, we can simply add the equations

$$y(3) = x + e(3)$$
$$y(4) = x + e(4)$$

to the two regression equations given for time instant t. Assume further that

$$\begin{aligned} E\big[e(k)\big] &= 0, \quad k = 1, \ldots, 4 \\ E\big[e(k)e(k+l)\big] &= 0, \quad l > 0 \\ E\big[e(1)^2\big] = E\big[e(3)^2\big] &= 1, \qquad E\big[e(2)^2\big] = E\big[e(4)^2\big] = 4 \end{aligned}$$

Hence, the error covariance matrix is given by the diagonal matrix R_{ee} with diagonal elements 1, 4, 1, and 4.

Given the measurements $y(1), \ldots, y(4)$, a "reasonable" estimate is given by

$$\widehat{x} = \frac{1}{4}y(1) + \frac{1}{4}y(2) + \frac{1}{4}y(3) + \frac{1}{4}y(4)$$

In this case

$$x - \widehat{x} =_{[y(k)=x+e(k)]} -\frac{1}{4}\sum_{k=1}^{4} e(k)$$

so that the bias and estimation variance are given by

$$\begin{aligned} E\big[x - \widehat{x}\big] &= -\frac{1}{4}\sum_{k=1}^{4} E\big[e(k)\big] =_{[E[e(.)]=0]} = 0 \\ E\big[(x - \widehat{x})^2\big] &= \frac{1}{16}(1 + 4 + 1 + 4) = \frac{10}{16} \end{aligned}$$

Hence, the estimate is unbiased, and the variance of the estimate is equal to 10/16. However, the "best" estimate is found from weighted least-squares estimation with

$$\Phi = [1\ 1\ 1\ 1]^T, \qquad W = R_{ee}^{-1}, \quad \text{and} \quad R_{ee} = \begin{bmatrix} 1 & 0 & 0 & 0 \\ 0 & 4 & 0 & 0 \\ 0 & 0 & 1 & 0 \\ 0 & 0 & 0 & 4 \end{bmatrix}$$

so that

$$\Phi^T W = [1\ \tfrac{1}{4}\ 1\ \tfrac{1}{4}], \qquad \Phi^T W \Phi = \frac{5}{2}$$

$$\Longrightarrow \quad (\Phi^T W \Phi)^{-1} \Phi^T W = [\tfrac{2}{5}\ \tfrac{1}{10}\ \tfrac{2}{5}\ \tfrac{1}{10}]$$

and thus,

$$\widehat{x} = \frac{4}{10} y(1) + \frac{1}{10} y(2) + \frac{4}{10} y(3) + \frac{1}{10} y(4)$$

In this case

$$\begin{aligned} x - \widehat{x} &= x - \left(\frac{4}{10} y(1) + \frac{1}{10} y(2) + \frac{4}{10} y(3) + \frac{1}{10} y(4) \right) \\ &= x - \left(\frac{4}{10}\big(x + e(1)\big) + \frac{1}{10}\big(x + e(2)\big) + \frac{4}{10}\big(x + e(3)\big) + \frac{1}{10}\big(x + e(4)\big) \right) \\ &= -\left(\frac{4}{10} e(1) + \frac{1}{10} e(2) + \frac{4}{10} e(3) + \frac{1}{10} e(4) \right) \end{aligned}$$

Consequently, given the independence of the errors, so that $E[e(k)e(k+l)] = 0$, $l > 0$, we have

$$E[x - \widehat{x}] = E\left[-\left(\frac{4}{10} e(1) + \frac{1}{10} e(2) + \frac{4}{10} e(3) + \frac{1}{10} e(4) \right) \right] = 0$$

$$\begin{aligned} E\big[(x - \widehat{x})^2\big] &= (0.4)^2 E\big[e(1)^2\big] + (0.1)^2 E\big[e(2)^2\big] + (0.4)^2 E\big[e(3)^2\big] \\ &\quad + (0.1)^2 E\big[e(1)^2\big] = \frac{4}{10} \end{aligned}$$

which yields, although not proven here, the minimum estimation error variance for all unbiased estimates.

From this last example we see that multiple outputs can be easily handled by just adding extra regression equations and subsequently performing a weighted least-squares estimation with a weighting matrix equal to the inverse of the error covariance matrix.

We conclude here with the statement, found in many textbooks on least-squares estimation and known as the Gauss–Markov theorem, that for the linear regression model (5.2) with mutually uncorrelated errors and constant variance the least-squares estimate given by (5.10) provides the smallest covariance of all *unbiased* linear estimators of the form $\widehat{\vartheta}_A = Ay$. This property, in addition to its simplicity, makes the least-squares estimate very popular. However, in general, least-squares estimation does not guarantee a minimum mean-square error (MSE) in the estimates. To see this, let us first present an expression of the MSE matrix for estimate $\widehat{\vartheta}$, where $E[\widehat{\vartheta}] = \overline{\vartheta}$. Using (5.24) and (5.28),

$$\begin{aligned}
\mathrm{MSE}\,\widehat{\vartheta} &= E\big[(\widehat{\vartheta} - \vartheta)(\widehat{\vartheta} - \vartheta)^T\big] \\
&= E\big[(\widehat{\vartheta} - \overline{\vartheta} + \overline{\vartheta} - \vartheta)(\widehat{\vartheta} - \overline{\vartheta} + \overline{\vartheta} - \vartheta)^T\big] \\
&= E\big[(\widehat{\vartheta} - \overline{\vartheta})(\widehat{\vartheta} - \overline{\vartheta})^T\big] + (\overline{\vartheta} - \vartheta)(\overline{\vartheta} - \vartheta)^T \\
&= \mathrm{Cov}\,\widehat{\vartheta} + bb^T
\end{aligned} \tag{5.34}$$

This matrix clearly emphasizes the trade-off between bias and covariance. Hence, finite bias may be worth exchanging for a reduction of the covariance matrix. The class of the so-called minimum mean-square estimators will not be described here in any detail. It suffices to say that reduction of the MSE of the estimate can be obtained by the *constrained least-squares* (CLS) estimate

$$\widehat{\vartheta}_R = \big(\Phi^T \Phi + K\big)^{-1} \Phi^T y \tag{5.35}$$

which is also known as a *regularization* or *smoothing* algorithm. The symmetric matrix K can take different forms, but the simplest is $K = kI$ with k a positive scalar. It can also been shown that this specific choice of K reduces ill-conditioning in least-squares problems.

Algorithm 5.2 Constrained least-squares estimation of ϑ in linear static systems

1. Given $y(t)$ and $\phi_j(t)$ for $t = 1, \ldots, N$ and $j = 1, \ldots, p$, define the N-dimensional vector $y := [y(1), \ldots, y(N)]^T$.
2. Form the $N \times p$ matrix Φ with elements $\Phi_{t_j} := \phi_j(t)$, where ϕ_j is the jth regressor.
3. For a specific choice of the symmetric matrix K, calculate from (5.35) the constrained least-squares estimate of the unknown p-dimensional parameter vector ϑ.

5.1.6 Identifiability

An essential question prior to the parameter estimation procedure is whether the unknown model parameters can be uniquely, albeit locally, estimated from the data. Let us demonstrate this issue by a simple example.

Example 5.13 Identifiability: Let a static system be described by

$$y(t) = (\alpha_1 + \alpha_2)u(t)$$

Notice then that, given measurements of $u(t)$ and $y(t)$, we can only estimate the sum $\alpha_1 + \alpha_2$. Consequently, we cannot uniquely estimate each individual parameter from the data. Both α_1 and α_2 are what we call *unidentifiable* parameters.

This question about the uniqueness of the estimates is the main issue in *identifiability analysis* and has received much attention in the literature. When the question only focuses on the case where the experiment and model structure, in principle, lead to unique parameter values and thus without regard to uncertainties, the analysis is indicated as *theoretical* or *structural* or *deterministic identifiability analysis*. Most of the tools for this type of analysis are restricted to rather simple problems with only a few unknowns and thus not further explored here.

An exception is given by the following numerical procedure. Recall that for the (weighted) least-squares case, the identification criterion is given by (5.12). On the basis of this criterion and in analogy with the definition of an identifiable structure given by [BK70], the following definition is given:

Definition 5.1 Assume that the measured output is generated by a system with parameter vector ϑ^*. The model structure is called *locally identifiable* if the criterion function $J_W(\vartheta)$ has a local minimum at $\vartheta = \vartheta^*$.

Notice that in this definition it is implicitly assumed that the model structure is a valid representation of the system under consideration. To study the theoretical identifiability properties of the model, data can be generated from a thought experiment. Assume therefore that the data has been generated by a regression model with parameter vector $\vartheta^* \in \mathbb{R}^p$, so that

$$y(t) = \widehat{y}(t; \vartheta^*), \quad t = 1, \ldots, N \tag{5.36}$$

The model is called locally identifiable in ϑ^* if $J_W(\vartheta)$ in the neighborhood of ϑ^* has a unique minimum which occurs at ϑ^*. Obviously the main disadvantage of this definition is that it only holds in the neighborhood of ϑ^* which must be specified by the user on the basis of prior knowledge of the parameter values. Therefore, in practice, often a number of points are evaluated to obtain some regional insight in the identifiability properties. From the conditions for a local minimum it can be easily derived that a sufficient condition for a model structure to be locally identifiable in ϑ^* is that the gradient $(\partial J_W(\vartheta)/\partial\vartheta_i)$ for $i = 1, \ldots, p$ is zero and the Hessian $(\partial^2 J_W(\vartheta)/\partial\vartheta_i\partial\vartheta_j)$, the $p \times p$ matrix containing the second derivatives, is positive definite for $\vartheta = \vartheta^*$.

This condition for positive definiteness is equal to the condition of full column rankness of the matrix Φ, which implies that the columns of Φ are independent. The test on full rankness can be easily performed by calculating the *singular values* of a matrix. The so-called *singular value decomposition* (SVD) technique (see

Appendix A) is based on decomposing the $N \times p$ regressor matrix Φ as follows:

$$\Phi = USV^T \tag{5.37}$$

In (5.37), U and V are orthogonal matrices of dimensions $N \times N$ and $p \times p$, respectively, such that $U^T U = I_N$ and $V^T V = I_p$. The $N \times p$ singular value matrix S has the following structure:

$$S = \begin{bmatrix} \sigma_1 & 0 & \cdots & 0 \\ 0 & \sigma_2 & \cdots & 0 \\ \vdots & \vdots & & \vdots \\ 0 & 0 & \cdots & \sigma_p \\ \hdashline & & 0_{(N-p)\times p} & \end{bmatrix} \tag{5.38}$$

where $0_{(N-p)\times p}$ denotes an $(N-p) \times p$ zero (or null) matrix. If the SVD of Φ is calculated (for details, see Appendix A) and $\sigma_1 \geq \cdots \geq \sigma_r > \sigma_{r+1} = \cdots = \sigma_p = 0$, then the rank of Φ is equal to r. Hence, there exists a clear link between the rank of a matrix and its singular values. Instead of demanding that $\sigma_{r+1} = 0$, in practice, the numerical rank is introduced where $\sigma_{r+1} < \varepsilon$ to account for numerical errors during the computation of the SVD. Let us illustrate this technique to the moving object example.

Example 5.14 Moving object: SVD of the regressor matrix Φ associated with a specific experiment, using MATLAB's function *svd*, gives

$$\Phi = \begin{bmatrix} 1 & 0.0 & 0 \\ 1 & 0.2 & 0.02 \\ 1 & 0.4 & 0.08 \\ 1 & 0.6 & 0.18 \\ 1 & 0.8 & 0.32 \\ 1 & 1.0 & 0.50 \end{bmatrix} = USV^T$$

with

$$U = \begin{bmatrix} 0.3051 & -0.6230 & 0.5833 & 0.0332 & -0.0913 & -0.4112 \\ 0.3405 & -0.4289 & -0.0851 & 0.1568 & 0.4209 & 0.7008 \\ 0.3785 & -0.2143 & -0.4269 & -0.6582 & -0.4397 & 0.0445 \\ 0.4191 & 0.0208 & -0.4420 & 0.7013 & -0.2930 & -0.2256 \\ 0.4623 & 0.2764 & -0.1304 & -0.2214 & 0.6781 & -0.4290 \\ 0.5082 & 0.5525 & 0.5079 & -0.0117 & -0.2750 & 0.3207 \end{bmatrix}$$

$$S = \begin{bmatrix} 2.8127 & 0 & 0 \\ 0 & 0.8177 & 0 \\ 0 & 0 & 0.1089 \\ 0 & 0 & 0 \\ 0 & 0 & 0 \\ 0 & 0 & 0 \end{bmatrix}$$

$$V = \begin{bmatrix} 0.8582 & -0.5094 & 0.0635 \\ 0.4796 & 0.7516 & -0.4529 \\ 0.1829 & 0.4191 & 0.8893 \end{bmatrix}$$

Usually, U is called the left singular vector matrix, and V the right singular vector matrix. From these results it can be concluded that, for $\varepsilon = 10^{-6}$, Φ has full rank, since the smallest singular value is significantly larger than 10^{-6}. Consequently, it is expected that the unknowns can be uniquely estimated from experimental data, because this full-rank condition implies that $(\Phi^T \Phi)^{-1}$ exists.

The effect of changing the time coordinates in the moving object example is illustrated in the next example.

Example 5.15 Moving object: Let the time start at 10 s rather than at time zero. Then,

$$\Phi = \begin{bmatrix} 1 & 10.0 & 50 \\ 1 & 10.2 & 50.02 \\ 1 & 10.4 & 54.08 \\ 1 & 10.6 & 56.18 \\ 1 & 10.8 & 58.32 \\ 1 & 11.0 & 60.50 \end{bmatrix}$$

with singular values

$$\sigma = \{137.8981,\ 0.8336,\ 0.0022\}$$

and right singular vector matrix

$$V = \begin{bmatrix} 0.0177 & -0.1872 & 0.9822 \\ 0.1865 & -0.9645 & -0.1872 \\ 0.9823 & 0.1865 & 0.0178 \end{bmatrix}$$

The resulting estimates are

$$\hat{\vartheta} = [428.0\ -319.1\ 55.4]^T$$

which substantially deviate from previous estimation results. Especially, x_0 and v_0 are badly estimated.

Let us analyze this result in some more detail. First, the smallest singular value is very small, indicating that some of the regressors are close to being linearly dependent. This result can also be directly seen from Φ in Example 5.14 by inspection of the first two columns. Notice that the second column is approximately 10 times column one. Consequently, bad estimates of x_0 and v_0 result. Secondly, let us premultiply the linear regression equation by U^T, so that

$$\begin{aligned} y^* &= U^T y = U^T \Phi\vartheta + U^T e \\ &= U^T U S V^T \vartheta + U^T e \\ &= S\vartheta^* + e^* \end{aligned} \tag{5.39}$$

where $\vartheta^* = V^T\vartheta$ and $e^* = U^T e$.

Notice from the orthogonality of U with $U^T U = I_N$ that $U^T = U^{-1}$ and thus $U U^T = I_N$. Then, given the transformed prediction error $\varepsilon^* = y^* - S\vartheta^* = U^T\varepsilon$, it follows that $J^* := (\varepsilon^*)^T \varepsilon^* = \varepsilon^T U U^T \varepsilon =_{[UU^T = I_N]} J$. Thus, it can be easily verified that the sum of squares is not altered by this transformation. The first term on the right-hand side of the linear regression model is transformed into

$$S\vartheta^* = \begin{bmatrix} \sigma_1\vartheta_1^* \\ \sigma_2\vartheta_2^* \\ \vdots \\ \sigma_p\vartheta_p^* \\ 0 \\ 0 \\ \vdots \\ 0 \end{bmatrix} \tag{5.40}$$

so that $J(\vartheta) = (y^* - S\vartheta^*)^T(y^* - S\vartheta^*)$ is minimized when $\widehat{\vartheta}_i^* = y_i^*/\sigma_i$ setting $\varepsilon_i^* = 0$ for $i = 1, \ldots, p$. Consequently, the parameter estimates can be readily obtained from $\widehat{\vartheta} = V\widehat{\vartheta}^*$, because V is an orthogonal matrix for which $V^T V = I$ and thus $(V^T)^{-1} = V$. In case $\sigma_i \approx 0$, the associated parameter ϑ_i^* can be chosen arbitrarily, because the complete term $\sigma_i\vartheta_i^*$ does not contribute too much to the sum of squares. Hence, a better choice is to reparameterize the model by setting the linear parameter combination $\vartheta_i^* = v_i^T\vartheta$ equal to zero, so that it does not affect the original parameters too much. This method is also known as the *truncated* least-squares method.

Algorithm 5.3 Truncated least-squares estimation of ϑ in linear static systems

1. Given $y(t)$ and $\phi_j(t)$ for $t = 1, \ldots, N$ and $j = 1, \ldots, p$, define the N-dimensional vector $y := [y(1), \ldots, y(N)]^T$.
2. Form the $N \times p$ matrix Φ with elements $\Phi_{t_j} := \phi_j(t)$, where ϕ_j is the jth regressor.
3. Calculate the SVD of Φ, using for example MATLAB's *svd*, which gives U, S, and V.

4. Premultiply y with U^T, leading to y^*.
5. For $i = 1, \ldots, p$ calculate the transformed estimates $\widehat{\vartheta}_i^* = y_i^*/\sigma_i$.
6. Calculate from $\widehat{\vartheta} = V\widehat{\vartheta}^*$ the truncated least-squares parameter estimate of the unknown p-dimensional parameter vector ϑ.

Again, let us apply this to the moving object example with shifted time axis.

Example 5.16 Moving object: Recall that an SVD of Φ with time starting at 10 s gives

$$S = \begin{bmatrix} 137.8981 & 0 & 0 \\ 0 & 0.8336 & 0 \\ 0 & 0 & 0.0022 \\ 0 & 0 & 0 \\ 0 & 0 & 0 \\ 0 & 0 & 0 \end{bmatrix}$$

and

$$V = \begin{bmatrix} 0.0177 & -0.1872 & 0.9822 \\ 0.1865 & -0.9645 & -0.1872 \\ 0.9823 & 0.1865 & 0.0178 \end{bmatrix}$$

so that the following estimates $\widehat{\vartheta}_i^* = y_i^*/\sigma_i$ for $i = 1, 2, 3$ are obtained:

$$\widehat{\vartheta}^* = [-2.44\ -237.96\ 481.09]^T$$

with $y^* = [-337.1\ -198.4\ 1.0\ -0.5\ 12.5\ 1.4]^T$. By setting $\widehat{\vartheta}_3^* = 0$, since $\sigma_3 \approx 0$, the sum of squares are increased from 157.86 to 158.96, and the following estimates are obtained from $V\widehat{\vartheta}^*$:

$$\widehat{\vartheta} = [-44.51\ -229.05\ 46.78]$$

which give reasonable predictions but are still unrealistic. Clearly, in this case the best solution is to shift the time coordinates 10 s to the left.

So far, the analysis has only been focussed on Φ, and as yet no output data has been incorporated. Identifiability analysis which includes measurement uncertainty in the output and numerical inaccuracy is called *practical identifiability analysis*. In practical identifiability analysis the analysis is completely focussed on the covariance matrix of the estimates. Let us illustrate this by an example.

Example 5.17 Moving object: Recall that the covariance matrix of the estimates in the original linear regression model was given by

$$\text{Cov}\,\widehat{\vartheta} = \begin{bmatrix} 43.20 & -155.0 & 234.9 \\ -155.0 & 956.1 & -1762 \\ 234.9 & -1762 & 3524 \end{bmatrix}$$

An SVD of this matrix gives:

$$\sigma = \{4438, 78.70, 6.651\}$$

$$U = V = \begin{bmatrix} 0.0635 & -0.5094 & 0.8582 \\ -0.4529 & 0.7516 & 0.4796 \\ 0.8893 & 0.4191 & 0.1829 \end{bmatrix}$$

where U is equal to V, because the covariance matrix is symmetric, and thus $\mathrm{Cov}\,\widehat{\vartheta} = VSV^T$. Consequently, with $V^T = V^{-1}$ because of the orthogonality of V, i.e., $V^T V = I_p$, $\mathrm{Cov}\,\widehat{\vartheta}\,V = VS$ defines an eigenvalue decomposition of $\mathrm{Cov}\,\widehat{\vartheta}$. Hence, singular value or eigenvalue decomposition of a covariance matrix will give the same result. For further analysis of this result, it should be mentioned that each of the singular values or eigenvalues is associated with a corresponding column in V. Each column in V defines a direction in the parameter space (see also Appendix B). Furthermore, a small singular value indicates a well-defined direction. Hence, since the third singular value is small, the parameter combination $0.8582x_0 + 0.4796v_0 + 0.1829a$ can be accurately estimated from the experimental data. This conclusion further implies that a specific combination of x_0 and v_0, due to their large contribution to this well-defined direction, can be estimated rather accurately. A similar conclusion can be drawn from our previous analysis of the estimates uncertainty.

Let us visualize the result in $\mathbb{R}^2$ for the parameters v_0 and a, neglecting the effect of x_0 on the output. Recall that, using (5.10),

$$\widehat{\vartheta} = [252\ 29.3]^T$$

The corresponding covariance matrix is given by

$$\mathrm{Cov}\,\widehat{\vartheta} = \begin{bmatrix} 352.9 & -811.2 \\ -811.2 & 1983 \end{bmatrix}$$

An SVD of this matrix gives:

$$\sigma = \{18.1, 2318\}$$

$$U = V = \begin{bmatrix} -0.9243 & -0.3816 \\ -0.3816 & 0.9243 \end{bmatrix}$$

The uncertainty ellipse (see Appendix B), which in this case is an isoline connecting points of equal objective function values (sum of squares), is presented in Fig. 5.2.

Notice from Fig. 5.2 that the uncertainty ellipse is rather thin in one direction. However, for a correct geometrical interpretation of the result, we must plot the ellipse with equally scaled axes, as in Fig. 5.3. Notice from this figure that the main axis of the uncertainty ellipse is more or less aligned with the y-axis. To be more specific, this main axis is described by the second column vector of V. In other words, the estimate of the acceleration a is most uncertain, as we concluded before. Consequently, the parameter combination $0.9243v_0 + 0.3816a$, with a large weight on v_0, can be most accurately estimated from the experimental data.

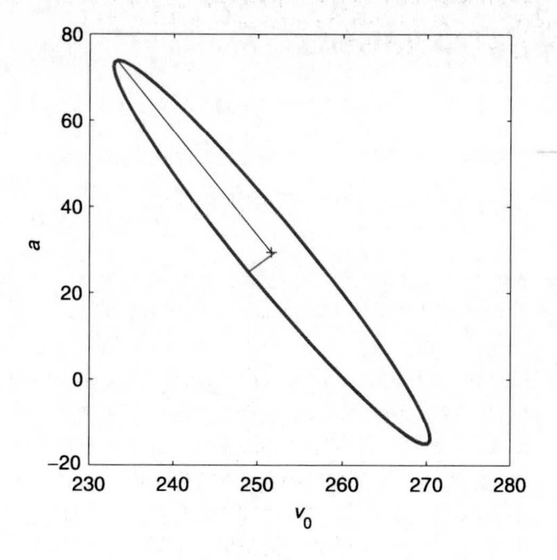

Fig. 5.2 Uncertainty ellipse of the estimates of the velocity (v_0) and the acceleration (a)

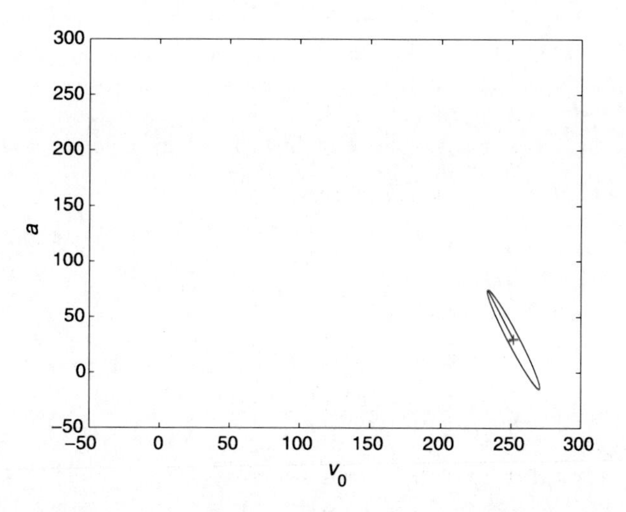

Fig. 5.3 Uncertainty ellipse of the estimates of the velocity (v_0) and the acceleration (a); equal scale plot

Notice that the identifiability analysis in the previous example does not evaluate the uncertainty with respect to the estimated value, indicating that x_0 is roughly equal to the standard deviation of its estimate, which is an indication of an inappropriate model structure for the given output data. Notice also the similarity between V and the right singular value matrix obtained from an SVD of the original regressor matrix Φ. This similarity can be verified using (5.32), which expresses the covariance matrix as a function of Φ and the property $(\Phi^T\Phi)V = VS^TS$ (see Appendix A for details).

In conclusion, for both practical and theoretical local identifiability studies using thought experiments, it suffices to evaluate the SVD of the regressor matrix Φ.

5.1.7 *Errors-in-variables Problem

Recall that in Example 5.9 the so-called errors-in-variables (EIV) problem has already been introduced as a result of noise in the explanatory variables. Applying ordinary least-squares estimation will in general lead to bias (see (5.26)). In this subsection, we will now introduce the so-called *Total Least-Squares* (TLS) method, which is able to properly solve this type of problems using SVD.

Before focusing on the TLS method, let us first introduce the norm of a vector $x \in \mathbb{R}^n$, denoted by $\|x\|$. A vector norm on $\mathbb{R}^n$ for $x, y \in \mathbb{R}^n$ satisfies the following properties:

$$\|x\| \geq 0 \quad \big(\|x\| = 0 \iff x = 0\big) \tag{5.41}$$

$$\|x + y\| \leq \|x\| + \|y\| \tag{5.42}$$

$$\|\alpha x\| = |\alpha|\|x\| \tag{5.43}$$

where $|\alpha|$ denotes the absolute value of the scalar $\alpha \in \mathbb{R}$. Many vector norms satisfy the properties of (5.41)–(5.43). Some frequently used norms, such as the 1-, 2-, and ∞-norm, are given by

$$\|x\|_1 = |x_1| + \cdots + |x_n| \tag{5.44}$$

$$\|x\|_2 = \left(x_1^2 + \cdots + x_n^2\right)^{\frac{1}{2}} \tag{5.45}$$

$$\|x\|_\infty = \max_{1 \leq i \leq n} |x_i| \tag{5.46}$$

where the subscripts on the double bar are used to denote a specific norm. Consequently, so far we have used the 2-norm or Euclidean norm to define the length of a vector. However, this idea of norms can be further extended to matrices $A, B \in \mathbb{R}^{m \times n}$ with the same kind of properties as presented above ((5.41)–(5.43)). In particular, in what follows, we will use the so-called *Frobenius* norm $\|\cdot\|_F$,

$$\|A\|_F = \sqrt{\sum_{i=1}^{m} \sum_{j=1}^{n} |a_{ij}|^2} \tag{5.47}$$

Given the norms of a vector and a matrix, the weighted least-squares problem could also be formulated as

$$\min_{y+e \in \operatorname{ran}(\Phi)} \left\| W(y - \Phi\vartheta) \right\|_2, \quad y \in \mathbb{R}^N \tag{5.48}$$

where $\operatorname{ran}(\Phi) = \{\widehat{y} \in \mathbb{R}^N : \widehat{y} = \Phi\vartheta \text{ for some } \vartheta \in \mathbb{R}^p\}$, the range of the matrix Φ (see Appendix A for details). If, however, errors are also present in the data matrix Φ, then it would be more natural to formulate the estimation problem as

$$\min_{y+e \in \operatorname{ran}(\Phi+Z)} \left\| D[Z, e]T \right\|_F \tag{5.49}$$

with $Z \in \mathbb{R}^{N \times p}$, a matrix containing the errors in Φ, and $e \in \mathbb{R}^N$. Furthermore, the nonsingular matrices $D = \text{diag}(D_{11}, \ldots, D_{NN})$ and $T = \text{diag}(T_{11}, \ldots, T_{pp}, T_{p+1,p+1})$ are added to weight the different errors. This estimation problem is referred to as the total least-squares (TLS) problem. For the multioutput case with observation matrix $Y \in \mathbb{R}^{N \times k}$ and observation noise matrix $E \in \mathbb{R}^{N \times k}$, (5.49) can be written as

$$\min_{\text{ran}(Y+E) \subseteq \text{ran}(\Phi+Z)} \left\| D[Z, E]T \right\|_F \tag{5.50}$$

If $[Z_0, E_0]$ solves (5.50), then any $\Theta \in \mathbb{R}^{p \times k}$ that satisfies $(\Phi + Z_0)\Theta = (Y + E_0)$ is said to be a TLS solution. The next question is: "how do we compute Θ, preferably in a direct (noniterative) way?" In what follows, we will only focus on the single output case, i.e., $k = 1$. Assume that $N \geq p + 1$ and let U, V, and S be obtained from an SVD of $[\Phi, y]$ with

$$U = [U_1 \; U_2], \qquad V = \begin{bmatrix} V_{11} & V_{12} \\ V_{21} & V_{22} \end{bmatrix}$$

$$S = \begin{bmatrix} S_1 & 0 \\ 0 & S_2 \end{bmatrix}$$

with $U_1 \in \mathbb{R}^{N \times (N-1)}$, $U_2 \in \mathbb{R}^{N \times 1}$, $V_{11}, S_1 \in \mathbb{R}^{p \times p}$, $V_{12} \in \mathbb{R}^{p \times 1}$, $V_{21} \in \mathbb{R}^{1 \times p}$, and $V_{22}, S_2 \in \mathbb{R}$. If $\sigma_p([\Phi, y]) > \sigma_{p+1}([\Phi, y])$, then the matrix $D[Z_0 \; e_0]T$ defined by

$$D[Z_0, e_0]T := -U_2 S_2 \left[V_{12}^T, V_{22}^T\right] \tag{5.51}$$

solves (5.49). If $T_1 = \text{diag}(T_{11}, \ldots, T_{pp})$ and $T_2 = T_{p+1,p+1}$, then the unique TLS solution is given by

$$\vartheta_{\text{TLS}} = -T_1 V_{12} V_{22}^{-1} T_2^{-1} \tag{5.52}$$

Algorithm 5.4 Total least-squares estimation of ϑ in linear static systems

1. Given $y(t)$ and $\phi_j(t)$ for $t = 1, \ldots, N$ and $j = 1, \ldots, p$, define the N-dimensional vector $y := [y(1), \ldots, y(N)]^T$.
2. Form the $N \times p$ matrix Φ with elements $\Phi_{t_j} := \phi_j(t)$, where ϕ_j is the jth regressor.
3. Define the weighting matrices D, T.
4. Calculate the SVD of $[\Phi, y]$, using, for example, MATLAB's *svd*, which gives U, S, and V.
5. Calculate from (5.52) the total least-squares estimate of the unknown p-dimensional parameter vector ϑ.
6. Calculate the noise-free regressors and residual vectors from (5.51).

Let us illustrate the TLS method to Example 5.3, but now without the estimation of the initial distance, which cannot be estimated accurately.

Example 5.18 Moving object: Without the estimation of the initial distance, the data matrices become

$$\Phi = \begin{bmatrix} 0.0 & 0 \\ 0.2 & 0.02 \\ 0.4 & 0.08 \\ 0.6 & 0.18 \\ 0.8 & 0.32 \\ 1.0 & 0.50 \end{bmatrix}, \qquad y = \begin{bmatrix} 3 \\ 59 \\ 98 \\ 151 \\ 218 \\ 264 \end{bmatrix}$$

Let $D = I_N$ and $T = I_{p+1}$. Then, the TLS algorithm computes the vector $\vartheta = \left[\begin{smallmatrix} v \\ a \end{smallmatrix}\right] \in \mathbb{R}^2$ such that $(\Phi + Z_0)\widehat{\vartheta} = (y + e_0)$ and $\|[Z_0, e_0]\|_F$ minimal.

The SVD of $[\Phi, y]$ gives $\sigma_1 = 391.3024 > \sigma_2 = 0.1479 > \sigma_3 = 0.0534$ with

$$U = \begin{bmatrix} -0.0077 & -0.0258 & -0.2185 & -0.1916 & -0.6406 & -0.7103 \\ -0.1508 & -0.4717 & -0.5369 & -0.0172 & -0.4025 & 0.5516 \\ -0.2504 & -0.5017 & 0.4377 & -0.6951 & 0.1023 & -0.0185 \\ -0.3859 & -0.3825 & 0.4476 & 0.6709 & -0.1990 & -0.1211 \\ -0.5571 & -0.1139 & -0.5070 & 0.0535 & 0.5475 & -0.3421 \\ -0.6747 & 0.6048 & 0.1226 & -0.1638 & -0.2790 & 0.2434 \end{bmatrix}$$

$$V = \begin{bmatrix} -0.0038 & -0.0734 & 0.9973 \\ -0.0016 & 0.9973 & 0.0734 \\ -1.0000 & -0.0013 & -0.0039 \end{bmatrix}$$

so that

$$\widehat{\vartheta}_{\text{TLS}} = \begin{bmatrix} 0.9973/0.0039 \\ 0.0734/0.0039 \end{bmatrix} = \begin{bmatrix} 256.2520 \\ 18.8637 \end{bmatrix}$$

The ordinary least-squares estimate is given by $\left[\begin{smallmatrix} 251.6304 \\ 29.3478 \end{smallmatrix}\right]$. Furthermore,

$$\begin{aligned} [Z_0, e_0] &= -U_2\sigma_3\left[V_{12}^T, V_{22}^T\right] \\ &= \begin{bmatrix} 0.0379 & 0.0028 & -0.0001 \\ -0.0294 & -0.0022 & 0.0001 \\ 0.0010 & 0.0001 & -0.0000 \\ 0.0065 & 0.0005 & -0.0000 \\ 0.0182 & 0.0013 & -0.0001 \\ -0.0130 & -0.0010 & 0.0001 \end{bmatrix} \end{aligned}$$

The model outputs related to $\widehat{\vartheta}_{\text{LS}}$ and $\widehat{\vartheta}_{\text{TLS}}$ can be seen in Fig. 5.4.

Notice that inclusion of the initial distance as an unknown parameter will add a *noise-free* column with ones in Φ.

So far, only the basic TLS method has been introduced. To close this section, it should be emphasized that in the last decade many modifications to handle, for instance, noise-free columns in Φ (as in the Moving object example), correlation

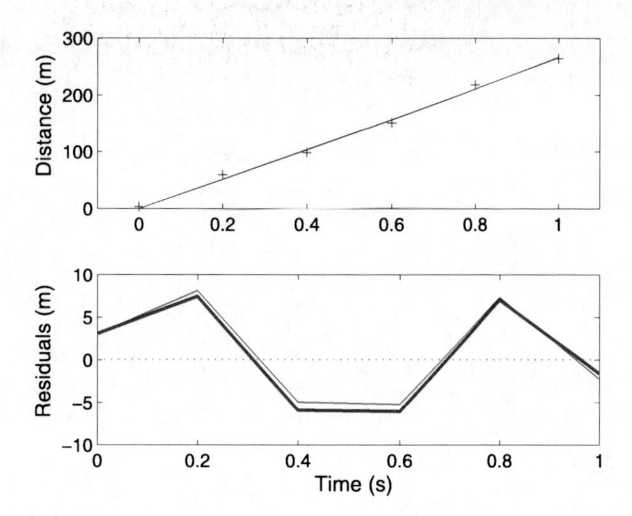

Fig. 5.4 TLS results (*solid line*) with measurements (+) (*top*), and residuals related to TLS (*bold line*) and OLS (*thin line*) results (*bottom*)

between rows or columns, and presence of bias due to nonlinearities in the data, have been proposed as well.

*5.1.8 *Bounded-noise Problem: Linear Case*

During the last two decades a growing amount of papers on so-called set-membership identification or parameter-bounding approaches has become available. The key problem in this bounded-noise identification is not to find a single vector of optimal parameter estimates, but a set of feasible parameter vectors that are consistent with a given model structure and data with bounded uncertainty. A bounded error characterization, as opposed to a statistical characterization in terms of mean, variances, or probability distributions, is favored when the central limit theorem (see Appendix B) is inapplicable, for example, in situations with small data sets or with heavily structured (modeling) errors. Typically, the set-membership approach has been applied for estimation of economic, ecological, and environmental systems, which were all characterized by small data sets. However, it has also been used in signal classification and fault detection in industrial applications.

Recall from Sect. 5.1.1 that a linear regression type of model can be represented as

$$y = \Phi\vartheta + e \tag{5.53}$$

In the set-membership context, the error or information uncertainty vector e is commonly assumed to be point-wise bounded, that is,

$$\|e\|_\infty \le \varepsilon \tag{5.54}$$

with constant error bound ε, a fixed positive number. Define the error set as

$$\Omega_e := \left\{ e \in \mathbb{R}^N : \|e\|_\infty \le \varepsilon \right\} \tag{5.55}$$

Hence, a measurement uncertainty set (MUS), containing all possible output vectors consistent with the observed output data and uncertainty characterization, is defined as

$$\Omega_y := \left\{\tilde{y} \in \mathbb{R}^N : \left\|y - \tilde{y}\right\|_\infty \leq \varepsilon\right\} \tag{5.56}$$

This set is a hypercube in $\mathbb{R}^N$. Let the set

$$\Omega_\vartheta := \left\{\vartheta \in \mathbb{R}^p : \|y - \Phi\vartheta\|_\infty \leq \varepsilon\right\} \tag{5.57}$$

define the feasible parameter set (FPS). Then, the set-membership estimation problem is to characterize this feasible parameter set, which is consistent with the model (5.53), the data (y), and uncertainty characterization (5.54).

For further analysis, the image set, which is a p-dimensional variety in the N-dimensional measurement space, is defined as follows:

$$\Omega_{\tilde{y}} := \left\{\tilde{y} \in \mathbb{R}^N : \tilde{y} = \Phi\vartheta;\, \vartheta \in \mathbb{R}^p\right\} \tag{5.58}$$

The image set related to the FPS, also called the feasible model output set, is then defined as

$$\Omega_{\widehat{y}} := \left\{\widehat{y} \in \mathbb{R}^N : \widehat{y} = \Phi\vartheta;\, \vartheta \in \Omega_\vartheta\right\} \tag{5.59}$$

$$= \Omega_{\tilde{y}} \cap \Omega_y \tag{5.60}$$

Let us illustrate the introduced sets by a simple example with two measurements and two unknown parameters. Furthermore, the example will also show some of the specific estimation problems in linear bounded-noise identification.

Example 5.19 Moving object (constant velocity): Consider an object moving in a straight line with constant velocity, so that $y(t) = \vartheta_1 + \vartheta_2 t$. The first three measurements are: $t(1) = 1$, $y(1) = 9$; $t(2) = 2$, $y(2) = 15$; and $t(3) = 3$, $y(3) = 19$ ([You84], p. 18). Assume that the error bound is given by $\varepsilon = 2$. Hence, when only one measurement at $t(1)$ is available, Ω_y is an interval, in this case $[7, 11]$. The parameter set Ω_ϑ is unbounded, that is, only bounded by a pair of bounds: $\vartheta_1 + \vartheta_2 = 7$ and $\vartheta_1 + \vartheta_2 = 11$ (see bold lines for $t = 1$ in Fig. 5.5). Consequently, the image set is equal to the real axis, and the feasible model output set is equal to the measurement uncertainty set $\Omega_{\hat{y}} = [7, 11]$.

When the second measurement at $t(2)$ becomes available, Ω_y becomes a square with center $[9\ 15]^T$ and edges with length 2ε in the measurement space. Consequently, in the parameter space another pair of bounds (bold lines for $t = 2$) is added, which, together with the bounds related to the first measurement, defines an exact solution to the parameter bounding estimation problem. Notice from Fig. 5.5 that after processing two measurements, Ω_ϑ becomes a convex set, in this case a parallelogram. The vertex set of Ω_ϑ, after two measurements, is given by

$$\left\{\begin{bmatrix}1\\6\end{bmatrix}, \begin{bmatrix}9\\2\end{bmatrix}, \begin{bmatrix}-3\\10\end{bmatrix}, \begin{bmatrix}5\\6\end{bmatrix}\right\}$$

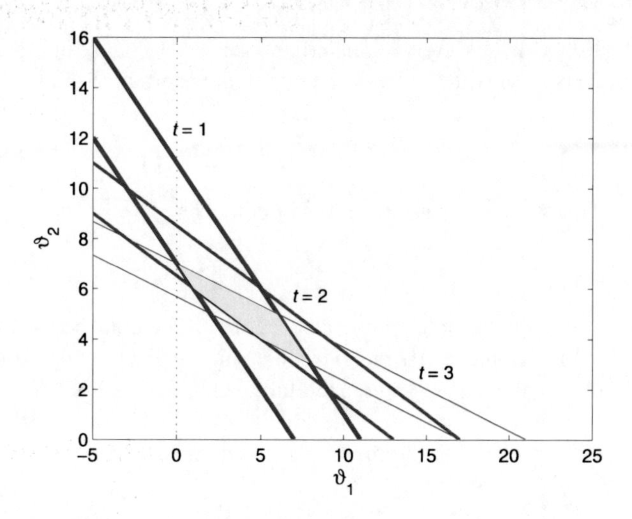

Fig. 5.5 Bounded-noise parameter estimation results

The image set is equal to $\mathbb{R}^2$, and again the feasible model output set is equal to the MUS. Furthermore, if prior knowledge requires $\vartheta_1 \geq 0$, a so-called polytope results. The vertex set of this polytope becomes

$$\left\{ \begin{bmatrix} 0 \\ 7 \end{bmatrix}, \begin{bmatrix} 0 \\ 8.5 \end{bmatrix}, \begin{bmatrix} 1 \\ 6 \end{bmatrix}, \begin{bmatrix} 9 \\ 2 \end{bmatrix}, \begin{bmatrix} 5 \\ 6 \end{bmatrix} \right\}$$

However, when the third measurement at $t(3) = 3$ becomes available, the feasible model output set will no longer be equal to the MUS, a box in $\mathbb{R}^3$ with center $[9\ 15\ 19]^T$, and the image set becomes a two-dimensional variety in $\mathbb{R}^3$. The measurements (*), error bounds on the measurements (−), and the possible feasible model outputs (shaded region) as functions of time are presented in Fig. 5.6. It can be seen from Fig. 5.6 that the feasible model outputs do not span the full region described by the bounded measurements.

The vertex set of Ω_ϑ, after three measurements, is given by

$$\left\{ \begin{bmatrix} 0 \\ 7 \end{bmatrix}, \begin{bmatrix} 1 \\ 6 \end{bmatrix}, \begin{bmatrix} 5 \\ 4 \end{bmatrix}, \begin{bmatrix} 8 \\ 3 \end{bmatrix}, \begin{bmatrix} 6 \\ 5 \end{bmatrix} \right\}$$

(see also the colored region in Fig. 5.5).

As shown in the example, it appears that at sample instant t, each measurement with its associated noise bounds defines two bounding surfaces in the parameter space, which bound a feasible parameter region ($\Omega_\vartheta(t)$). Hence, each parameter vector situated within this region is consistent with the uncertain measurement. Consequently, the intersection of these individual regions will provide an exact characterization of Ω_ϑ, that is,

$$\Omega_\vartheta := \bigcap_{t=1}^{N} \Omega_\vartheta(t) \tag{5.61}$$

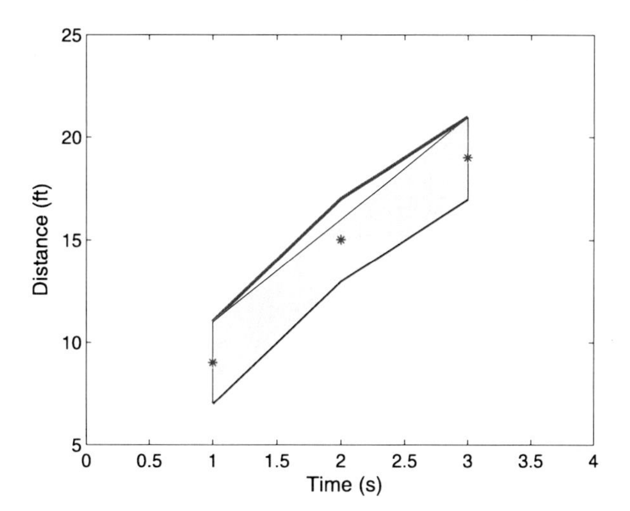

Fig. 5.6 Bounded-noise model output results

When the model is linear in the parameters, the feasible set becomes a polytope. The complexity of the resulting polytope depends on the number of data and especially on the parameter dimensionality. Efficient algorithms have been developed to solve the problem for models with limited number of parameters.

Instead of trying to find an exact characterization, we could also try to encapsulate the solution set by a set with lower complexity, as orthotopes (hypercubes), parallelotopes, or ellipsoids. In fact, for orthotopic bounding, supporting hyperplanes are found by solving a couple of LP problems. Define therefore the individual parameter uncertainty interval

$$b_i := \Big[\min_{\vartheta \in \Omega_\vartheta} \vartheta_i, \max_{\vartheta \in \Omega_\vartheta} \vartheta_i\Big] \quad \text{for } i = 1, \ldots, p \tag{5.62}$$

The resulting orthotopic outer-bounding set, which can thus be found by solving $2p$ LP problems with $2N$ constraints, becomes

$$\mathscr{B} = b_1 \times b_2 \times \cdots \times b_p \tag{5.63}$$

In spite of the fact that the resulting orthotope aligned with the coordinate axes provides minimum uncertainty intervals, the approximation can be very rough if the exact solution set shows parameter interactions. Therefore, for the linear case, it has been suggested to solve the resulting LP problems on a rotated basis. Alternatively, $\Omega_\vartheta(t)$ can also be approximated by an ellipsoid, that is,

$$\mathscr{E}(t) = \big\{\vartheta \in \mathbb{R}^p : \big(\vartheta - m(t)\big)^T P(t)^{-1}\big(\vartheta - m(t)\big) \leq 1\big\} \tag{5.64}$$

where m is the center of the ellipsoid, and the $p \times p$ matrix P defines the orientation and size of the ellipsoid. However, the intersection of the individual ellipsoids, as in (5.61), will in general not lead to an ellipsoid. Consequently, an ellipsoidal outer-bounding step is needed, which in a sequential version of the algorithm is performed after each update.

Finally, also projection set algorithms have been proposed for solving the set-membership estimation problem approximately. Essentially, in these algorithms the measurement uncertainty set (5.56) is projected onto the subspace $\mathrm{ran}(\Phi)$, the range of regressor matrix Φ. In particular, the specific case of so-called ℓ_2-projection (least-squares) under ∞-norm bounded noise has been analyzed. Here, the projection set is found by orthogonal projection of the vertices of Ω_y onto the image set $\Omega_{\tilde{y}}$. A particular result is obtained when a weighting is introduced. Under a specific choice of the weights a minimum-volume weighted least-squares set, a parallelotope in $\mathbb{R}^p$, can be found. This set can be computed rather efficiently when the data is processed sequentially.

5.2 Nonlinear Static Systems

5.2.1 Nonlinear Regression

The linear regression case associated with linear static system estimation problems can be easily extended to the nonlinear case. Consider therefore the following nonlinear regression model:

$$y = f(\Phi, \vartheta) + e \tag{5.65}$$

where $y = [y(1), y(2), \ldots, y(N)]^T$ is the measured output vector, $e = [e(1), e(2), \ldots, e(N)]^T$ is the prediction error vector, and $f(\Phi, \vartheta)$ is a vector function relating the explanatory variables to the output. Again, ϑ denotes the unknown parameter vector. Essentially, ϑ contains all the unknowns that have to be estimated from the data. As in previous sections, we will mainly focus on parameter estimation. On the basis of this nonlinear regression model, predicted values, with t as an explanatory variable, can be obtained from the following predictor:

$$\widehat{y}(t, \vartheta) = f(\Phi, \vartheta; t) \tag{5.66}$$

Let us illustrate this by an example.

Example 5.20 Nitrification experiment: The maximal oxygen demand rate $rs_{\max}(t)$ in a nitrification experiment can be expressed as

$$rs_{\max}(t) = rs_{\max}(0)e^{-bt} + \mu_{\max} B\left[1 - e^{-bt}\right]/b$$

with nitrogen load $B = 0.281$ kg N/m^{3}day. The unknown parameters are: b, the death rate of the nitrifying biomass, and $\mu_{\max}$, the maximal growth rate of the nitrifying biomass. Hence, given N measurements of $rs_{\max}$, we define

$$y := \left[rs_{\max}(0)\ rs_{\max}(1) \cdots rs_{\max}(N)\right]^T$$

and

$$f(\Phi, \vartheta) := r_{S\,\max}(0)e^{-bt} + \mu_{\max} B\left[1 - e^{-bt}\right]/b$$

with $\vartheta := [b\ \mu_{\max}]^T$ and explanatory variables t and B (fixed in this experiment).

For application in further analyzes, the sensitivity matrix $X(\vartheta) \in \mathbb{R}^{N\times p}$ is introduced. This sensitivity matrix is given by

$$X(\vartheta) = \begin{bmatrix} \frac{\partial \widehat{y}(1;\vartheta)}{\partial \vartheta_1} & \frac{\partial \widehat{y}(1;\vartheta)}{\partial \vartheta_2} & \cdots & \frac{\partial \widehat{y}(1;\vartheta)}{\partial \vartheta_p} \\ \frac{\partial \widehat{y}(2;\vartheta)}{\partial \vartheta_1} & \frac{\partial \widehat{y}(2;\vartheta)}{\partial \vartheta_2} & & \frac{\partial \widehat{y}(2;\vartheta)}{\partial \vartheta_p} \\ \vdots & & & \vdots \\ \frac{\partial \widehat{y}(N;\vartheta)}{\partial \vartheta_1} & \frac{\partial \widehat{y}(N;\vartheta)}{\partial \vartheta_2} & \cdots & \frac{\partial \widehat{y}(N;\vartheta)}{\partial \vartheta_p} \end{bmatrix} \tag{5.67}$$

which contains partial differential coefficients of the model output with respect to the unknown parameters. This $N \times p$ matrix, which expresses the sensitivities of $\widehat{y}(t;\vartheta)$ with respect to ϑ, is also indicated as the Jacobi matrix of f with respect to ϑ. Notice that in the linear case where $f(\Phi, \vartheta) = \Phi\vartheta$, the sensitivity matrix is equal to Φ. Hence, the sensitivity matrix can also be considered as a local regressor matrix at the point ϑ.

Example 5.21 Nitrification experiment: The sensitivity vectors at a specific time instant for the previous nonlinear regression model, denoted as $\psi(t, \vartheta)$, with explicit reference to the explanatory variable t, is given by

$$\begin{aligned} \psi(t, \vartheta) &= \left[\frac{\partial f(\Phi,\vartheta;t)}{\partial \vartheta_1}\ \frac{\partial f(\Phi,\vartheta;t)}{\partial \vartheta_2}\right]^T \\ &= \begin{bmatrix} -r_{S\,\max}(0)te^{-bt} - \frac{\mu_{\max} B}{b^2} + \frac{t\mu_{\max} B}{b}e^{-bt} + \frac{\mu_{\max} B}{b^2}e^{-bt} \\ \frac{B}{b} - \frac{B}{b}e^{-bt} \end{bmatrix} \end{aligned}$$

Hence, for a given experiment, including initial guesses of $\mu_{\max}$ and b, the sensitivity matrix $X(\vartheta) = [\psi(1, \vartheta), \ldots, \psi(N, \vartheta)]^T$ can be easily evaluated for the different sample instants.

5.2.2 Nonlinear Least-squares Estimation

As in the linear case, we can try to minimize the sum of squares of the prediction errors, that is, for the nonlinear regression model (5.65)

$$\begin{aligned} J(\vartheta) &= \varepsilon^T \varepsilon \\ &= \left(y - f(\Phi, \vartheta)\right)^T \left(y - f(\Phi, \vartheta)\right) \end{aligned} \tag{5.68}$$

Again, ϑ is chosen such that the gradient of J with respect to ϑ is zero, that is,

$$\frac{\partial J(\vartheta)}{\partial \vartheta} = -2\left(\frac{\partial f(\Phi,\vartheta)}{\partial \vartheta}\right)^T \big(y - f(\Phi,\vartheta)\big) = 0 \tag{5.69}$$

where $\frac{\partial f(\Phi,\vartheta)}{\partial \vartheta} = X(\vartheta,t)$ is the sensitivity matrix. Hence,

$$\left(\frac{\partial f(\Phi,\vartheta)}{\partial \vartheta}\right)^T f(\Phi,\vartheta) = \left(\frac{\partial f(\Phi,\vartheta)}{\partial \vartheta}\right)^T y \tag{5.70}$$

which represents the generalized normal equations. Substituting $\Phi\vartheta = f(\Phi,\vartheta)$ and $\Phi = \frac{\partial f(\Phi,\vartheta)}{\partial \vartheta}$ into (5.70), which holds for the linear case, gives (5.9). Due to the dependence of the sensitivity matrix with respect to ϑ, the solution to (5.70) has to be found iteratively by numerical procedures.

5.2.3 Iterative Solutions

In essence, numerical minimization of the nonlinear function $J(\vartheta)$ is based on iterative updates of the estimates according to

$$\widehat{\vartheta}^{(i+1)} = \widehat{\vartheta}^{(i)} + \alpha^{(i)} s^{(i)} \tag{5.71}$$

where i is the iteration index, $\alpha^{(i)}$ the step size, and $s^{(i)}$ the search direction at the ith iteration. In the literature many minimization methods have been proposed, but essentially they can be classified as:

- zeroth-order methods that only use function values,
- first-order methods that use function values and gradients,
- second-order methods that use function values, gradients, and second derivatives.

Typical examples of zeroth- to second-order search methods are the *simplex method*, the *steepest-descent method*, and the *Gauss–Newton method*. The class of well-known *Newton methods*, belonging to the third class, uses

$$s^{(i)} = -\big[J''\big(\widehat{\vartheta}^{(i)}\big)\big]^{-1} J'\big(\widehat{\vartheta}^{(i)}\big) \tag{5.72}$$

which originates from the Newton–Raphson formula for finding a root of the function $J'(\vartheta) = \frac{\partial J(\vartheta)}{\partial \vartheta}$. The problem here is how to determine the Hessian $J''(\cdot)$, a matrix of second derivatives. Methods that, in each ith iteration, use an approximation of $J''(\cdot)$, which in what follows is denoted by the matrix $R^{(i)}$, are called *quasi-Newton methods*.

A general family of search routines is thus given by

$$\widehat{\vartheta}^{(i+1)} = \widehat{\vartheta}^{(i)} - \alpha^{(i)}\big[R^{(i)}\big]^{-1} J'\big(\widehat{\vartheta}^{(i)}\big) \tag{5.73}$$

where $R^{(i)}$ is a $p \times p$ matrix that modifies the gradient $J'(\cdot)$ and $\alpha^{(i)}$ is chosen such that at each iteration step the function value decreases. The choice of $\alpha^{(i)}$ generally results from a line search procedure. Sometimes it is chosen as a constant or as a prespecified decreasing function of i.

The simplest choice of $R^{(i)}$ is

$$R^{(i)} = I \tag{5.74}$$

which is the case in the so-called *gradient* or steepest-descent methods. It appears that this method is not very effective near the optimum. From (5.72) it can be directly verified that

$$R^{(i)} = \left[J''\left(\widehat{\vartheta}^{(i)}\right)\right] \tag{5.75}$$

leads to the Newton methods. A reasonable approximation of $J''(\cdot)$ is given by

$$J''\left(\widehat{\vartheta}^{(i)}\right) \simeq 2X\left(\widehat{\vartheta}^{(i)}\right)^T X\left(\widehat{\vartheta}^{(i)}\right) \tag{5.76}$$

which for

$$R^{(i)} = 2X\left(\widehat{\vartheta}^{(i)}\right)^T X\left(\widehat{\vartheta}^{(i)}\right) \tag{5.77}$$

gives the so-called *Gauss–Newton methods*. Substituting (5.69) and (5.76) into (5.73) gives

$$\widehat{\vartheta}^{(i+1)} = \widehat{\vartheta}^{(i)} + \alpha^{(i)}\left[X\left(\widehat{\vartheta}^{(i)}\right)^T X\left(\widehat{\vartheta}^{(i)}\right)\right]^{-1} X\left(\widehat{\vartheta}^{(i)}\right)^T \left(y - f\left(\Phi, \widehat{\vartheta}^{(i)}\right)\right) \tag{5.78}$$

This Gauss–Newton algorithm has been a starting point for many search routines. For instance, in the widely applied Levenberg–Marquardt procedure, the following approximation is chosen:

$$R^{(i)} = 2X\left(\widehat{\vartheta}^{(i)}\right)^T X\left(\widehat{\vartheta}^{(i)}\right) + \delta I \tag{5.79}$$

where δ is a small positive scalar. This procedure basically incorporates a regularization technique. In (5.79), the extension δI is introduced to prevent singularities of $R^{(i)}$.

However, in addition to the choice of a specific search routine, and because of its iterative character, we also have to specify a stopping criterion. Typically, choices of a stopping criterion, with $\delta, \varepsilon \in \mathbb{R}$ small, are: (i) $\|\vartheta^{(i)} - \vartheta^{(i-1)}\| \leq \delta$, (ii) $J(\vartheta^{(i)}) - J(\vartheta^{(i-1)})| \leq \varepsilon$, or (iii) the maximum number of iterations.

To summarize, the nonlinear least-squares estimation algorithm, for a fixed number of iterations M, is given in the following.

Algorithm 5.5 Nonlinear least-squares estimation of ϑ in linear static systems

1. Given $y(t)$ and $f(\Phi, \vartheta)$ for $t = 1, \ldots, N$, define the N-dimensional vector $y := [y(1), \ldots, y(N)]^T$.
2. Choose the estimation method and related tuning parameters, as, for example, step size (α) and regularization parameter (δ).

Table 5.2 Data nitrification experiment

Time t (d)	$r_{S\,\max}$ (kg/m^3 d)
0	0.268
2	0.305
4	0.347
7	0.399
8	0.499
10	0.504
14	0.431
23	0.735
27	0.809
35	0.930

3. Specify the initial guess $\widehat{\vartheta}^{(0)}$.
4. In the case of the Gauss–Newton method:

$$\begin{aligned}&\text{for } i = 0 : M\\ &\quad \text{calculate } X(\widehat{\vartheta}^{(i)}) \text{ from (5.67)}\\ &\quad \widehat{\vartheta}^{(i+1)} = \widehat{\vartheta}^{(i)} + \alpha^{(i)}\left[X(\widehat{\vartheta}^{(i)})^T X(\widehat{\vartheta}^{(i)})\right]^{-1} X(\widehat{\vartheta}^{(i)})^T \left(y - f(\Phi, \widehat{\vartheta}^{(i)})\right)\\ &\text{end.}\end{aligned}$$

The main bottle-neck in all the numerical minimization procedures is that in general no global optimum can be guaranteed. Furthermore, these iterative procedures can be very time-consuming if the problem is not well posed. In the next section, when applicable, model reparameterization is suggested as an alternative. Let us first apply the Gauss–Newton method to the parameter estimation problem related to the nitrification experiment.

Example 5.22 Nitrification experiment: From the nitrification experiment the following data, as presented in Table 5.2, became available.

We suggest the following initial parameter guesses: $b^{(0)} = 0.01$ and $\mu_{\max}^{(0)} = 0.1$. Let us then investigate the first iteration in the estimation of $\vartheta = [b\ \mu_{\max}]^T$ using the Gauss–Newton method with $\alpha^{(i)} = 1$, so that

$$\widehat{\vartheta}^{(i+1)} = \widehat{\vartheta}^{(i)} + \left[X(\widehat{\vartheta}^{(i)})^T X(\widehat{\vartheta}^{(i)})\right]^{-1} X(\widehat{\vartheta}^{(i)})^T \left(y - f(\Phi, \widehat{\vartheta}^{(i)})\right)$$

Recall that, given the initial guesses and the N-load $B = 0.281$ kg N/m^3 d, the sensitivity vectors at each time instant can be calculated from

$$\psi(t, \widehat{\vartheta}^{(0)}) = \begin{bmatrix} -r_{S\,\max}(0) t e^{-b^{(0)} t} - \frac{\mu_{\max}^{(0)} B}{(b^{(0)})^2} + \frac{t \mu_{\max}^{(0)} B}{b^{(0)}} e^{-b^{(0)} t} + \frac{\mu_{\max}^{(0)} B}{(b^{(0)})^2} e^{-b^{(0)} t} \\ \frac{B}{b^{(0)}} - \frac{B}{b^{(0)}} e^{-b^{(0)} t} \end{bmatrix}$$

Hence,

$$X(\widehat{\vartheta}^{(0)}) = \begin{bmatrix} 0 & 0 \\ -0.581 & 0.556 \\ -1.249 & 1.102 \\ -2.406 & 1.900 \\ -2.832 & 2.160 \\ -3.740 & 2.674 \\ -5.772 & 3.671 \\ -11.283 & 5.774 \\ -14.097 & 6.649 \\ -20.287 & 8.298 \end{bmatrix}$$

and

$$(y - f(\Phi, \widehat{\vartheta}^{(0)})) = \begin{bmatrix} 0 \\ -0.013 \\ -0.021 \\ -0.041 \\ 0.036 \\ -0.006 \\ -0.169 \\ -0.055 \\ -0.061 \\ -0.089 \end{bmatrix}$$

Consequently,

$$[X(\widehat{\vartheta}^{(0)})^T X(\widehat{\vartheta}^{(0)})] = \begin{bmatrix} 800.5639 & -370.7911 \\ -370.7911 & 176.8326 \end{bmatrix}$$

$$\Longrightarrow \quad [X(\widehat{\vartheta}^{(0)})^T X(\widehat{\vartheta}^{(0)})]^{-1} = \begin{bmatrix} 0.0433 & 0.0909 \\ 0.0909 & 0.1962 \end{bmatrix}$$

and

$$X(\widehat{\vartheta}^{(0)})^T (y - f(\Phi, \widehat{\vartheta}^{(0)})) = \begin{bmatrix} 4.3048 \\ -2.1249 \end{bmatrix}$$

Thus, after one iteration we obtain the following estimates:

$$\widehat{\vartheta}^{(1)} = \begin{bmatrix} 0.0035 \\ 0.0743 \end{bmatrix}$$

5.2.4 Accuracy

In the analysis of the estimation uncertainty for the nonlinear case, mainly two approaches prevail: the Monte Carlo approach and first-order variance propagation analysis. The Monte Carlo approach essentially evaluates the nonlinear mapping from random samples of output vector $y^{(k)}$ to the parameter estimates $\widehat{\vartheta}^{(k)}$, where $k = 1, \dots, M$ is the sample number. Hence, the probability distributions of $y(t)$ for $t = 1, \dots, N$ have to be specified, and an appropriate sampling scheme has to be selected. Usually one probability distribution is chosen, so that for one run, N samples from this distribution have to be drawn using, for instance, a Monte Carlo (random) sampling scheme. The resulting estimates are then evaluated with respect to mean value and variance, and sometimes the complete distribution of the estimates is recovered. Clearly, this approach is rather computationally consuming and thus not well suited for practical cases with complex models.

Therefore, in practice and for deterministic regressors, one usually applies the following expression:

$$\operatorname{Cov}\widehat{\vartheta}^* = \widehat{\sigma}_\varepsilon^2 \big[X\big(\widehat{\vartheta}^*\big)^T X\big(\widehat{\vartheta}^*\big)\big]^{-1} \tag{5.80}$$

which results from first-order variance propagation analysis by linearization of the vector function $f(\cdot,\cdot)$ in the optimum $\widehat{\vartheta}^*$. Clearly the covariance matrix is a function of the estimate $\widehat{\vartheta}^*$ and thus represents only local properties. Notice the similarity between (5.32) and (5.80), where the deterministic matrix Φ has been substituted by the sensitivity matrix $X(\widehat{\vartheta}^*)$.

Example 5.23 Nitrification experiment: On the basis of previous estimation results, with

$$\widehat{\vartheta}^{(1)} = \begin{bmatrix} 0.0035 \\ 0.0743 \end{bmatrix}$$

we obtain

$$X\big(\widehat{\vartheta}^{(1)}\big) = \begin{bmatrix} 0 & 0 \\ -0.574 & 0.560 \\ -1.223 & 1.116 \\ -2.334 & 1.943 \\ -2.741 & 2.217 \\ -3.609 & 2.762 \\ -5.555 & 3.840 \\ -10.928 & 6.212 \\ -13.739 & 7.243 \\ -20.105 & 9.262 \end{bmatrix}$$

$$\Longrightarrow \quad \big[X\big(\widehat{\vartheta}^{(1)}\big)^T X\big(\widehat{\vartheta}^{(1)}\big)\big]^{-1} = \begin{bmatrix} 0.0561 & 0.1064 \\ 0.1064 & 0.2065 \end{bmatrix}$$

We find $\widehat{\sigma}_\varepsilon^2 = 0.0026$ and thus

$$\operatorname{Cov}\widehat{\vartheta}^{(1)} = 10^{-3} \cdot \begin{bmatrix} 0.1438 & 0.2727 \\ 0.2727 & 0.5293 \end{bmatrix}$$

From this covariance matrix the standard deviations are calculated as: $\sigma_b = 0.012\ \text{d}^{-1}$ and $\sigma_{\mu_{\max}} = 0.023\ \text{d}^{-1}$, indicating that no reliable estimate of b can be found from this experiment, since the standard deviation is larger than the estimated value.

5.2.5 *Model Reparameterization: Static Case*

Model reparameterization is useful in those cases where, for instance, the effect of numerical errors in the optimization step becomes an important issue. It has been shown in Sect. 5.1.6 that for linear relationships, SVD is a useful tool for analysis of model structures. For nonlinear static relationships as presented in this section, however, no general tool is available. Nevertheless, we can try to reparameterize the nonlinear model structure such that numerical errors, as well as local minima in numerical minimization studies, can be avoided to a large extent. However, the question is: "how should we reparameterize?" Some feasible solutions to the model reparameterization problem will be illustrated by the next examples.

Example 5.24 Pendulum experiment: The pendulum experiment is a simple experiment to estimate the local gravitational constant g, since

$$T = 2\pi \sqrt{\frac{l}{g}}$$

Herein, T is the period of the pendulum, which is the time needed for a complete cycle, and l is the length of the pendulum. However, the unknown parameter g is nonlinearly related to T, which is measured. Thus, a general approach would be to use the Gauss–Newton algorithm (5.78), with all its drawbacks. Several approaches exist to reparameterize the nonlinear relationship. For instance, define $\tilde{g} := g^{-\frac{1}{2}}$, so that the linear regression $T = 2\pi\sqrt{l}\,\tilde{g}$ results. Another approach is to square both sides of the equation, $T^2 = 4\pi^2 l/g$, and take the inverses, so that $1/T^2 = 1\ 4\pi^2 lg$. This result is again a linear regression. We could also directly evaluate the rational relationship such that $2\pi\sqrt{l} = T\sqrt{g} =_{[\tilde{g}:=\sqrt{g}]}= T\tilde{g}$. Finally, taking the natural logarithm of both sides will result in $\ln T = \ln 2\pi + \frac{1}{2}\ln l - \frac{1}{2}\ln g$. Thus, with $\tilde{g} := \ln g$, $\ln T - \ln 2\pi - \frac{1}{2}\ln l = \frac{1}{2}\tilde{g}$.

Consequently, as illustrated above, a model reparameterization step from a static nonlinear relationship to a linear regression is not unique. Moreover, going from a nonlinear relationship to a linear regression will, in general lead to error distortion, that is, the initially assumed probability density function of the measurement error may significantly change. The effect of error distortion on the estimates, in terms of bias, will not be further evaluated here. For details on bias in nonlinear estimation, see [Box71].

Example 5.25 Membrane bioreactor fouling: As suggested by [OWG04], the following relationship represents the changes in transmembrane pressure (TMP) during the first period of filtration operation in a membrane bioreactor.

$$\Delta P = \frac{\Delta P_0}{1 - \alpha \Delta P_0 t^2/2}$$

where ΔP and ΔP_0 are the TMP and initial TMP, respectively, t is the time after the start of a new filtration operation, and α is an unknown parameter that combines a couple of physically interpretable parameters. The underlying hypothesis of this relationship is that the open surface of a membrane after, for instance, a cleaning step, is reduced due to a successive blocking of membrane pores. Notice that α is in a nonlinear way related to the measured TMP (ΔP). Reparameterization leads to

$$\Delta P - \Delta P \alpha \Delta P_0 t^2/2 = \Delta P_0$$
$$\Longrightarrow \quad \Delta P - \Delta P_0 = \Delta P \Delta P_0 t^2/2\alpha$$

which is a linear regression. As ΔP is measured and thus corrupted with noise, an errors-in-variables (EIV) problem results. For possible solutions to EIV problems, see Sect. 5.1.7.

Example 5.26 Respiration rate experiment: For the estimation of the maximum degradation rate of a substrate (μ) and the corresponding half saturation constant (K_S), a respiration rate experiment using a respirometer can be conducted. The following relationship between the respiration rate and the unknown parameters holds:

$$r = \mu \frac{S}{K_S + S}$$

where r is the respiration rate, and S is the substrate concentration. This nonlinear relationship between $\vartheta := [\mu \ \ K_S]^T$ and the measured respiration rate r can be reparameterized to

$$rK_S + rS = \mu S$$
$$\Longrightarrow \quad rS = [S \quad -r] \begin{bmatrix} \mu \\ K_S \end{bmatrix}$$

Hence, we obtain a linear regression. Since typically both r and S contain measurement errors, as in the previous example, an EIV problem results.

In the next example, experimental data will be used in the model reparameterization procedure.

Example 5.27 Nitrification experiment: Recall that from the nitrification experiment the following estimates after one iteration have been obtained:

$$\widehat{b} = 0.0035 \pm 0.012; \qquad \widehat{\mu}_{\max} = 0.0743 \pm 0.023$$

As mentioned before, the estimate of the death rate b is unreliable, and therefore this parameter can be set to zero. In other words, there is no clear evidence that the data supports the prior idea of incorporating the death process in the model. Consequently, the model is modified to

$$\begin{aligned} r_{S\max}(t) &= \lim_{b\to 0} r_{S\max}(0)e^{-bt} + \mu_{\max} B\left[1 - e^{-bt}\right]/b \\ &= r_{S\max}(0) + \lim_{b\to 0} \mu_{\max} B\left[1 - e^{-bt}\right]/b \\ &=_{[L'H\hat{o}pital]} r_{S\max}(0) + \mu_{\max} Bt \end{aligned}$$

which appears to be linear in the parameter $\mu_{\max}$. Hence, $\mu_{\max}$ can be easily found by applying the ordinary least-squares algorithm, which gives $\widehat{\mu}_{\max} = 0.0688\ \mathrm{d}^{-1}$. Notice then that for $t \to \infty$, this linear relationship does not give a reliable prediction unlike the nonlinear model with the limit given by $\frac{\mu_{\max} B}{b}$. From this we conclude that due to the reparameterization, $b \to 0$, a much simpler estimation problem results, but the applicability region of the resulting linear model is limited and in fact has been dictated by the finite experimental data.

Generally, parameters that appear nonlinearly in the model output are estimated by nonlinear least-squares (NLS) optimization algorithms. As an alternative, for nonlinear static models with a so-called rational structure in inputs and parameters, in this section a method has been illustrated to re-parameterize the model such that the model becomes linear in its new parameters (see [DK09, KD09] for details). In addition to this, on the basis of an evaluation of prior estimation results, physically based model reduction techniques may also be applied, which in the last example again led to a reparameterized model that is linear in the parameters. Consequently, in all these cases of nonlinear-in-the-parameter models, the new parameters can be estimated by direct least-squares methods.

5.2.6 **Maximum Likelihood Estimation*

Let $p(\vartheta)$ and $p(e)$ denote the probability density functions (pdf) of ϑ and e, respectively. The *conditional* pdf of the parameter vector ϑ, given the observation vector y, is denoted by $p(\vartheta|y)$ and also called the *a posteriori* pdf , while $p(\vartheta)$ is called the *a priori* pdf of ϑ. The well-known Bayes' rule is given by

$$p(\vartheta|y) = \frac{p(y|\vartheta)p(\vartheta)}{p(y)} \tag{5.81}$$

relating a posteriori pdf's to a priori pdf's.

Let us assume that a given set of experimental single-output data can be modeled as the nonlinear regression

$$y(t) = \widehat{y}(t,\vartheta) + e(t), \quad t = 1,\ldots,N \tag{5.82}$$

as in (5.65)–(5.66). Then, given (5.82),

$$p(y|\vartheta) = p(e)|_{e=y-\widehat{y}(t,\vartheta)} \tag{5.83}$$

Consequently, if the pdf's $p(\vartheta)$ and $p(e)$ are known, it is possible to calculate $p(\vartheta|y)$, where $p(y)$ is just a number once the measurements y have been taken. Since by definition $\int p(\vartheta|y)\,\mathrm{d}\vartheta = 1$, it is not necessary to calculate $p(y)$; it simply becomes a scaling factor. Hence, given the observation vector y, $p(\vartheta|y)$ provides complete information about ϑ and can thus be used to define an estimate of ϑ. For instance, taking the maximum of $p(\vartheta|y)$ results in the well-known *maximum a posteriori* (MAP) estimator. In general, analytical solutions to this specific problem are not available, except for some very simple cases. Hence, we have to rely on demanding numerical solutions associated to these so-called Bayesian estimation problems. The problem becomes much simpler when we assume that ϑ is completely unknown and thus $\vartheta \in [-\infty, \infty]$. Since this assumption on ϑ does not affect $p(\vartheta|y)$, the so-called maximum likelihood (ML) estimation theory focuses on the likelihood function $p(y|\vartheta)$, where y is a vector with realized measurements. Hence, formally speaking, a likelihood function is a conditional probability function considered as a function of its second argument, with its first argument fixed. In that sense, a likelihood function can be thought a "reversed" version of conditional probability density function. Consequently, a likelihood function allows us to estimate unknown parameters based on known outcomes.

If (5.82) holds and if we assume that the measurement errors $e(t)$, $t = 1, \ldots, N$, are independent, homoscedastic (also known as homogeneous in variance, i.e., all $e(t)$ have the same variance), zero-mean, and Gaussian distributed, in short, $e(t) \sim N(0, R_{ee})$ with $R_{ee} = \mathrm{Cov}\, e$ and in what follows denoted by R, then

$$p(y|\vartheta) = \frac{1}{(2\pi)^{N/2}|R|^{1/2}} \exp\left(-\frac{1}{2}\left[y - \widehat{y}(t,\vartheta)\right]^T R^{-1}\left[y - \widehat{y}(t,\vartheta)\right]\right) \tag{5.84}$$

with $|R|$ the determinant of the covariance matrix R. The parameter vector ϑ is found by maximizing (5.84). In practice, and especially when Gaussian noise is considered, it is always more convenient to work with the logarithm of the likelihood function, called the log-likelihood, $L(\vartheta, R) = -\ln p(y|\vartheta, R)$ given by

$$L(\vartheta, R) = \frac{N}{2}\ln 2\pi|R| + \frac{1}{2}\left[y - \widehat{y}(t,\vartheta)\right]^T R^{-1}\left[y - \widehat{y}(t,\vartheta)\right] \tag{5.85}$$

Hence, under our assumptions, the ML estimator is given by

$$\left(\widehat{\vartheta}, \widehat{R}\right) = \arg\min_{\vartheta, R} L(\vartheta, R) \tag{5.86}$$

However, the objective function derived from (5.85) depends on the assumptions made on the covariance matrix R.

1. If R is known and thus the first term on the righthand-side of (5.85) is constant, the ML estimator corresponds to the so-called Gauss–Markov estimator, which

Table 5.3 Process data

t	1	2
$y_1(t)$	1	2
$y_2(t)$	3	1

minimizes the objective function

$$J(\vartheta) = \big[y - \widehat{y}(t, \vartheta)\big]^T R^{-1}\big[y - \widehat{y}(t, \vartheta)\big] \tag{5.87}$$

2. If $R = aI_p$ with a positive real number a and the $(p \times p)$ identity matrix I_p, the ML estimator corresponds to the ordinary least-squares estimator, which minimizes

$$J(\vartheta) = \big[y - \widehat{y}(t, \vartheta)\big]^T \big[y - \widehat{y}(t, \vartheta)\big] \tag{5.88}$$

The ML estimate of R is given by

$$\widehat{R} = \frac{J(\widehat{\vartheta})}{N} I_p \tag{5.89}$$

3. If R is completely unknown, the ML estimator minimizes

$$J(\vartheta) = \ln\big[\det\big[y - \widehat{y}(t, \vartheta)\big]^T \big[y - \widehat{y}(t, \vartheta)\big]\big] \tag{5.90}$$

In this case, the ML estimate of R is given by

$$\widehat{R} = \frac{1}{N}\big[y - \widehat{y}(t, \vartheta)\big]\big[y - \widehat{y}(t, \vartheta)\big]^T \tag{5.91}$$

Let us illustrate the ML estimation theory by an example.

Example 5.28 Constant process state: Let for the system of Example 5.7 with

$$\begin{bmatrix} y_1(t) \\ y_2(t) \end{bmatrix} = \begin{bmatrix} 1 \\ 1 \end{bmatrix} x + \begin{bmatrix} e_1(t) \\ e_2(t) \end{bmatrix}$$

and thus $\vartheta = x$, the following measurements be given in Table 5.3.

In the case that $R = \left[\begin{smallmatrix} 1 & 0 \\ 0 & 4 \end{smallmatrix}\right]$ is known, the Gauss–Markov estimate is given by $\widehat{\vartheta} = 1.90$.

Assuming that R is proportional to the identity matrix, the ML estimates are $\widehat{\vartheta} = 1.75$ and $\widehat{R} = 0.6875$.

If we consider R to be completely unknown, $\widehat{\vartheta} = \arg\min_{\vartheta} \ln(4x^2 - 14x + 15) = 1.75$, where we have expressed the determinant of (5.90) directly in terms of the

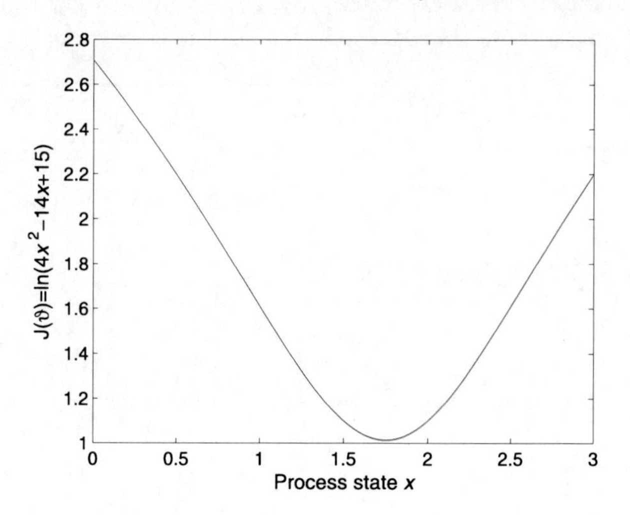

Fig. 5.7 Objective function values as a function of x

unknown x. Thus,

$$\begin{aligned}
&\det\left[y-\widehat{y}(t,\vartheta)\right]^T\left[y-\widehat{y}(t,\vartheta)\right] \\
&= \left|\left[\,1-x\ \ 2-x\,\right]\begin{bmatrix}1-x\\2-x\end{bmatrix}+\left[\,3-x\ \ 1-x\,\right]\begin{bmatrix}3-x\\1-x\end{bmatrix}\right| \\
&= 4x^2-14x+15
\end{aligned}$$

The unknown process state can also be found graphically, as in Fig. 5.7.

The ML estimate of R is given by $\widehat{R}=\left[\begin{smallmatrix}1.0625 & -0.5625\\ -0.5625 & 0.3125\end{smallmatrix}\right]$. Consequently, knowledge of the covariance matrix significantly affects the estimates of ϑ and R.

Typically, the uncertainty in the ML parameter estimates is evaluated via the computation of the Fisher information matrix (FIM). The FIM is given by

$$F\left(\vartheta^*\right)=-E\left[\frac{\partial^2}{\partial\vartheta\,\partial\vartheta^T}\ln p(y|\vartheta)\right]_{\vartheta=\vartheta^*} \tag{5.92}$$

with ϑ^* the true, but usually unknown, parameter vector. Under a number of technical assumptions, the covariance matrix of the parameter estimates $\operatorname{Cov}\widehat{\vartheta}$ satisfies the following inequality:

$$\operatorname{Cov}\widehat{\vartheta}\geq F^{-1}\left(\vartheta^*\right) \tag{5.93}$$

which is known as the Cramér–Rao inequality. In practice, most often $F^{-1}(\vartheta^*)$ is approximated by $F^{-1}(\widehat{\vartheta})$. However, notice from (5.92) that the likelihood function must be known or at least partially known. If, for instance, the measurement errors obey the rather strict assumptions presented in the beginning of this subsection, relatively simple analytical expressions for $F(\vartheta^*)$ can be obtained. In general, the likelihood function is unknown, especially for limited data sets, and thus in practice we most often rely on expressions like (5.80).

Recall that, given a set of experimental data and a nonlinear regression model, the maximum likelihood method leads to model parameter estimates that maximize the likelihood function. The merit of maximum likelihood estimation is that it provides a unified framework to estimation, which is well defined in the case of normal distributions. However, in practice, often complex problems with nonnormal or with unknown distributions occur. In such cases the maximum-likelihood estimators may be unsuitable or may not even exist. Hence, the application of maximum likelihood estimators is rather limited in practice.

5.2.7 *Bounded-noise Problem: Nonlinear Case

Let us extend the ideas given in Sect. 5.1.8 to the nonlinear set-membership identification problem, which frequently occurs in practice. Recall that the set-membership approach is in particular useful in the case of small data sets. Instead of the linear regression model (5.53), in this section we consider the following nonlinear regression type of model, as given by (5.65). Thus,

$$y = f(\Phi, \vartheta) + e \tag{5.94}$$

where $f(\Phi, \vartheta)$ is a nonlinear vector function mapping the unknown parameter vector $\vartheta \in \mathbb{R}^p$ into a noise-free model output $\widehat{y}(\vartheta)$. Again, the error vector e is assumed to be point-wise bounded with constant error bound ε and similar sets, as in Sect. 5.1.8, can be defined. However, in what follows, we use $f(\Phi, \vartheta)$, in short $f(\vartheta)$, instead of $\Phi\vartheta$. Let us illustrate this by a simple example with two measurements and two unknown parameters. Furthermore, the example will also show some of the specific estimation problems in nonlinear bound-based identification.

Example 5.29 Sinusoidal model: Suppose that $f(\vartheta)$ is given by $f(\vartheta) = \sin(\vartheta_1 t) + \vartheta_2$ and the measurements are: $t(1) = 1$, $y(1) = 1.0$ and $t(2) = 3$, $y(2) = 0.5$ with error bound $\varepsilon = 0.5$. Hence, when only one measurement at $t(1)$ is available, Ω_y is an interval, in this case $[0.5, 1.5]$, and Ω_ϑ is an unbounded set that is only bounded by a pair of bounds, $\sin(\vartheta_1) + \vartheta_2 = 0.5$ and $\sin(\vartheta_1) + \vartheta_2 = 1.5$ (see Fig. 5.8). Consequently, the image set is equal to the real axis, and the feasible model output set is equal to the measurement uncertainty set.

When the second measurement at $t(2)$ becomes available, Ω_y becomes a square with center $[1\ 0.5]^T$ and edges with unit length in the measurement space. Consequently, in the parameter space another pair of bounds is added, which, together with the bounds related to the first measurement, define an exact solution to the parameter bounding estimation problem. Notice from Fig. 5.8 that Ω_ϑ (dotted regions) becomes a nonconnected set with nonconvex subsets. Furthermore, prior knowledge restricts ϑ_1 to the interval $[0, 2\pi]$. The image set is equal to $\{\tilde{y} \in \mathbb{R}^2 : \tilde{y} = [\vartheta_{11} + \vartheta_2 \vartheta_{12} + \vartheta_2]^T; \vartheta_{11}, \vartheta_{12} \in [-1, 1], \vartheta_2 \in \mathbb{R}\}$, a strip in $\mathbb{R}^2$, and again the feasible model output set is equal to the MUS (see Fig. 5.9). However,

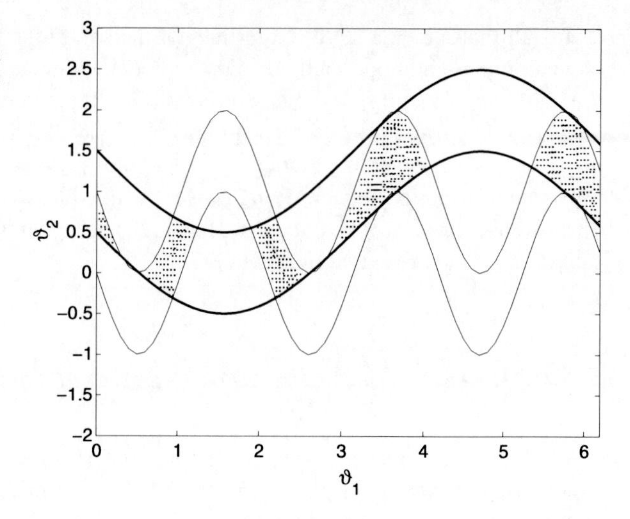

Fig. 5.8 Bounded-noise results of sinusoidal model in the parameter space after two measurements, for $t(1)$ (*bold lines*) and $t(2)$ (*thin lines*)

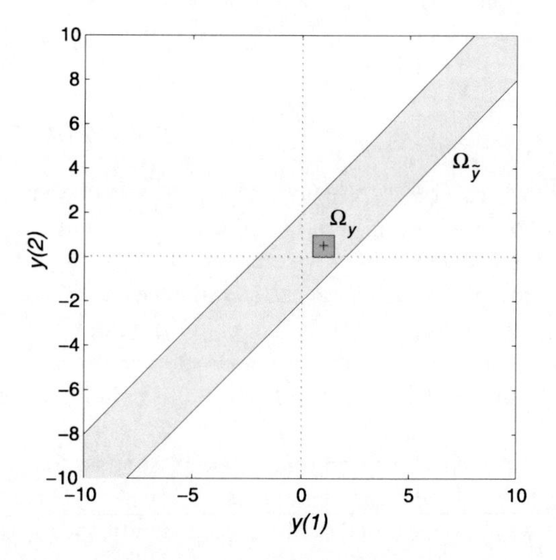

Fig. 5.9 Bounded-noise results of sinusoidal model in the measurement space after two measurements

when a third measurement becomes available, the feasible model output set generally is not be equal to the MUS, and the image set becomes a two-dimensional variety in $\mathbb{R}^3$.

In addition to intersection, encapsulation, and projection approaches, as presented in Sect. 5.1.8, but now slightly modified for the nonlinear case, a fourth class of algorithms can be introduced. This class of algorithms suitable for approximately solving the nonlinear set-membership estimation problem consists of algorithms that in a discrete way approximate the bounding surfaces (either by random or deterministic search) of the FPS and algorithms which provide inner/outer approximations of the FPS. Two important algorithms from this class will be briefly introduced.

Table 5.4 Exponential model data

t	0.0	0.2	0.4	0.6	0.8
$y(t)$	3.4	2.3	1.7	1.2	0.9

Especially for lower dimensional problems where $f(\vartheta)$ is an explicit function of the model parameters, the so-called SIVIA (set inversion via interval analysis) algorithm is superior, because it can inner and outer bound the solution set FPS by a pavement of boxes. Theoretically, the set enclosure can be made as accurate as we wish, but, as expected, the number of boxes (which is proportional to the computing time) increases quickly when more accuracy is required.

The second algorithm point-wise approximates the FPS by proper sampling of the parameter space and thus a finite solution set results. Especially for model prediction, the inner approximation, using a parameter space sampling method, is well suited. In this algorithm each feasible and unfalsified parameter vector that obeys the definition of the FPS can be directly used in the model prediction step. Furthermore, in this indirect type of algorithm, $f(\vartheta)$ can have a very general structure; it can be simply the result of a dynamic simulation. It can also deal with nonconnected sets. Thus, estimation problems related to nonlinear state-space models can also be handled. However, the main disadvantage is its computational inefficiency, which becomes clearly visible in higher-dimensional parameter estimation problems. In the literature this problem is partially compensated by using adaptively rotated bases, as in the Monte Carlo Set-Membership algorithm, or by step-wise decreasing the error bound. A rotation based on the eigenvalue decomposition of the dispersion matrix related to the finite feasible parameter vector set found in a previous iteration appears to be rather effective (see [Kee90] for details). The application of adaptively rotated bases will be illustrated in the following example. Another disadvantage of this algorithm, in addition to its computational inefficiency, is that we cannot easily give an idea of the accuracy of the finite solution set.

Example 5.30 Exponential model: Consider the following exponential model, which, for instance, can be interpreted as the impulse response of the first-order LTI system

$$y(t) = \mu_1 e^{-\nu_1 t} + e(t) \tag{5.95}$$

Let the measurements presented in Table 5.4 be available. The time variable error bound $\varepsilon(t)$ is assumed to be equal to $0.1|y(t)| + 0.5$ (see also Fig. 5.10).

For this two-parameter case, the feasible parameter or exact solution set can be represented graphically (see Fig. 5.11). In this figure, each line fulfills the constraint

$$\mu_1 e^{-\nu_1 t} = y(t) \pm \varepsilon(t) \quad \text{for } t = 0, 0.2, \ldots, 0.8 \tag{5.96}$$

From 5000 randomly chosen parameter vectors within the region with vertex set,

$$\left\{ \begin{bmatrix} 2.56 \\ -0.84 \end{bmatrix}, \begin{bmatrix} 4.24 \\ 1.68 \end{bmatrix}, \begin{bmatrix} 4.24 \\ 4.97 \end{bmatrix}, \begin{bmatrix} 2.56 \\ 2.44 \end{bmatrix} \right\}$$

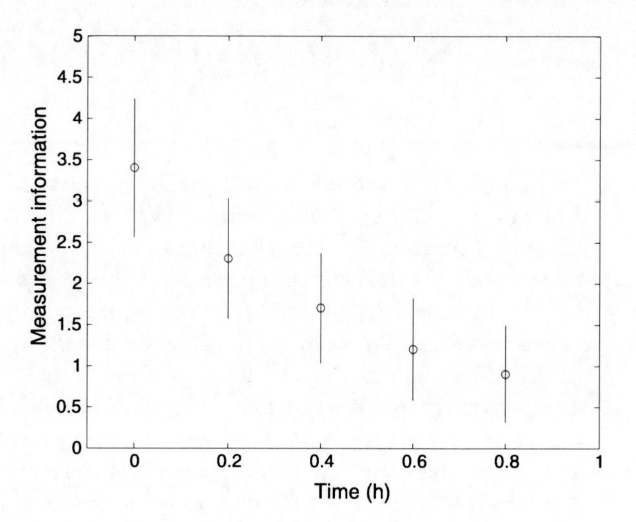

Fig. 5.10 Measured data with bounded uncertainty

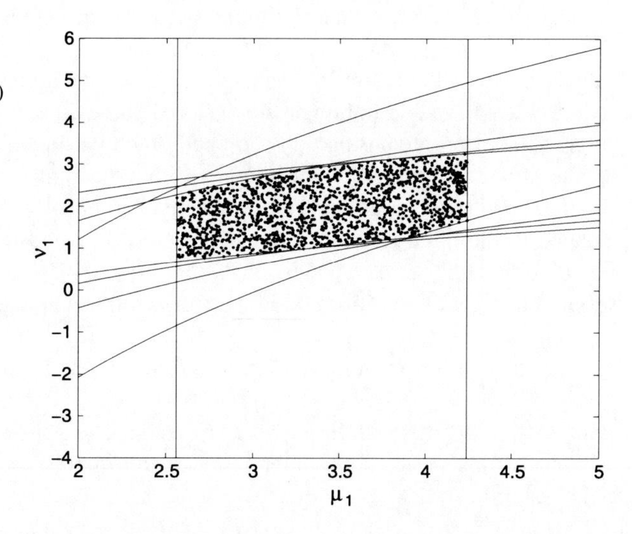

Fig. 5.11 Exact solution set using intersection and discrete approximation set ($\cdot$) after 5000 trials

as a result of the intersection of $\Omega_\vartheta(t(1))$ and $\Omega_\vartheta(t(2))$, 1692 (i.e., the efficiency of 34%) parameter vectors appear to be feasible (presented as dots in Fig. 5.11). The efficiency of feasible hits can be significantly increased (up to 70%) when, for instance, after 1000 samples, the orientation and size of the approximate feasible parameter set are analyzed, and subsequently a new sampling strategy based on a rotated basis with projected intervals is applied. A typical example of the point-wise discrete approximation, including the rotation step, is presented in Fig. 5.12.

Solving (5.62) and (5.63) for this specific case leads to the box

$$\mathscr{B} = [2.56,\ 4.24] \times [0.68,\ 3.27] \tag{5.97}$$

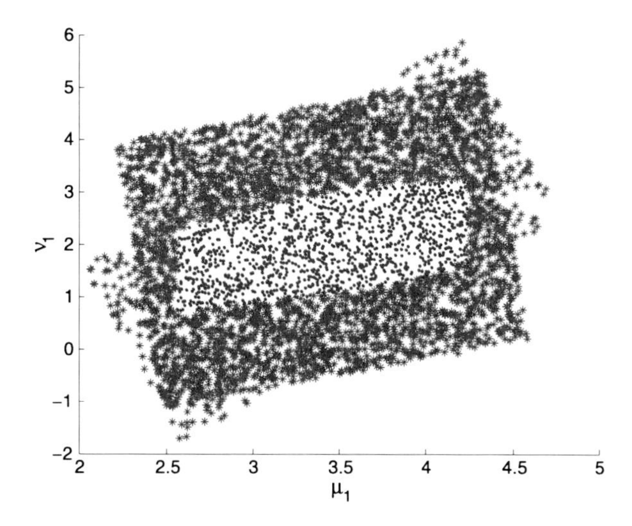

Fig. 5.12 Discrete approximation set (·) and unfeasible parameter set (*) after 5×1000 trials

For general nonlinear static estimation problems under bounded noise, an interval-based algorithm, like SIVIA, could be a good choice, even for a relatively large number of parameters, if the required accuracy is not too high. For particular polynomial problems, however, a specific signomial programming method may be a good alternative. As illustrated by Example 5.29, a point-wise discrete approximation algorithm, using an appropriate sampling and updating strategy, can be applied for the estimation of general nonlinear (dynamic) simulation models.

5.3 Historical Notes and References

The material in this chapter originates from the work of Gauss on least-squares estimation, which, as he claims, started in 1795. The book of Sorenson [Sor80], Chap. 1, gives a nice historical perspective of estimation theory in general. Since the work of Gauss, many articles and books have appeared on linear regression and the least-squares method, see, for instance, [DS98, MPV06] for linear regression issues and [Bar74, GVL89, Bjo96, Ips09] for solving linear and nonlinear least-squares problems.

Identifiability of model structures has been a subject in many articles, using the Laplace transform, Taylor series expansion, and the exhaustive modeling or similarity transformation approach, for linear and nonlinear systems. For the class of linear models, we refer, for instance, to [BK70, GW74, NW82, Wal82, vS94, vdH98, ADSC98, PC07] and for nonlinear model structures to [VGR89, WP90, DvdH96, MRCW01, ECCG02, CGCE03, SAD03, PH05].

Errors-in-variables (EIV) estimation problems are covered in many books and articles. The first solutions of the dynamic EIV identification problem have been proposed by Koopmans [Koo37] and Levin [Lev64]. For other references to the identification of dynamic systems using Maximum Likelihood techniques, Frisch scheme-based algorithms, Instrumental Variable (IV) methods or Total Least Squares

(TLS), and other least-squares methods, see [Lev64, GVL80, You84, And85, SD98, MWDM02, SSM02, KMVH03, VHMVS07, HSZ07, Söd07, Söd08, HS09], to mention a few.

The bounded-noise problem, which is in literature also referred to as unknown-but-bounded or set-membership identification, has initially been tackled by Schweppe [Sch73], Kurzhanski [Kur77], Chernousko and Melikyan [CM78] and, in particular for parameter estimation problems, by Milanese and Belforte [MB82] and Fogel and Huang [FH82]. Since then, many papers and books have appeared on this subject. For overviews, we refer to [Nor87, MV91a, MNPLE96, Wal03, Nor03, Kee03]. The use of least-squares techniques to solve the set-membership estimation problem has been emphasized by [Mil95, Kee97]. The first link between support vector machines, popular in statistical learning, and nonlinear set-membership identification has been published by [KS04].

5.4 Problems

Problem 5.1 For the determination of the unknown parameters in the linear growth model

$$y(t) = y_0 + \mu t$$

where $y(t)$ is the crop height, y_0 the initial crop height, and μ the growth rate, a number of experiments are performed.

(a) Using least-squares estimation, determine the coefficients y_0 and μ, and the associated estimation errors, when for $t = 1$, the measured output is equal to 3. Explain your result.
(b) As (a), but now including a second measurement which for $t = 2$ gives a measured output of 5. Explain your result.
(c) Idem, if the next experimental results are $t = 3$, $y(t) = 7$; $t = 4$, $y(t) = 13$; and $t = 10$, $y(t) = 21$. Explain your result.

Problem 5.2 Consider the moving object example (Example 5.2).

(a) Repeat the steps that lead to the residuals (e).
(b) Calculate the bias (b) in the estimates of the three unknown parameters (see (5.26)). What do you conclude from this?
(c) Calculate the variance of the residuals and use this estimate of the variance for the calculation of the accuracy in the estimates of x_0, v, and a. What do you conclude with respect to the accuracy in the estimates?

Problem 5.3 Let the following (normalized) data from an experiment investigating the effect of the feed rate on the substrate concentration in a reactor be given (see Problem 4.3 and Table 5.5):

Table 5.5 Normalized data chemical reactor

Time	1	2	3	4	5	6	7	8	9	10	11
$u(t)$ (m^3/s)	1	1	1	1	1	−1	−1	−1	−1	−1	−1
$y(t)$ (kg/m^3)	0	0.13	0.09	0.10	0.10	0.10	−0.17	−0.08	−0.11	−0.10	−0.10

Table 5.6 Compartmental model data

t (s)	0	0.5	0.75	1.25	1.75	2.25
y (m)	0	90	115	85	55	40

(a) Assume furthermore that this process can be described by the following linear regression model:

$$y(t) = g(0)u(t) + g(1)u(t-1) + g(2)u(t-2)$$

Determine the least-squares estimates of the impulse response coefficients $g(0)$, $g(1)$, and $g(2)$ from this data using all information available.

(b) Calculate the residuals and plot them. Interpret your result.

(c) To evaluate the uncertainty in the estimates, calculate the covariance matrix and give the estimation variances for each of the coefficients. Interpret your result in terms of accuracy and reliability of the estimates.

Problem 5.4 For the estimation of unknown parameters in nonlinear relationships from given experimental data, the MATLAB function *lsqnonlin* can be used. In the modeling of biological systems, so-called compartmental models are frequently used. For the linear case and as a result of an impulsive input, a multiexponential response model will appear. An example of such a model is

$$y(t) = c\left(e^{\lambda_1 t} + e^{\lambda_2 t}\right)$$

The unknown parameters are c, λ_1, and λ_2. These can be estimated from the following measurements in Table 5.6,

(a) Plot the measurements and interpret the result.

(b) Examine the MATLAB function *lsqnonlin* and try the given examples.

(c) Estimate the three unknown parameters (c, λ_1, and λ_2) in the given model (*NB*: use the function *myfun* to calculate the residuals on the basis of the given model and data).

(d) Estimate the Jacobi matrix (J, see *help lsqnonlin*) as well and determine the covariance matrix related to the parameter estimation errors.

(e) Perform an eigenvalue decomposition of the covariance matrix and evaluate the result in terms of parameter sensitivities.

Problem 5.5 Step responses are frequently used to obtain a first indication of the process dynamics of low-order processes. In the following we will investigate the estimation uncertainty properties as a function of the sampling strategy (frequency). Consider, for simplicity, the step response of a first-order system without time delay,

$$y(t) = K\left(1 - e^{-\alpha t}\right)$$

Calculate $y(t)$ for $K = 2$, $\alpha = 0.1$, and $t = 0, 0.1, 0.2, \ldots, 100$ and add normally distributed noise with a variance of 0.1 to it (store this data set).

(a) Given the generated data set, estimate the parameters K and α using a nonlinear least-squares method (MATLAB: *lsqnonlin*). Store these results together with the associated covariance matrix, its determinant, and the norm of the residuals in a table.
(b) Repeat (a), but now for $t = 0, 1, 2, \ldots, 100$, i.e., resample your stored data set, and add the results to the table.
(c) Repeat (a), but now for $t = 0, 10, 20, \ldots, 100$, i.e., resample your data set at an even lower sampling frequency, and again add the results to the table. Explain your results.
(d) Let us now focus on the effect of nonequidistant sampling. Determine from the model equation the parameter sensitivities $\partial y/\partial K$ and $\partial y/\partial \alpha$ and plot these as a function of time. Interpret the results.
(e) Suppose that we are mainly interested in the estimation of the time constant α. Considering the parameter sensitivities obtained in (d), select 11 "optimal" sampling instants and motivate your choice.
(f) Repeat (a), but now for the 11 sampling instants of (e), and add the results to the table. Explain your results.

Chapter 6
Dynamic Systems Identification

6.1 Linear Dynamic Systems

6.1.1 Transfer Function Models

In Part I the transfer function model representation for linear time-invariant systems has already been introduced. In what follows, however, as in Chap. 5, the model structure will include a noise term to account for the misfit between output measurements and model output. In this chapter we will consider several parameterizations of transfer function-noise model structures describing the dynamic system behavior in discrete time.

Let us start with the simplest structure, the convolution model structure where $G(q)$ is replaced by $B(q)$ for reasons that will become clear later and extended with a noise term, represented by

$$\begin{aligned} y(t) &= b_1 u(t-1) + b_2 u(t-2) + \cdots + e(t) \\ &= B(q)u(t) + e(t) \end{aligned} \tag{6.1}$$

with $B(q) = \sum_{k=1}^{\infty} b_k q^{-k} = b_1 q^{-1} + b_2 q^{-2} + \cdots$, a polynomial in the backward shift operator q^{-1} (see Appendix E), and a white noise error term $e(t)$. In what follows, it is assumed that a real system is not strictly causal, which means that the actual input $u(t)$ cannot have a direct effect on the output $y(t)$. Therefore, the polynomial starts with $k = 1$. This structure is also called an *IIR* (Infinite Impulse Response) model structure. In practice, however, it mostly suffices to take just n_b terms, so that $B(q) = \sum_{k=1}^{n_b} b_k q^{-k} = b_1 q^{-1} + b_2 q^{-2} + \cdots + b_{n_b} q^{-n_b}$. This structure is then called a *FIR* (Finite Impulse Response) model structure.

Another simple input–output relationship, introduced in Chap. 1 and extended with a noise term, is given by the linear difference equation

$$\begin{aligned} y(t) + a_1 y(t-1) + \cdots + a_{n_a} y(t-n_a) &= b_1 u(t-1) \\ &\quad + \cdots + b_{n_b} u(t-n_b) + e(t) \end{aligned} \tag{6.2}$$

K.J. Keesman, *System Identification*,
Advanced Textbooks in Control and Signal Processing,
DOI 10.1007/978-0-85729-522-4_6,

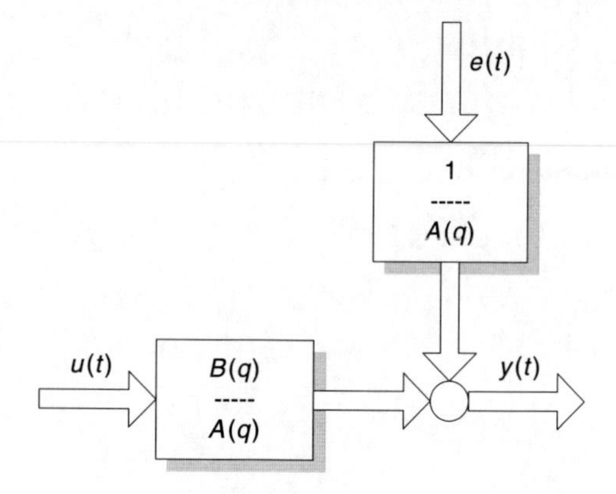

Fig. 6.1 ARX model structure

Since $e(t)$ enters as a direct error in the difference equation, (6.2) is also called an *equation error model* structure. Rewriting (6.2) in transfer-function form gives

$$A(q)y(t) = B(q)u(t) + e(t) \tag{6.3}$$

where $A(q) = \sum_{k=0}^{n_a} a_k q^{-k} = a_0 + a_1 q^{-1} + a_2 q^{-2} + \cdots + a_{n_a} q^{-n_a}$ with $a_0 = 1$, and again $B(q) = \sum_{k=1}^{n_b} b_k q^{-k} = b_1 q^{-1} + b_2 q^{-2} + \cdots + b_{n_b} q^{-n_b}$. Notice that (6.3) has an AutoRegressive part $A(q)y(t)$ and an eXogenous part $B(q)u(t)$. Therefore, this model structure is also indicated as an ARX model, which can be rewritten in explicit form as

$$y(t) = \frac{B(q)}{A(q)} u(t) + \frac{1}{A(q)} e(t) \tag{6.4}$$

(see also Fig. 6.1 for the signal flows). More specifically, ARX model structures are also denoted as $ARX(n_a, n_b, n_k)$, where n_k indicates the number of sampling intervals related to dead time. Consequently, in case of dead time $b_1 = \cdots = b_{n_k} = 0$.

A special case is obtained when $n_a = 0$, which reduces the ARX to an FIR model structure.

A further extension is obtained when the error term is modeled as a moving average of white noise, that is,

$$\begin{aligned} y(t) + a_1 y(t-1) + \cdots + a_{n_a} y(t - n_a) = b_1 u(t-1) & \\ + \cdots + b_{n_b} u(t - n_b) & \\ + e(t) + c_1 e(t-1) + \cdots & \\ + c_{n_c} e(t - n_c) & \end{aligned} \tag{6.5}$$

Due to the moving average part, (6.6) will be called an ARMAX model structure . Rewriting (6.5) in transfer-function form gives

$$y(t) = \frac{B(q)}{A(q)} u(t) + \frac{C(q)}{A(q)} e(t) \tag{6.6}$$

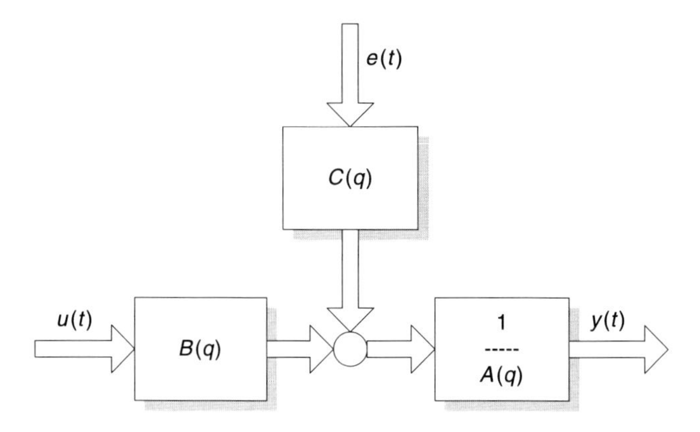

Fig. 6.2 Equation error model family

with $A(q)$ and $B(q)$ as defined before, and with $C(q) := \sum_{k=0}^{n_c} c_k q^{-k} = c_0 + c_1 q^{-1} + c_2 q^{-2} + \cdots + c_{n_c} q^{-n_c}$, $c_0 = 1$. This ARMAX model structure is very popular in controller design procedures. In the case of systems with slow disturbances, the so-called ARIMA(X) is often used, where I stands for integrated. In model structures of this type, the output $y(t)$ is replaced by $\Delta y(t) = y(t) - y(t-1)$; this extension will not be further discussed here.

So far the equation-error has played an important role, leading to transfer function models with a common polynomial A in the denominators (see Fig. 6.2).

However, if it is imposed that the linear difference equation is error-free, but that the noise consists of white measurement noise only, then we obtain the following description:

$$\xi(t) + f_1 \xi(t-1) + \cdots + f_{n_f} \xi(t - n_f) = b_1 u(t-1) + \cdots + b_{n_b} u(t - n_b) \tag{6.7}$$

$$y(t) = \xi(t) + e(t) \tag{6.8}$$

where $\xi(t)$ is the noise-free output of the dynamic system, and $F(q)$ is defined as $F(q) := \sum_{k=0}^{n_f} f_k q^{-k} = 1 + f_1 q^{-1} + f_2 q^{-2} + \cdots + f_{n_f} q^{-n_f}$. We can rewrite this so-called *output-error model* structure as

$$y(t) = \frac{B(q)}{F(q)} u(t) + e(t) \tag{6.9}$$

(see Fig. 6.3).

The last model structure we will discuss in this subsection is the so-called *Box–Jenkins model* structure. This model structure is a natural extension of the output-error model structure. In this structure the output error is modeled as an ARMA model, so that

$$y(t) = \frac{B(q)}{F(q)} u(t) + \frac{C(q)}{D(q)} e(t) \tag{6.10}$$

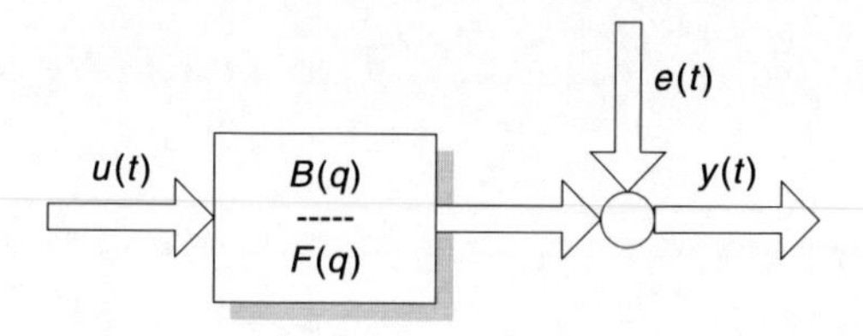

Fig. 6.3 Output error model structure

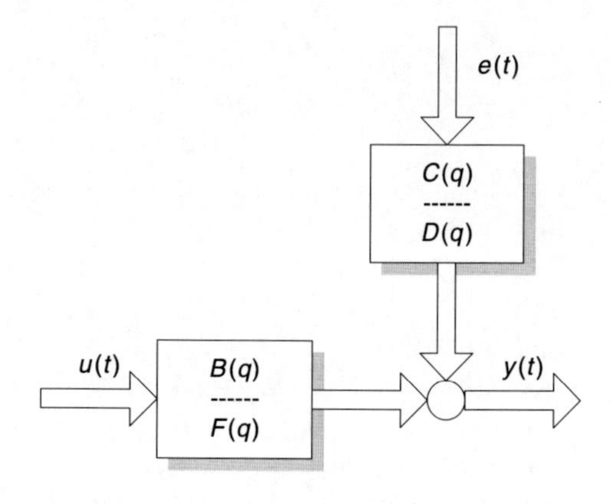

Fig. 6.4 Box–Jenkins model structure

with polynomial $D(q) = \sum_{k=0}^{n_d} d_k q^{-k} = d_0 + d_1 q^{-1} + d_2 q^{-2} + \cdots + d_{n_d} q^{-n_d}$, $d_0 = 1$. The signals flows in the Box–Jenkins model structure are presented in Fig. 6.4.

From these results the following generalized model structure can be derived:

$$A(q)y(t) = \frac{q^{-n_k} B(q)}{F(q)} u(t) + \frac{C(q)}{D(q)} e(t) \tag{6.11}$$

with appropriate polynomials and where $n_k \geq 0$ is the number of time delay intervals, plus one default delay in the definition of $B(q)$. It should be mentioned here that, to avoid over-parameterization, in applications either $A(q)$ or $F(q)$ is set equal to one. In what follows, we will represent the whole class of transfer function models as

$$y(t) = G(q)u(t) + H(q)e(t) \tag{6.12}$$

where $G(q) = \sum_{k=1}^{\infty} g(k) q^{-k}$ and $H(q) = 1 + \sum_{k=1}^{\infty} h(k) q^{-k}$. Notice here that both G and H are not only simple polynomials, but in general a ratio of polynomials, more commonly referred to as rational transfer functions. It can be easily verified by long division that, in general, the rational transfer functions $G(q) = \frac{B(q)}{F(q)}$ and $H(q) = \frac{C(q)}{D(q)}$ lead to infinite impulse response functions. Furthermore, for subsequent analyzes, we introduce the filtered white noise term $v(t) = H(q)e(t)$.

It should be mentioned here that in practical situations, the input–output data is usually pretreated by removing off-sets, drifts, trends, etc., since the class of transfer function models, represented by (6.12), does not describe nonstationary effects. A natural way to remove off-set, for instance, is to subtract sample means from both the input and output data. Drifts and trends can be removed by high-pass filtering of the data, a subject which will not be further treated here.

6.1.2 Equation Error Identification

In the previous subsection several transfer function model structures have been introduced. However, as yet, no attention has been paid to the estimation of the model parameters in these structures from input–output data. In this and the next two sections, therefore, the focus will be on estimation algorithms for the different model structures.

Notice that for the estimation of the unknown coefficients $b_1, b_2, \dots, b_{n_b}$ in the FIR model structure from input–output data, the model output can be rewritten as the linear regression

$$\widehat{y}(t, \vartheta) = \phi(t)^T \vartheta \tag{6.13}$$

with $\phi(t)^T = [u(t-1), u(t-2), \dots, u(t-n_b)]$ and $\vartheta = [b_1, b_2, \dots, b_{n_b}]^T$. Let the inputs $u(0), u(1), \dots, u(N)$ and corresponding outputs $y(0), y(1), \dots, y(N)$ be recorded with $N \gg n_b$. Then, in vector-matrix notation the output vector is defined as $y := [y(n_b), \dots, y(N)]^T$, and the regressor matrix is

$$\Phi = \begin{bmatrix} u(n_b-1) & u(n_b-2) & \cdots & u(0) \\ u(n_b) & u(n_b-1) & \cdots & u(1) \\ u(n_b+1) & \vdots & & \vdots \\ \vdots & & & \\ u(N-1) & u(N-2) & \cdots & u(N-n_b) \end{bmatrix} \tag{6.14}$$

Hence, unlike the methods presented in Part I, the impulse response coefficients can also be estimated from the data by using the ordinary least-squares method. It can then be shown that for the same observations, the Wiener–Hopf equation approach (see Sect. 4.2.2) gives the same estimates as those obtained from the least-squares method.

The model output of an ARX model structure can also be rewritten as a linear regression,

$$\widehat{y}(t, \vartheta) = \phi(t)^T \vartheta \tag{6.15}$$

with $\phi(t)^T = [-y(t-1), -y(t-2), \dots, -y(t-n_a), u(t-1), u(t-2), \dots, u(t-n_b)]$ and $\vartheta = [a_1, a_2, \dots, a_{n_a}, b_1, b_2, \dots, b_{n_b}]^T$. Let again the inputs $u(0)$, $u(1), \dots, u(N)$ and corresponding outputs $y(0), y(1), \dots, y(N)$ be recorded with $N \gg \max(n_a, n_b)$. In vector-matrix notation the output vector is defined as $y :=$

$[y(\max(n_a, n_b)), \ldots, y(N)]^T$, and the regressor matrix, for $n_a \geq n_b$, is

$$\Phi = \begin{bmatrix} -y(n_a-1) & \cdots & -y(0) & u(n_a-1) & \cdots & u(n_a-n_b) \\ -y(n_a) & & -y(1) & u(n_a) & \cdots & u(n_a-n_b+1) \\ -y(n_a+1) & & \vdots & & \vdots & \\ \vdots & & & & & \\ -y(N-1) & \cdots & -y(N-n_a) & u(N-1) & \cdots & u(N-n_b) \end{bmatrix} \tag{6.16}$$

Similarly, the regressor matrix for $n_b > n_a$ can be formed. Consequently, the unknown parameters $a_1, a_2, \ldots, a_{n_a}, b_1, b_2, \ldots, b_{n_b}$ can be directly found from input–output data using ordinary least-squares estimation. In order to avoid unwanted side-effects in the estimates due to off-sets and trends in the data, it is advisable to remove the mean from both the input and output data and to detrend the data to remove nonstationary behavior. In the following algorithms, it is always assumed that preprocessed input–output data for $t = 1, \ldots, N$, with N large enough to avoid practical identifiability problems, is available.

Algorithm 6.1 Identification of ARX model parameters from input–output data

1. Specify an ARX model structure in terms of n_a and n_b.
2. Define the vector $y := [y(n_a), \ldots, y(N)]^T$ and the matrix Φ, as in (6.16), for $n_a \geq n_b$.
3. Calculate from (5.10) the least-squares estimate of the unknown $(n_a + n_b)$-dimensional parameter vector ϑ.

Example 6.1 Heating system: In an identification experiment of the heating system the following inputs and outputs are measured (see Fig. 6.5). The input signal is a Random Binary Signal (RBS) around zero with $p_0 = 0.2$, $N = 1000$, and sampling interval $T_s = 0.08$ s. The output signal is pretreated by subtracting its mean value and discarding the first 100 output samples to eliminate the start-up effects.

Let us suppose that the system can be described by an ARX(1, 1, 1) model, where the arguments indicate the number of autoregressive and exogenous terms, and the number of sampling intervals related to the dead-time. Hence, the model is

$$y(t) = -a_1 y(t-1) + b_1 u(t-2) + e(t)$$

which can be written in vector-matrix form with $y = [y(2), y(3), \ldots, y(902)]^T$, $\vartheta = [a_1, b_1]^T$, and

$$\Phi = \begin{bmatrix} -y(1) & u(0) \\ -y(2) & u(1) \\ -y(3) & \\ \vdots & \\ -y(901) & u(900) \end{bmatrix}$$

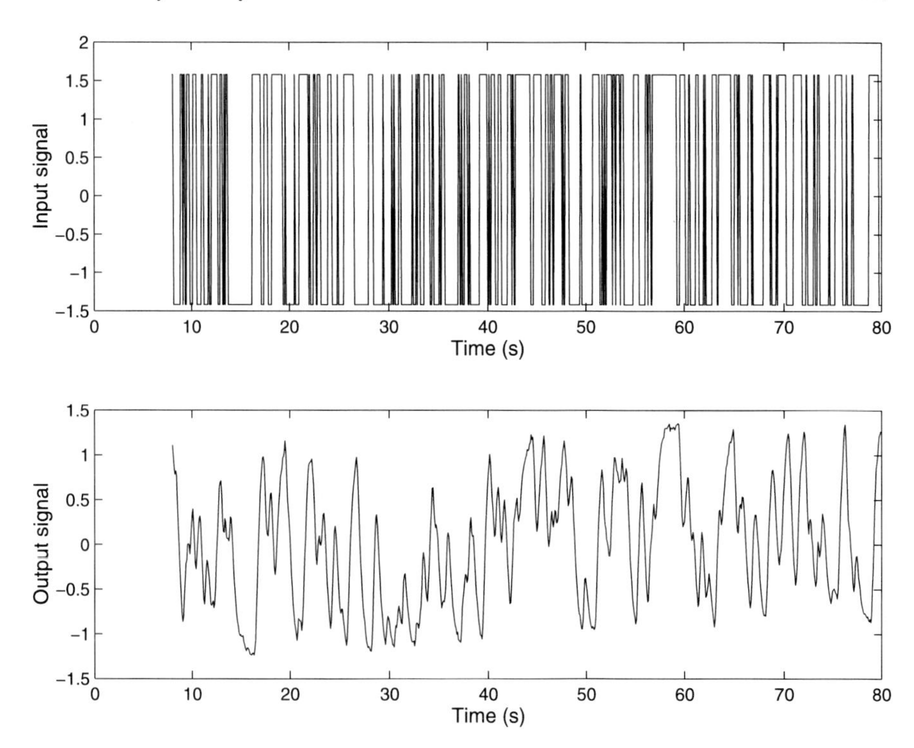

Fig. 6.5 Input–output data from identification experiment

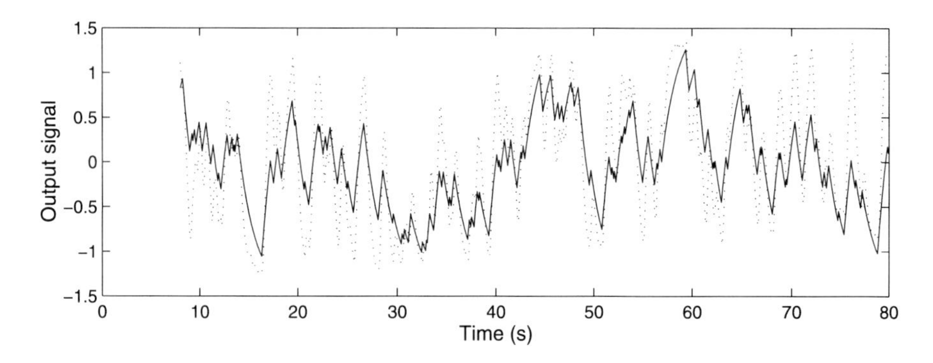

Fig. 6.6 Measured (*dotted line*) and predicted (*solid line*) output

The estimates can be simply found by applying MATLAB's function *arx*, which gives $\widehat{\vartheta} = [-0.9558\ \ 0.0467]^T$ and associated standard deviations 0.0066 and 0.0029. Comparison between the predicted model output and the measured output reveals that the model predictions are inaccurate (see Fig. 6.6). Hence, other model structures must be tried and evaluated.

Let us now try to obtain a linear regression for the ARMAX model output. However, a problem directly appears, since the past error terms $e(t-1)$, $e(t-2), \ldots, e(t-n_c)$ are unknown. As a solution to this, it is common practice to substitute the error terms by the prediction errors $\varepsilon(t-1,\vartheta), \varepsilon(t-2,\vartheta), \ldots,$ $\varepsilon(t-n_c,\vartheta)$, where $\varepsilon(t,\vartheta) = y(t) - \widehat{y}(t,\vartheta)$ and $\vartheta = [a_1, \ldots, a_{n_a}, b_1, \ldots, b_{n_b},$ $c_1, \ldots, c_{n_c}]^T$. The prediction errors, however, depend on the parameter values of ϑ, so that no true linear regression can be obtained. If we introduce the vector $\phi(t,\vartheta)^T = [-y(t-1), -y(t-2), \ldots, -y(t-n_a), u(t-1), u(t-2), \ldots,$ $u(t-n_b), \varepsilon(t-1,\vartheta), \varepsilon(t-2,\vartheta), \ldots, \varepsilon(t-n_c,\vartheta)]$. Then, the model output can be written as

$$\widehat{y}(t,\vartheta) = \phi(t,\vartheta)^T\vartheta \tag{6.17}$$

which is sometimes called a *pseudo-linear regression* because of the nonlinear effect of ϑ on the model output. Clearly, no direct methods exist, and thus an iterative solution method has to be used. Let again the inputs $u(0), u(1), \ldots, u(N)$ and corresponding outputs $y(0), y(1), \ldots, y(N)$ be recorded with $N \gg \max(n_a, n_b, n_c)$. In vector-matrix notation the output vector is defined as $y := [y(\max(n_a, n_b, n_c)), \ldots,$ $y(N)]^T$, and the regressor matrix at iteration i, for $n_a \geq n_b, n_c$, is

$$\Phi^{(i)} = \begin{bmatrix} y(n_a-1) & \cdots & y(0) & u(n_a-1) & \cdots & u(n_a-n_b) & \varepsilon(n_a-1,\widehat{\vartheta}^{(i-1)}) & \cdots & \varepsilon(n_a-n_c,\widehat{\vartheta}^{(i-1)}) \\ y(n_a) & & y(1) & u(n_a) & \cdots & u(n_a-n_b+1) & \varepsilon(n_a,\widehat{\vartheta}^{(i-1)}) & & \varepsilon(n_a-n_c+1,\widehat{\vartheta}^{(i-1)}) \\ y(n_a+1) & & \vdots & \vdots & & \vdots & \vdots & & \vdots \\ \vdots & & & & & & \vdots & & \vdots \\ y(N-1) & \cdots & y(N-n_a) & u(N-1) & \cdots & u(N-n_b) & \varepsilon(N-1,\widehat{\vartheta}^{(i-1)}) & \cdots & \varepsilon(N-n_c,\widehat{\vartheta}^{(i-1)}) \end{bmatrix} \tag{6.18}$$

Usually, for $i=0$, an ordinary least-squares solution is used. Subsequently, this solution provides prediction errors, which are used in the next step. In the successive steps, new estimates and including $c_1, \ldots, c_{n_c}$ are found. These estimates determine new prediction errors. This procedure is repeated a number of times until the estimates converge or the maximum number of iterations is reached. This iterative method is called the *extended least-squares* method.

Algorithm 6.2 Identification of ARMAX model parameters from input–output data

1. Specify an ARMAX model structure in terms of n_a, n_b, and n_c.
2. Define the vector $y := [y(n_a), \ldots, y(N)]^T$ and the matrix $\Phi^{(i)}$ with $i=0$, as in (6.16), for $n_a \geq n_b, n_c$.

3. Calculate from (5.10) the least-squares estimate of the unknown $(n_a + n_b)$-dimensional parameter vector $\vartheta^{(0)}$.
4. Calculate the prediction errors $\varepsilon(t-1, \vartheta^{(0)}), \varepsilon(t-2, \vartheta^{(0)}), \ldots, \varepsilon(t-n_a, \vartheta^{(0)})$, where $\varepsilon(t, \vartheta^{(0)}) = y(t) - \widehat{y}(t, \vartheta^{(0)})$ and $\vartheta^{(0)} = [a_1, \ldots, a_{n_a}, b_1, \ldots, b_{n_b}]^T$
5. Given the prediction errors from an ordinary least-squares estimation of the ARX-parameters, execute subsequently the following loop, with a fixed number M of iterations:

for $i = 1 : M$

define $\Phi^{(i)}$, as in (6.18)

calculate from (5.10) the least-squares estimate of the unknown

$(n_a + n_b + n_c)$-dimensional parameter vector $\vartheta^{(i)}$

calculate $\varepsilon\big(t-1, \vartheta^{(i)}\big), \varepsilon\big(t-2, \vartheta^{(i)}\big), \ldots, \varepsilon\big(t-n_c, \vartheta^{(i)}\big)$

end.

6.1.3 Output Error Identification

As for an ARMAX model structure , the estimation of the output error (OE) model parameters $b_1, b_2, \ldots, b_{n_b}, f_1, f_2, \ldots, f_{n_f}$ cannot be performed directly, since the noise-free output $\xi(t, \vartheta)$ is not observed and is a function of the unknown parameters. However, using the predicted values,

$$\widehat{y}(t, \vartheta) = \frac{B(q)}{F(q)} u(t) = \xi(t, \vartheta) \tag{6.19}$$

the regressor vector can be defined as $\phi(t, \vartheta)^T := (u(t-1), u(t-2), \ldots, u(t-n_b), -\xi(t-1, \vartheta), -\xi(t-2, \vartheta), \ldots, -\xi(t-n_f, \vartheta))$ with $\vartheta = (b_1, b_2, \ldots, b_{n_b}, f_1, f_2, \ldots, f_{n_f})^T$. Consequently,

$$\widehat{y}(t, \vartheta) = \phi(t, \vartheta)^T \vartheta \tag{6.20}$$

which again is a pseudo-linear regression, and which again requires an iterative solution.

However, let us first investigate the effect of substituting (6.8) in (6.7) so that

$$y(t) + f_1 y(t-1) + \cdots = b_1 u(t-1) + \cdots + e(t) + f_1 e(t-1) + \cdots \tag{6.21}$$

and, by (6.9),

$$F(q)y(t) = B(q)u(t) + v(t) \tag{6.22}$$

where the noise term $v(t) = F(q)e(t)$ is a moving average of $n_f + 1$ successive samples of the original white noise sequence $\{e\}$. Hence, as can be seen from the next example, the sequence $\{v\}$ is generally autocorrelated even if $\{e\}$ is not.

Example 6.2 Output error model: Suppose that a system is described by the first-order discrete-time model with a time delay of one sample interval, that is, $n_k = 1$:

$$y(t) = \frac{b_1 q^{-2}}{1 + f_1 q^{-1}} u(t) + e(t)$$

so that

$$y(t) = -f_1 y(t-1) + b_1 u(t-2) + v(t)$$

where $v(t) = e(t) + f_1 e(t-1)$. Suppose further that $\{e\}$ is a zero-mean white noise sequence with constant variance σ^2. Then,

$$\begin{aligned} r_{vv}(0) &= E\left[\{e(t) + f_1 e(t-1)\}^2\right] \\ &= E\left[e^2(t)\right] + f_1^2 E\left[e^2(t-1)\right] \\ &= (1 + f_1^2)\sigma^2 \\ r_{vv}(1) &= E\left[\{e(t) + f_1 e(t-1)\}\{e(t+1) + f_1 e(t)\}\right] \\ &= E\left[f_1 e^2(t)\right] \\ &= f_1 \sigma^2 \\ &= r_{vv}(-1) \\ r_{vv}(l) &= E\left[\{e(t) + f_1 e(t-1)\}\{e(t+l) + f_1 e(t-1+l)\}\right] \\ &= 0 \\ &= r_{vv}(-l) \quad \forall l \geq 2 \end{aligned}$$

and thus $\{v\}$ is autocorrelated. Similarly,

$$\begin{aligned} r_{vy}(0) &= E\left[\{e(t) + f_1 e(t-1)\}\right. \\ &\qquad \left.\times \{-f_1 y(t-1) + b_1 u(t-2) + e(t) + f_1 e(t-1)\}\right] \\ &= E\left[e^2(t)\right] + f_1^2 E\left[e^2(t-1)\right] \\ &= (1 + f_1^2)\sigma^2 \\ r_{vy}(1) &= f_1 \sigma^2 \\ r_{vy}(l) &= 0 \quad \forall l \geq 2 \end{aligned}$$

The consequence of writing the output error model as in (6.22) is that $\{v\}$ is autocorrelated and that this leads to correlation between $v(t)$ and one or more regressors $y(t-1), \ldots, y(t-n_f)$. Thus ordinary least-squares estimation for models of this type leads to bias, since $b = E[(\Phi^T \Phi)^{-1} \Phi^T v]$ is in general not equal to zero due to the dependence between Φ and v.

Substitution of $\widehat{y}(t, \vartheta) = \xi(t, \vartheta)$ for $y(t)$, which avoids this correlation between error and regressors but requires an iterative solution, leads to the so-called *Instrumental Variable* (IV) methods. Let the inputs $u(0), u(1), \ldots, u(N)$ and corresponding outputs $y(0), y(1), \ldots, y(N)$ be recorded with $N \gg \max(n_f, n_b)$. In vector-matrix notation the output vector for the case $n_k = 0$ is defined as $y := [y(\max(n_f, n_b)), \ldots, y(N)]^T$, and for $n_f \geq n_b$, the instrumental variable matrix at iteration i is given by

$$Z^{(i)} = \begin{bmatrix} \xi(n_f - 1, \widehat{\vartheta}^{(i-1)}) & \cdots & \xi(0, \widehat{\vartheta}^{(i-1)}) & u(n_f - 1) & \cdots & u(n_f - n_b) \\ \xi(n_f), \widehat{\vartheta}^{(i-1)} & & \xi(1, \widehat{\vartheta}^{(i-1)}) & u(n_f) & \cdots & u(n_f - n_b + 1) \\ \xi(n_f + 1, \widehat{\vartheta}^{(i-1)}) & & \vdots & \vdots & & \vdots \\ \vdots & & \vdots & & & \\ \xi(N - 1, \widehat{\vartheta}^{(i-1)}) & \cdots & \xi(N - n_f, \widehat{\vartheta}^{(i-1)}) & u(N - 1) & \cdots & u(N - n_b) \end{bmatrix} \tag{6.23}$$

where the error-correlated regressors have been replaced by so-called instrumental variables not correlated with the error, but with a large correlation with respect to the original regressors. The *instrumental variable estimate*, which in general is unbiased, is found from

$$\widehat{\vartheta}_{\mathrm{IV}} = \left[Z^T \Phi\right]^{-1} Z^T y \tag{6.24}$$

where Z has to be evaluated at each iteration. The constant regressor matrix Φ is defined as

$$\Phi := \begin{bmatrix} y(n_f - 1) & \cdots & y(0) & u(n_f - 1) & \cdots & u(n_f - n_b) \\ y(n_f) & & y(1) & u(n_f) & \cdots & u(n_f - n_b + 1) \\ y(n_f + 1) & & \vdots & \vdots & & \vdots \\ \vdots & & & & & \\ y(N - 1) & \cdots & y(N - n_f) & u(N - 1) & \cdots & u(N - n_b) \end{bmatrix} \tag{6.25}$$

As in the case of an ARMAX model structure, usually for $i = 0$, an ordinary least-squares method is applied, so that $Z^{(0)} = \Phi$. The resulting least-squares estimate $\widehat{\vartheta}^{(0)}$ is then used to generate $\xi(t, \widehat{\vartheta}^{(0)})$, which will appear in the matrix $Z^{(1)}$ (see (6.23)). These steps are repeated until convergence of the estimates occurs. In general, only a limited number of iterations is needed.

Algorithm 6.3 Identification of OE model parameters from input–output data

1. Specify an OE model structure in terms of n_b and n_f.
2. Define the vector $y := [y(n_f), \ldots, y(N)]^T$ and the matrix Φ, as in (6.25), for $n_f \geq n_b$.
3. Calculate from (5.10) the biased least-squares estimate of the unknown $(n_f + n_b)$-dimensional parameter vector $\vartheta^{(0)}$.
4. Calculate the instrumental variables $\xi(0, \vartheta^{(0)}), \xi(1, \vartheta^{(0)}), \ldots, \xi(N-1, \vartheta^{(0)})$, where $\xi(t, \vartheta^{(0)}) = \widehat{y}(t, \vartheta^{(0)})$ and $\vartheta^{(0)} = [f_1, \ldots, f_{n_f}, b_1, \ldots, b_{n_b}]^T$.
5. Given the biased least-squares estimates of the OE model parameters, execute subsequently the following loop, with a fixed number M of iterations:

for $i = 1 : M$

define $Z^{(i)}$, $\Phi^{(i)}$, as in (6.23)–(6.25)

calculate from (5.10) the least-squares estimate of the unknown

$(n_f + n_b)$-dimensional parameter vector $\vartheta^{(i)}$

calculate $\xi\big(0, \vartheta^{(i)}\big), \xi\big(1, \vartheta^{(i)}\big), \ldots, \xi\big(N-1, \vartheta^{(i)}\big)$

end.

Example 6.3 Output error model: Consider again the first-order output error model with time delay,

$$y(t) = -f_1 y(t-1) + b_1 u(t-2) + v(t)$$

Then, for given inputs, $u(0), u(1), \ldots, u(N)$, and corresponding outputs $y(0), y(1), \ldots, y(N)$, the output vector is defined as $y := [y(2), \ldots, y(N)]^T$, and

$$\Phi = \begin{bmatrix} y(1) & u(0) \\ y(2) & u(1) \\ \vdots & \vdots \\ y(N-1) & u(N-2) \end{bmatrix}$$

From this the least-squares estimate $\widehat{\vartheta^{(0)}}^{(1)} = (f_1^{(1)}, b_1^{(1)})^T = [\Phi^T \Phi]^{-1} \Phi^T y$ can be simply found. In the next iteration, the matrix $Z^{(1)}$ is defined as

$$Z^{(1)} := \begin{bmatrix} \widehat{y}(1, \widehat{\vartheta}^{(0)}) & u(0) \\ \widehat{y}(2, \widehat{\vartheta}^{(0)}) & u(1) \\ \vdots & \vdots \\ \widehat{y}(N-1, \widehat{\vartheta}^{(0)}) & u(N-2) \end{bmatrix}$$

where $\widehat{y}(t, \widehat{\vartheta}^{(0)}) = -f_1^{(1)} \widehat{y}(t-1, \widehat{\vartheta}^{(0)}) + b_1^{(1)} u(t-2)$ and $\widehat{y}(0, \widehat{\vartheta}^{(0)}) = y(0)$. Notice that for the case $n_k = 0$, the row dimension of $\Phi^{(0)}$ would be equal to N. However, the row dimension of $\Phi^{(i)}$, with $i > 0$, would be equal to $N-1$, since $\widehat{y}(0, \widehat{\vartheta}^{(i)})$ in the matrix $Z^{(i)}$ cannot be evaluated for the given model structure.

As an alternative to the introduction of instrumental variables, as in (6.24), we can also try to whiten the error term $v(t)$ with covariance matrix R_{vv}, such that the covariance matrix of the whitened equation error becomes of the form $\sigma^2 I$. One way to do this is by premultiplication of the terms in the regression equation (5.2) by an $N \times N$ matrix Q, so that

$$\begin{aligned} y' &= Qy = Q(\Phi\vartheta + v) \\ &= \Phi'\vartheta + v' \end{aligned} \tag{6.26}$$

where $\Phi' = Q\Phi$ and $v' = Qv$. It can be easily verified that $E[v'] = 0$ and, by (5.28), that $\operatorname{Cov} v' = E[Qvv^T Q^T] = QR_{vv}Q^T$. Recall that a covariance matrix is positive definite and symmetric. Then, the Choleski decomposition of R_{vv} gives $R_{vv} = LL^T$ with L a lower triangular matrix, which can be considered as the matrix square root of R_{vv}. Notice that Q is unspecified so far. If Q is then chosen to be equal to L^{-1} so that

$$Q^{-1}Q^{-T} = R_{vv} \tag{6.27}$$

the covariance matrix of the N-dimensional vector v' becomes equal to $QQ^{-1}Q^{-T}Q^T = I$. Hence, $\{v'\}$ is a mutually uncorrelated sequence with constant variance and zero mean. The ordinary least-squares estimate of the filtered equation (6.26), which is unbiased, is given by

$$\begin{aligned} \widehat{\vartheta} &= \left[\Phi'^T\Phi'\right]^{-1}\Phi'^T y' \\ &= \left[\Phi^T Q^T Q\Phi\right]^{-1}\Phi^T Q^T Q y \\ &= \left[\Phi^T R_{vv}^{-1}\Phi\right]^{-1}\Phi^T R_{vv}^{-1} y \end{aligned} \tag{6.28}$$

since $(Q^{-1}Q^{-T})^{-1} = Q^T Q = R_{vv}^{-1}$. This estimate is called the *Markov estimate* or *generalized least-squares estimate*. The corresponding covariance matrix of the estimates is given by

$$\operatorname{Cov}\widehat{\vartheta} = E\left[\left[\Phi'^T\Phi'\right]^{-1}\right] = E\left[\left[\Phi^T R_{vv}^{-1}\Phi\right]^{-1}\right] \tag{6.29}$$

In practice, however, R_{vv} is never known in advance, and thus it has to be estimated from the data, as is illustrated by the following example. But let us first present the algorithm.

Algorithm 6.4 Identification of OE model parameters from input–output data using the generalized least-squares method

1. Specify an OE model structure in terms of n_b and n_f.
2. Given the OE model structure, derive the autocorrelation function $r_{vv}(l)$ for $l = 0, 1, \ldots$ analytically, as in Example 6.2.
3. Given the autocorrelation function of v, form the corresponding autocorrelation matrix R_{vv}.

4. Define the vector $y := [y(n_f), \dots, y(N)]^T$ and the matrix Φ, as in (6.16) with $n_a = n_f$, for $n_f \geq n_b$.
5. Calculate from (5.10) the biased least-squares estimate of the unknown $(n_f + n_b)$-dimensional parameter vector $\vartheta^{(0)}$.
6. Given the biased least-squares estimates of the OE model parameters $f_1^{(0)}, \dots, f_{n_f}^{(0)}$, execute subsequently the following loop, with a fixed number M of iterations:

for $i = 1 : M$

calculate $R_{vv}(i)$, as a function of $f_1^{(i-1)}, \dots, f_{n_f}^{(i-1)}$

calculate from (6.28) the weighted least-squares estimate of the

unknown $(n_f + n_b)$-dimensional parameter vector $\vartheta^{(i)}$

end.

Example 6.4 Output error model: Recall that for the given output error model structure

$$y(t) = -f_1 y(t-1) + b_1 u(t-2) + v(t)$$

the autocorrelation function of v is given by

$$r_{vv}(0) = (1 + f_1^2)\sigma^2$$
$$r_{vv}(\pm 1) = f_1 \sigma^2$$
$$r_{vv}(\pm l) = 0, \quad \forall l > 1$$

Hence,

$$\mathrm{Cov}\,\widehat{v} = \widehat{\sigma}_\varepsilon^2 \begin{bmatrix} (1+\widehat{f}_1^2) & \widehat{f}_1 & 0 & \cdots & 0 \\ \widehat{f}_1 & (1+\widehat{f}_1^2) & \widehat{f}_1 & & \vdots \\ 0 & \widehat{f}_1 & & & 0 \\ \vdots & & & & \widehat{f}_1 \\ 0 & \cdots & 0 & \widehat{f}_1 & (1+\widehat{f}_1^2) \end{bmatrix} \simeq R_{vv}$$

and thus R_{vv} can be approximated at each iteration on the basis of estimates of the autoregressive parameters and of the error variance.

For large data sets, this implementation is unattractive, since at each iteration an $N \times N$ matrix has to be inverted. In the following section, a more convenient implementation using low-order linear filters is introduced. In conclusion, for output error model structures, nonlinear regressions between model predictions and parameters result, which asks for iterative estimation procedures as Markov estimation or Instrumental Variable methods.

6.1.4 Prediction Error Identification

The results of the previous section can be generalized by considering the equation and output errors as prediction errors, more specifically as l-steps-ahead prediction errors with $l = 1$ and ∞, respectively. Using the generalized transfer function model (6.12), the following expression of the error $e(t)$ is found:

$$e(t) = -H^{-1}(q)G(q)u(t) + H^{-1}(q)y(t) \tag{6.30}$$

where $H^{-1}(q) = 1/H(q)$. Let us evaluate this expression for some common model structures.

Example 6.5 MA process: Suppose that

$$v(t) = e(t) + ce(t-1)$$

that is, $G(q) = 0$ and $H(q) = 1 + cq^{-1}$, and that $v(t)$ is observed. Then, by long division we find

$$\begin{aligned} H^{-1}(q) &= \frac{1}{1+cq^{-1}} \\ &= 1 - cq^{-1} + c^2q^{-2} - c^3q^{-3} + \cdots \\ &= \sum_{k=0}^{\infty}(-c)^k q^{-k} \end{aligned}$$

and thus,

$$e(t) = \sum_{k=0}^{\infty}(-c)^k v(t-k)$$

Example 6.6 AR process: Suppose that

$$v(t) + av(t-1) = e(t)$$

that is, $G(q) = 0$ and $H(q) = 1/(1 + aq^{-1})$, and that $v(t)$ is observed. Then,

$$H^{-1}(q) = 1 + aq^{-1}$$

and thus,

$$e(t) = v(t) + av(t-1)$$

For further analysis of the prediction error, let us consider a one-step-ahead prediction of the noise term $v(t) = H(q)e(t) = \sum_{k=0}^{\infty} h(k)e(t-k)$ given measurements of $v(s)$ for $s \le t-1$. Then, under the assumption that $H(q)$ is monic, i.e.,

$h_0 = 1$, and $\{e\}$ is a mutually uncorrelated sequence with zero-mean,

$$\begin{aligned}\widehat{v}(t|t-1) &= E\big[v(t|t-1)\big] \\ &= E\big[e(t)\big] + E\left[\sum_{k=1}^{\infty} h(k)e(t-k)\right] \\ &= \sum_{k=1}^{\infty} h(k)e(t-k)\end{aligned} \tag{6.31}$$

In terms of the rational transfer function H, we obtain

$$\begin{aligned}\widehat{v}(t|t-1) &= \big[H(q) - 1\big]e(t) \\ &= \frac{H(q)-1}{H(q)}v(t) = \big[1 - H^{-1}(q)\big]v(t)\end{aligned} \tag{6.32}$$

Example 6.7 MA process: Suppose that

$$v(t) = e(t) + ce(t-1)$$

Then, with $H(q) = 1 + cq^{-1}$ and after long division,

$$\widehat{v}(t|t-1) = \frac{cq^{-1}}{1+cq^{-1}}v(t) = -\sum_{k=1}^{\infty}(-c)^k v(t-k)$$

Example 6.8 AR process: Suppose that

$$v(t) + av(t-1) = e(t)$$

Then, with $H(q) = 1/(1 + aq^{-1})$,

$$\widehat{v}(t|t-1) = \big[1 - \big(1 + aq^{-1}\big)\big]v(t) = av(t-1)$$

The one-step-ahead prediction of the model output $y(t) = G(q)u(t) + v(t)$, given measurements of $u(s)$ and $y(s)$, and thus of $v(s)$ as well, for $s \leq t-1$, is found from

$$\begin{aligned}\widehat{y}(t|t-1) &= G(q)u(t) + \widehat{v}(t|t-1) \\ &= G(q)u(t) + \big[1 - H^{-1}(q)\big]v(t) \\ &= G(q)u(t) + \big[1 - H^{-1}(q)\big]\big[y(t) - G(q)u(t)\big] \\ &= H^{-1}(q)G(q)u(t) + \big[1 - H^{-1}(q)\big]y(t)\end{aligned} \tag{6.33}$$

From this we find

$$\begin{aligned}
y(t) - \widehat{y}(t|t-1) &= -H^{-1}(q)G(q)u(t) + H^{-1}(q)y(t) \\
&= H^{-1}(q)\big[y(t) - G(q)u(t)\big] \\
&= H^{-1}(q)v(t) \\
&= e(t)
\end{aligned} \tag{6.34}$$

so that $e(t)$ is the one-step-ahead prediction error that represents that part of the output $y(t)$ that cannot be predicted from past data. Hence, a realization of the error sequence $\{e\}$ is found from an evaluation of past prediction errors.

Suppose now that $v(s)$ has been observed for $s \leq t$, so that $e(t)$ is known. In order to derive the l-steps-ahead prediction of $v(t+l)$, we need to write the rational polynomial $H(q)$ as

$$H(q) = \overline{H}_l(q) + q^{-l}\widetilde{H}_l(q) \tag{6.35}$$

where $\overline{H}_l(q) = \sum_{k=0}^{l-1} h(k)q^{-k}$ and $\widetilde{H}_l(q) = \sum_{k=l}^{\infty} h(k)q^{-k+l}$. Consequently, $v(t+l)$ is split up in an unknown part including the error terms $e(t+l)$, $e(t+l-1), \ldots, e(t+1)$ and a known part, that is,

$$\begin{aligned}
v(t+l) &= \sum_{k=0}^{\infty} h(k)e(t+l-k) \\
&= \sum_{k=0}^{l-1} h(k)e(t+l-k) + \sum_{k=l}^{\infty} h(k)e(t+l-k)
\end{aligned} \tag{6.36}$$

The l-steps-ahead prediction of $v(t+l)$ is then given by

$$\begin{aligned}
\widehat{v}(t+l|t) &= \sum_{k=l}^{\infty} h(k)e(t+l-k) = \widetilde{H}_l(q)e(t) \\
&= \widetilde{H}_l(q)H^{-1}(q)v(t)
\end{aligned} \tag{6.37}$$

Let $y(-\infty), \ldots, y(t)$ and $u(-\infty), \ldots, u(t)$ be measured. Then

$$\begin{aligned}
\widehat{y}(t+l|t) &= G(q)u(t+l) + \widehat{v}(t+l|t) \\
&= G(q)u(t+l) + \widetilde{H}_l(q)H^{-1}(q)v(t) \\
&= G(q)u(t+l) + \widetilde{H}_l(q)H^{-1}(q)\big[y(t) - G(q)u(t)\big]
\end{aligned} \tag{6.38}$$

If we define $W_l(q) := 1 - q^{-l}\widetilde{H}_l(q)H^{-1}(q)$, then by (6.35), $W_l(q) = \overline{H}_l(q)H^{-1}(q)$, and after some manipulation we find

$$\widehat{y}(t+l|t) = W_l(q)G(q)u(t+l) + \widetilde{H}_l(q)H^{-1}(q)y(t) \tag{6.39}$$

or, after setting $t := t + l$, i.e., $\widehat{y}(t|t-l) = q^{-l}\widehat{y}(t+l|t)$,

$$\widehat{y}(t|t-l) = W_l(q)G(q)u(t) + \left[1 - W_l(q)\right]y(t) \tag{6.40}$$

The prediction errors associated with (6.38) are then given by

$$\begin{aligned} e(t+l|t) &= y(t+l) - \widehat{y}(t+l|t) \\ &= -W_l(q)G(q)u(t+l) + \left[q^l - \widetilde{H}_l(q)H^{-1}(q)\right]y(t) \\ &= W_l(q)\left[y(t+l) - G(q)u(t+l)\right] \\ &= W_l(q)H(q)e(t+l) \\ &= \overline{H}_l(q)e(t+l) \end{aligned} \tag{6.41}$$

Recall from (6.35) that $\overline{H}_l(q)$ is a polynomial of order $k-1$, so that $e_l(t+l)$ is a moving average of $e(t+l), \ldots, e(t+1)$. Hence, even if $e(t)$ is a white noise sequence, the more-steps-ahead prediction $e_l(t+l)$ is in general not.

For the following, it is important to notice from (6.33) that the predictor (6.40) is the one-step-ahead predictor of the model

$$y(t) = G(q)u(t) + W_l^{-1}(q)e(t) \tag{6.42}$$

where the last term represents some filtered noise.

In order to allow a large class of identification problems to be cast in the prediction-error framework, the prediction-error sequence $\{\varepsilon(t,\vartheta)\}$ is filtered by a stable linear filter $L(q)$ such that

$$\varepsilon_F(t,\vartheta) = L(q)\varepsilon(t,\vartheta), \quad t = 1, \ldots, N \tag{6.43}$$

where it is emphasized that ε is a function of both t and ϑ, which is especially important to realize when applying iterative solution procedures. A large class of *prediction-error identification methods* will try to minimize the following objective function:

$$J(\vartheta) := \sum_{t=1}^{N} \varepsilon_F^2(t,\vartheta) \tag{6.44}$$

The high- or low-frequency disturbances, which are thought to be unimportant for the identification results, can thus be removed from the error sequence by the filter L. From this point of view, the filter acts like frequency weighting. Notice, furthermore that

$$\varepsilon_F(t,\vartheta) = \left[L^{-1}(q)H(q,\vartheta)\right]^{-1}\left[y(t) - G(q,\vartheta)u(t)\right] \tag{6.45}$$

In [Lju87], p. 200, Ljung noticed then that "the effect of pre-filtering is thus identical to changing the noise model from $H(q,\vartheta)$ to $L^{-1}(q)H(q,\vartheta)$." From these results it can be deduced that for l-steps-ahead prediction-error identification, the filter $L(q)$ must be chosen identical to $\overline{H}_l(q)$ to minimize the sum of squares of the l-steps-ahead prediction-errors. Hence, using (6.41), we arrive at the following result for $l=1$: $\varepsilon_F(t,\vartheta)=\varepsilon_l(t|t-l,\vartheta)=\varepsilon(t,\vartheta)$ because $H(q)$ is considered to be monic, which implies that $\overline{H}_{l=1}(q)=1$ and thus $L(q)=1$. Since $H(q)$ in general is a low-pass filter, the one-step-ahead prediction error method implies high-pass filtering of the error sequence $\{y(t)-G(q,\vartheta)u(t)\}$. Notice furthermore that, since for an ARX model structure $G(q)=B(q)/A(q)$ and $H(q)=1/A(q)$, we have $\varepsilon_F(t,\vartheta)=L(q)[A(q,\vartheta)y(t)-B(q,\vartheta)u(t)]$. Hence, an equation-error method, which thus minimizes the sum of squares of the sequence $\{A(q)y(t)-B(q)u(t)\}$ with $L(q)=1$, minimizes the one-step-ahead prediction errors. On the other hand, for $l=\infty$, $\varepsilon_F(t,\vartheta)=\varepsilon_{l=\infty}(t|t-l,\vartheta)=H(q)\varepsilon(t,\vartheta)$, since $\overline{H}_{l=\infty}(q)=H(q)$, and thus $\varepsilon_F(t,\vartheta)=y(t)-G(q,\vartheta)u(t)$, which is the output-error. Consequently, an output-error method tends to minimize the ∞-steps-ahead prediction errors.

If the predictor is linear and time-invariant, the filtering of the prediction error $\varepsilon(t,\vartheta)$ is identical to first filter the input–output data and then apply the predictor. Let us apply this to an output-error estimation problem. If we rewrite the output-error model structure (6.9) as $F(q)e(t)=F(q)y(t)-B(q)u(t)$, then from this we obtain the following nonlinear expression of the prediction error in ϑ:

$$F(q,\vartheta)\varepsilon(t,\vartheta)=F(q,\vartheta)y(t)-B(q,\vartheta)u(t) \tag{6.46}$$

An iterative solution of the estimate $\widehat{\vartheta}$ is then found via the prediction error evaluation

$$\begin{aligned}\varepsilon\big(t,\widehat{\vartheta}^{(i)}\big)&=F\big(q,\widehat{\vartheta}^{(i)}\big)\big[F\big(q,\widehat{\vartheta}^{(i-1)}\big)\big]^{-1}y(t)\\&\quad-B\big(q,\widehat{\vartheta}^{(i)}\big)\big[F\big(q,\widehat{\vartheta}^{(i-1)}\big)\big]^{-1}u(t)\\&=F\big(q,\widehat{\vartheta}^{(i)}\big)\widetilde{y}(t)-B\big(q,\widehat{\vartheta}^{(i)}\big)\widetilde{u}(t)\end{aligned} \tag{6.47}$$

which allows at each iteration an unbiased least-squares estimation. In fact, this approach is an effective alternative, based on "noise whitening," to the previously described Markov estimation method. Clearly, this idea of repeated prefiltering of the input–output data to obtain a white error sequence can also be applied to other prediction-error identification problems.

Algorithm 6.5 Identification of OE model parameters from input–output data using prefiltering

1. Specify an OE model structure in terms of n_b and n_f.
2. Define the vector $y:=[y(n_f),\dots,y(N)]^T$ and the matrix Φ, as in (6.16) with $n_a=n_f$, for $n_f\geq n_b$.
3. Calculate from (5.10) the biased least-squares estimate of the unknown (n_f+n_b)-dimensional parameter vector $\vartheta^{(0)}$.

Table 6.1 Random process data

$x(t)$	1	2	3	4	5
$y(t)$	5.2	5.3	5.1	4.5	5.0

4. Given the biased least-squares estimates of the OE model parameters $f_1^{(0)}, \dots, f_{n_f}^{(0)}$, execute subsequently the following loop, with a fixed number M of iterations:

for $i = 1 : M$

evaluate $F(q, \widehat{\vartheta}^{(i-1)})$, as a function of $f_1^{(i-1)}, \dots, f_{n_f}^{(i-1)}$

prefilter both $y(t)$ and $u(t)$ with $[F(q, \widehat{\vartheta}^{(i-1)})]^{-1}$

calculate the ordinary least-squares estimate of the

unknown $(n_f + n_b)$-dimensional parameter vector $\vartheta^{(i)}$

end.

6.1.5 Model Structure Identification

So far, it has been assumed that the model structure is a priori given. However, in practice this is never fully the case; the input–output data may suggest an other structure than that obtained from prior system knowledge. A most natural way is to suggest a number of structures and evaluate its performance. At first instance, it appears to be a good idea to use the objective function value to discriminate between structures. Let us illustrate the consequence of this in the next example.

Example 6.9 Random process: Consider the following measurements, shown in Table 6.1, which originate from a random process.

A very simple model that approximately describes the data in Table 6.1 is given by

$$y(t) = \vartheta_0 + e(t)$$

with $\vartheta_0 = 5.02$ and the sum of squared prediction errors (see (5.4)) $\varepsilon^T \varepsilon = 0.388$. It can be easily verified that the alternative model

$$y(t) = \vartheta_0 + \vartheta_1 x(t) + \vartheta_2 x^2(t) + \vartheta_3 x^3(t) + \vartheta_4 x^4(t) + e(t)$$

with $\vartheta_0 = 6.4997$, $\vartheta_1 = -2.9661$, $\vartheta_2 = 2.2830$, $\vartheta_3 = -0.68325$, and $\vartheta_4 = 0.06666$ exactly describes the data. Hence, the objective function value (sum of squares) is equal to zero. However, model predictions outside the range, unlike the predictions from the first model $\widehat{y}(t) = 5.02$, become unstable. For instance, for $x(t) = 6$, $\widehat{y}(t) = 9.7$, and for $x(t) = 10$, $\widehat{y}(t) = 188$, so that for large values of x, the predicted output tends to infinity.

From this example it becomes clear that evaluation of the objective function values only is not a good idea, because despite the perfect fit, bad prediction models may result. We call these models *overparameterized*, since they fit the noise rather than the underlying process dynamics.

However, there is a more fundamental problem: given some data, there will always be an infinite number of models that fit the data equally well. Thus, without making additional assumptions, there is no reason to prefer one model over another. The additional assumptions may be expressed in terms of probabilities, evidential support, falsifiability, or minimum description length. Within a system identification context, model selection aims at choosing a model of optimal complexity for the given (finite) data. Many model selection procedures employ some form of parsimony. If a set of models fit the data equally well, the simplest model is preferred. Therefore, in addition to a measure of the misfit, a measure of model complexity has been introduced. The Akaike information criterion (AIC), for instance, provides a trade-off between the model complexity and the goodness of fit to the experimental data. The AIC is given by

$$\mathrm{AIC} = -2\log L + 2d_{\mathcal{M}} \tag{6.48}$$

where $\log L$ is the maximum log-likelihood, and $d_{\mathcal{M}}$ is the number of parameters in the model. The model with the lowest AIC should be preferred. The AIC is grounded in the concept of entropy. In fact, it quantifies a relative measure of the information loss when a model is used to describe a data set. It should be noted that the AIC is not a test of the model in the sense of hypothesis testing. It provides a test between models and is thus one of the tools for model selection.

Akaike's Final Prediction Error criterion (FPE) provides a measure of model quality for the case where the model is tested on a different data set. Hence, the model prediction quality is explicitly tested, as in our previous example. According to Akaike's theory, the most accurate model has the smallest FPE, where the Final Prediction Error is defined as

$$J_{\mathrm{FPE}}(\mathcal{M}) := \frac{1 + d_{\mathcal{M}}/N}{1 - d_{\mathcal{M}}/N}\,\frac{1}{N}\sum_{t=1}^{N}\frac{1}{2}\varepsilon^2(t,\hat{\vartheta}) \tag{6.49}$$

As before, it combines the model complexity and goodness of fit for a specific model $\mathcal{M}(\hat{\vartheta})$. In this criterion the model complexity is represented by the dimension of the model parameter vector ($d_{\mathcal{M}}$). The factor $\frac{1+d_{\mathcal{M}}/N}{1-d_{\mathcal{M}}/N}\frac{1}{N} = \frac{1+d_{\mathcal{M}}/N}{N-d_{\mathcal{M}}}$ can be interpreted as a corrected inverse of the degrees of freedom, $N - d_{\mathcal{M}}$, see also (5.33) for a comparison. The term $\frac{1}{N}\sum_{t=1}^{N}\frac{1}{2}\varepsilon^2(t,\hat{\vartheta})$, used in MATLAB's *System Identification Toolbox*, will be further indicated as the loss function and is clearly related to the least-squares objective function (5.3).

Example 6.10 Heating system: In Example 6.1 it appeared that an ARX(1, 1, 1) model structure was not appropriate to describe the data. Let us therefore evaluate a number of candidate ARX models. Define, with the help of MATLAB's function

Table 6.2 Model structure identification results

n_a	n_b	n_k	Loss function ($\times 10^{-4}$)	J_{FPE} ($\times 10^{-4}$)	$d_{\mathcal{M}}$
1	1	4	54.074	54.314	2
2	1	3	24.157	24.318	3
2	2	3	13.886	14.010	4
3	2	3	12.788	12.931	5
3	3	3	11.809	11.967	6
4	3	3	11.628	11.810	7

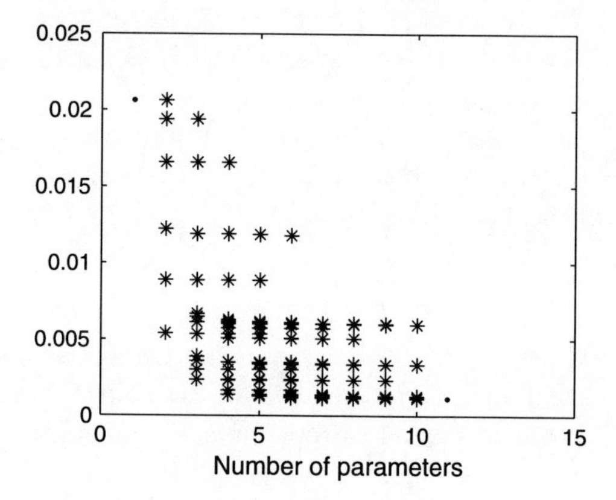

Fig. 6.7 FPE function values (*stars*) with corresponding number of parameters in ARX model

struc, a matrix of candidate structures ARX(1:5, 1:5, 0:5). Then, using *arxstruc* and *selstruc* for a given input–output data set, Fig. 6.7 will result.

From Fig. 6.7 it can be concluded that the FPE function value decreases with the number of model parameters or, in other words, with the model complexity. This decrease is caused by an increase in the degrees of freedom. A next step is to find the optimal combination of autoregressive and exogenous parameters and time delays. A natural way to find this is to look for the "knee" in the curve and then to evaluate all possible combinations for a specific number of parameters. The result of this for $d_{\mathcal{M}}$ ranging from two to seven is presented in Table 6.2.

On the basis of these results, a good choice would be an ARX(2, 2, 3) model structure, because more complex model structures will not significantly increase the model performance as measured by the values of the loss function and J_{FPE}. This result is further confirmed from an analysis using the unexplained output variance (in %), which is the variance of the ARX model prediction error. In other words, the unexplained output variance represents the portion of the model output not explained by the model. The results, based on the unexplained output variance and also leading to an ARX(2, 2, 3) model structure, are presented in Fig. 6.8.

It is important to mention here that the model performance is evaluated on the same data set that has been used for parameter estimation. Hence, so far no independent measure of model performance has been used.

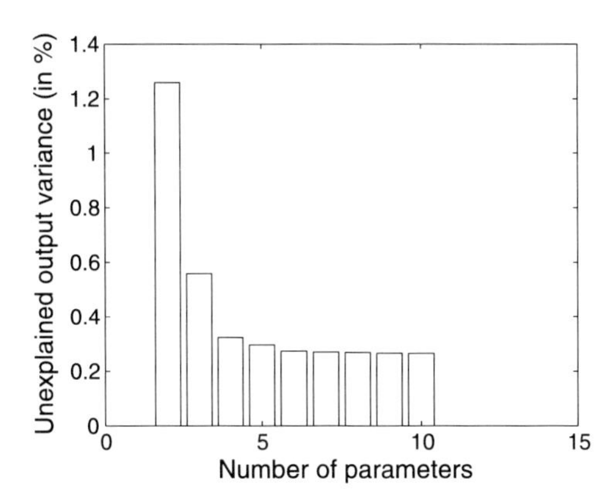

Fig. 6.8 Unexplained output variance with corresponding number of parameters in ARX model

Consequently, from these examples it appears that the comparison of model structures should essentially be based on *cross-validation*, where the identified model structures are confronted with fresh data, which has not been used for parameter estimation. This cross-validation prevents overparameterized models to a large extent, because the noise in the cross-validation data set will be different from the noise in the identification data set. Therefore, in practice, most often the data set is split up into an identification/calibration and a validation set. A further treatment of the model validation step will be found in Part IV.

6.1.6 *Subspace Identification

Subspace identification methods aim at directly estimating the system matrices A, B, C, D in a state-space model structure from noisy input–output data. It should be emphasized that these methods do not need an a priori specification of the structure of the system matrices. Hence, all the entries in the matrices follow from the input–output data.

In the following, it will be shown that subspace identification is a direct (noniterative) estimation method, which is also indicated in short as 4SID, i.e., Subspace State-Space System IDentification. The basic idea behind subspace identification starts from a given noisy unit impulse response realization of an LTI system and results in a minimal (data-based) state-space realization. Let us illustrate this in some more detail for the noise-free case.

Recall that for a discrete-time dynamic system, the output can be expressed by the convolution sum

$$\begin{aligned} y(t) &= \sum_{k=0}^{t} g(t-k)u(k) \\ &= \sum_{k=0}^{t} g(t)u(t-k), \quad t \in \mathbb{Z}^+ \end{aligned} \tag{6.50}$$

Alternatively, the system can also be described by the discrete-time state-space model (see Sect. 1.2.2)

$$\begin{aligned} x(t+1) &= Ax(t) + Bu(t) \\ y(t) &= Cx(t) + Du(t) \end{aligned} \tag{6.51}$$

with $x \in \mathbb{R}^n$. The goal is now to determine the matrices A, B, C, D (taking into account that these matrices are equivalent up to a linear transformation, that is, $\tilde{A} = SAS^{-1}$, $\tilde{B} = SB$, $\tilde{C} = CS^{-1}$, $\tilde{D} = D$). The following relationship between (6.50) and (6.51) exists:

$$g(t) = \begin{cases} 0, & t < 0 \\ D, & t = 0 \\ CA^{t-1}B, & t > 0 \end{cases} \tag{6.52}$$

This relationship allows us to construct the so-called Hankel matrix H on the basis of the given impulse response, where this matrix can be factorized as follows:

$$H = \Gamma_{n+1}\Omega_{n+1} \tag{6.53}$$

with

$$H = \begin{bmatrix} g(1) & g(2) & \cdots & g(n+1) \\ g(2) & g(3) & \cdots & g(n+2) \\ \vdots & \vdots & & \vdots \\ g(n+1) & g(n+2) & \cdots & g(2n+1) \end{bmatrix}$$

In fact (see (6.53)), the Hankel matrix can be factorized so that

$$\Gamma_n = \begin{bmatrix} C \\ CA \\ \vdots \\ CA^{n-1} \end{bmatrix} \tag{6.54}$$

which is known as the observability matrix, and

$$\Omega_n = \begin{bmatrix} B & AB & \cdots & A^{n-1}B \end{bmatrix} \tag{6.55}$$

the controllability matrix.

As the ranks of the observability and controllability matrix are equal to n, the rank of H is also n. This fact forms the basis for the estimation of the system matrices A, B, C, and D. In particular, separation of the Hankel matrix H in terms of an observability and controllability matrix (6.53) such that the upmost rows and the leftmost columns of the factors result in C (see (6.54)) and B (see (6.55)), seems to be an appropriate choice. Furthermore, from the observability matrix we can use the following relationship:

$$\Gamma_{2:n+1} = \Gamma_{1:n}A \tag{6.56}$$

where the matrices $\Gamma_{2:n+1}$ and $\Gamma_{1:n}$ have been derived by deleting the first and last row of Γ_{n+1}, respectively. The matrix A can then be simply found from

$$A = \Gamma_{1:n}^{+} \Gamma_{2:n+1} \tag{6.57}$$

where $\Gamma_{1:n}^{+} = (\Gamma_{1:n}^{T} \Gamma_{1:n})^{-1} \Gamma_{1:n}^{T}$, and is called the Moore–Penrose pseudo-inverse of $\Gamma_{1:n}$. The Moore–Penrose pseudo-inverse has also been used in the derivation of the ordinary least-squares estimator via the normal equations (see (5.9)–(5.10)). Finally, the matrix D is equal to $g(0)$. Consequently, given a unit impulse response, the system matrices A, B, C, and D can be found. Let us illustrate this by an example of a second-order process.

Example 6.11 Second-order process: Let a process be described in discrete-time by

$$\begin{aligned} x_1(t+1) &= x_2(t) \\ x_2(t+1) &= -\alpha_1 x_1(t) - \alpha_2 x_2(t) + \beta_1 u(t) \\ y(t) &= x_1(t), \quad t \in \mathbb{Z}^+ \end{aligned}$$

with $\alpha_1 = \alpha_2 = \beta_1 = 1$ and sampling interval of 1. Consequently, the system matrices are given by

$$A = \begin{bmatrix} 0 & 1 \\ -1 & -1 \end{bmatrix}, \qquad B = \begin{bmatrix} 0 \\ 1 \end{bmatrix}, \qquad C = [1\ 0], \qquad D = 0$$

The first eight elements of the unit impulse response, starting at $t = 0$, are

$$[0\ 0\ 1\ -1\ 0\ 1\ -1\ 0]$$

The Hankel matrix H is then given by

$$H = \begin{bmatrix} 0 & 1 & -1 \\ 1 & -1 & 0 \\ -1 & 0 & 1 \end{bmatrix} = \begin{bmatrix} 1 & 0 \\ 0 & 1 \\ -1 & -1 \end{bmatrix} \begin{bmatrix} 0 & 1 & -1 \\ 1 & -1 & 0 \end{bmatrix}$$

where the upmost row of the first matrix on the right-hand side (Γ_{n+1}) is C, and the leftmost column of the second matrix (Ω_{n+1}) is B. Furthermore, given

$$\Gamma_{n+1} = \begin{bmatrix} 1 & 0 \\ 0 & 1 \\ -1 & -1 \end{bmatrix}$$

A can be found from

$$A = \begin{bmatrix} 1 & 0 \\ 0 & 1 \end{bmatrix}^{+} \begin{bmatrix} 0 & 1 \\ -1 & -1 \end{bmatrix} = \begin{bmatrix} 0 & 1 \\ -1 & -1 \end{bmatrix}$$

Clearly, the key problem now is how to factorize H appropriately for some noisy input–output data set. Consider, therefore, the following state-space model of an LTI discrete-time system:

$$\begin{aligned} x(t+1) &= Ax(t) + Bu^o(t) + w(t) \\ y^o(t) &= Cx(t) + Du^o(t) \end{aligned} \tag{6.58}$$

where u^o and y^o are noise-free input and output signals. The state vector x is corrupted with an additional system noise term $w(t)$. Let

$$\begin{aligned} u(t) &= u^o(t) + z(t) \\ y(t) &= y^o(t) + v(t) \end{aligned} \tag{6.59}$$

where all the errors are assumed to be white. Substituting (6.59) into (6.58) gives

$$\begin{aligned} x(t+1) &= Ax(t) + Bu^o(t) + Bz(t) + w(t) \\ y(t) &= Cx(t) + Du^o(t) + Dz(t) + v(t) \end{aligned} \tag{6.60}$$

Let $Ke(t) = Bz(t) + w(t)$ and $e(t) = Dz(t) + v(t)$ with $e(t)$ white. Compose then the following column vectors of length $m-1$ and filled with future values from t to $t+m-1$:

$$Y(t) = \left[y(t)\ y(t+1) \cdots y(t+m-1) \right] \tag{6.61}$$

$$U(t) = \left[u(t)\ u(t+1) \cdots u(t+m-1) \right] \tag{6.62}$$

$$E(t) = \left[e(t)\ e(t+1) \cdots e(t+m-1) \right] \tag{6.63}$$

$$Z(t) = \left[z(t)\ z(t+1) \cdots z(t+m-1) \right] \tag{6.64}$$

Consequently,

$$Y(t) = \Gamma_m x(t) + H_m^u U(t) - H_m^u Z(t) + H_m^e E(t) \tag{6.65}$$

with

$$\Gamma_m = \begin{bmatrix} C \\ CA \\ \vdots \\ CA^{m-1} \end{bmatrix}$$

$$H_m^u = \begin{bmatrix} D & 0 & \cdots & 0 \\ CB & D & & \vdots \\ \vdots & \cdots & \ddots & \\ CA^{m-2}B & CA^{m-3}B & \cdots & D \end{bmatrix}$$

$$H_m^e = \begin{bmatrix} 1 & 0 & \cdots & 0 \\ CK & 1 & & \vdots \\ \vdots & \cdots & \ddots & \\ CA^{m-2}K & CA^{m-3}K & \cdots & 1 \end{bmatrix}$$

This relationship can be even further expanded. Let therefore,

$$Y = \left[Y(1)\; Y(2) \cdots Y(N) \right] \tag{6.66}$$

$$U = \left[U(1)\; U(2) \cdots U(N) \right] \tag{6.67}$$

$$E = \left[E(1)\; E(2) \cdots E(N) \right] \tag{6.68}$$

$$Z = \left[Z(1)\; Z(2) \cdots Z(N) \right] \tag{6.69}$$

$$X = \left[x(1)\; x(2) \cdots x(N) \right] \tag{6.70}$$

so that

$$Y = \Gamma_m X + H_m^u U - H_m^u Z + H_m^e E \tag{6.71}$$

with $Y \in \mathbb{R}^{(m-1)\times N}$, $\Gamma_m \in \mathbb{R}^{(m-1)\times n}$, $X \in \mathbb{R}^{n\times N}$, etc.

As mentioned before, subspace identification methods aim at estimating the observability matrix Γ_m from which the system matrices A, B, C, and D can be estimated. Basically, three subspace identification approaches can be distinguished, that is,

- output error approach, where $Z = 0$ and $K = 0$
- simultaneous output/state error approach, where $Z = 0$ and $K \neq 0$
- simultaneous output/state/input error approach

In what follows, we will only focus on the simplest case, that is, the output error approach. The basic output error 4SID method calculates the RQ factorization of the matrix $\frac{1}{\sqrt{N}}[U^T\; Y^T]^T$, that is,

$$\frac{1}{\sqrt{N}} \begin{bmatrix} U \\ Y \end{bmatrix} = \begin{bmatrix} R_{11} & 0 \\ R_{21} & R_{22} \end{bmatrix} \begin{bmatrix} Q_1 \\ Q_2 \end{bmatrix} \tag{6.72}$$

where R is a lower triangular matrix, and Q an orthogonal matrix with $Q^T Q = I$. Notice that the RQ factorization can be easily found from the well-known QR factorization by taking the transpose of both sides of (6.72), so that $\frac{1}{\sqrt{N}}[U\; Y] = QR$ with R an upper triangular matrix. The system order is determined via the SVD of R_{22}, namely

$$\begin{aligned} R_{22} &= USV^T \\ &= \left[\overline{U}_s\; \overline{U}_n \right] \begin{bmatrix} S_s & 0 \\ 0 & S_n \end{bmatrix} \begin{bmatrix} V_s^T \\ V_n^T \end{bmatrix} \\ &\approx \overline{U}_s S_s V_s^T \end{aligned} \tag{6.73}$$

Table 6.3 Second-order process data

t	0	1	2	3	4	5	6	7	8	9	10	11	12	13	14	15
u	1	1	−1	−1	1	1	1	−1	−1	−1	−1	−1	−1	−1	1	1
y	0	0	1	0	−2	1	2	−2	1	0	−2	1	0	−2	1	0

Via the separation of S into S_s and S_n, where S_s contains the dominant singular values, a separation between signal and noise is made. The observation matrix is now calculated from

$$\widehat{\Gamma}_m = \overline{U}_s S_s^{\frac{1}{2}} \tag{6.74}$$

from which A, B, C, and D can be found after a least-squares step and by inspection. The next example will further illustrate the output error 4SID procedure.

Example 6.12 Second-order process: Let the following noise-free input–output data, as presented in Table 6.3, be given. Using MATLAB's *N4SID*, we obtain

$$A = \begin{bmatrix} -0.5 & 0.866 \\ 0.866 & -0.5 \end{bmatrix}, \qquad B = \begin{bmatrix} -1.411 \\ 0.095373 \end{bmatrix}$$

$$C = \begin{bmatrix} 0.055064 & 0.81464 \end{bmatrix}, \qquad D = 0$$

$$K = \begin{bmatrix} 0 \\ 0 \end{bmatrix}, \qquad x(0) = \begin{bmatrix} -6.2009 \\ -6.313 \end{bmatrix} \times 10^{-16}$$

with $J_{\text{FPE}} = 3.99 \times 10^{-31}$ and loss function value of 1.81×10^{-31}. Notice the small error in $\widehat{x}(0)$. Furthermore, the estimated transfer function is given by

$$\widehat{G}(q) = \frac{-6.43 \times 10^{-7} q^{-2} + q^{-1}}{q^{-2} + q^{-1} + 1}$$

which, apart from a very small extra term in the numerator, coincides with the exact transfer function of the system, presented in Example 6.11.

Clearly, in subspace identification the Hankel matrix, for a prespecified prediction horizon m, plays a key role.

6.1.7 *Linear Parameter-varying Model Identification

In this subsection, the identification of discrete-time linear parameter-varying (LPV) models of nonlinear or time-varying systems is considered. We assume that inputs, outputs, and the scheduling parameters, which can be interpreted as the "set-point" of the system, can be directly measured. Furthermore, some form of the functional dependence of the system parameters on the scheduling parameters is known. Although these models are introduced to describe nonlinear or time-varying systems, it will be shown in the following that the model can be written as a linear regression. Recall that a model in linear regression form, as introduced in Chap. 5, allows us to use direct least-squares estimation methods. However, in this section, we will only show how to arrive at a linear regression, since from this the next estimation step becomes rather trivial. Finally, we will demonstrate the identification of a linear parameter-varying model by an example.

Consider the (noise-free) discrete-time LPV model

$$y(t) = G(p, q; \vartheta)u(t) \tag{6.75}$$

where p is a measured time-varying scheduling parameter, and ϑ contains the unknowns of the functional dependence between the system parameters and the scheduling parameters. In what follows, we assume that $G(p, q; \vartheta)$ is of the form

$$G(p, q; \vartheta) = \frac{B(p, q)}{A(p, q)} \tag{6.76}$$

where $B(p, q) = b_0(p) + b_1(p)q^{-1} + \cdots + b_{n_b}(p)q^{-n_b}$ and $A(p, q) = 1 + a_1(p)q^{-1} + \cdots + a_{n_a}(p)q^{-n_a}$. Hence, these polynomials contain $n = n_a + n_b + 1$ unknowns. Furthermore, we assume that $p = p(t)$ is a function of t with $t \in (Z)^+$. To be more specific, we assume that the functions a_i and b_i are linear combinations of the known fixed basis functions $f_1, \ldots, f_M$, so that, for example,

$$a_1(p) = a_1^1 f_1(p) + \cdots + a_1^M f_M(p) \tag{6.77}$$

with constant real numbers a_1^j. Thus, the problem is to find the parameters a_i^j with $i = 1, \ldots, n_a$ and b_i^j with $i = 0, \ldots, n_b$, $j = 1, \ldots, M$ from input–output data. As yet, we are free to choose the basis functions. However, in what follows, we choose these functions as powers of p, that is, $f_j(p) = p^j$.

Consequently, the system parameter functions become

$$\begin{aligned} a_i(p) &= a_i^1 + a_i^2 p + \cdots + a_i^M p^{M-1} \\ b_i(p) &= b_i^1 + b_i^2 p + \cdots + b_i^M p^{M-1} \end{aligned} \tag{6.78}$$

Obviously, many other choices are possible. For a direct estimation of the system parameters, it will be very helpful if we can write the model as a linear regression.

Therefore, we define

$$\Theta := \begin{bmatrix} a_1^1 & \cdots & a_1^M \\ a_2^1 & \cdots & a_2^M \\ \vdots & & \vdots \\ a_{n_a}^1 & \cdots & a_{n_a}^M \\ b_0^1 & \cdots & b_0^M \\ \vdots & & \vdots \\ b_{n_b}^1 & \cdots & b_{n_b}^M \end{bmatrix} \tag{6.79}$$

In addition to this, we define the extended regressor Ψ,

$$\Psi := \begin{bmatrix} -y(t-1) \\ \vdots \\ -y(t-n_a) \\ u(t) \\ \vdots \\ u(t-n_b) \end{bmatrix} \begin{bmatrix} 1 \; p(t) \cdots p(t)^{M-1} \end{bmatrix} \tag{6.80}$$

Given these definitions, we obtain the following regression:

$$y(t) = \langle \Theta, \Psi(t) \rangle = \mathrm{Tr}\big(\Theta^T \Psi(t)\big) \tag{6.81}$$

where $\langle \cdot, \cdot \rangle$ denotes the inner product of matrices (see Appendix A). This result can be verified as follows:

$$y(t) = \left\langle \begin{bmatrix} a_1^1 & \cdots & a_1^M \\ a_2^1 & \cdots & a_2^M \\ \vdots & & \vdots \\ a_{n_a}^1 & \cdots & a_{n_a}^M \\ b_0^1 & \cdots & b_0^M \\ \vdots & & \vdots \\ b_{n_b}^1 & \cdots & b_{n_b}^M \end{bmatrix}, \begin{bmatrix} -y(t-1) & \cdots & -y(t-1)p(t)^{M-1} \\ -y(t-2) & \cdots & -y(t-2)p(t)^{M-1} \\ \vdots & & \vdots \\ u(t) & \cdots & u(t)p(t)^{M-1} \\ \vdots & & \vdots \\ u(t-n_b) & \cdots & u(t-n_b)p(t)^{M-1} \end{bmatrix} \right\rangle$$

$$= \mathrm{Tr}\left(\begin{bmatrix} a_1^1 & \cdots & a_{n_a}^1 & b_0^1 & \cdots & b_{n_b}^1 \\ a_1^2 & \cdots & a_{n_a}^2 & \cdots & & \\ \vdots & & & & & \vdots \\ a_1^M & \cdots & & & & b_{n_b}^M \end{bmatrix} \right.$$

$$\times \begin{bmatrix} -y(t-1) & \cdots & -y(t-1)p(t)^{M-1} \\ -y(t-2) & \cdots & -y(t-2)p(t)^{M-1} \\ \vdots & & \vdots \\ u(t) & \cdots & u(t)p(t)^{M-1} \\ \vdots & & \vdots \\ u(t-n_b) & \cdots & u(t-n_b)p(t)^{M-1} \end{bmatrix} \Bigg)$$

$$= \left[a_1^1 \cdots a_{n_a}^1 \; b_0^1 \cdots b_{n_b}^1 \right] \begin{bmatrix} -y(t-1) \\ -y(t-2) \\ \vdots \\ u(t-n_b) \end{bmatrix}$$

$$+ \left[a_1^2 \cdots a_{n_a}^2 \; b_0^2 \cdots b_{n_b}^2 \right] \begin{bmatrix} -p(t)y(t-1) \\ -p(t)y(t-2) \\ \vdots \\ p(t)u(t-n_b) \end{bmatrix} + \cdots$$

$$+ \left[a_1^M \cdots a_{n_a}^M \; b_0^M \cdots b_{n_b}^M \right] \begin{bmatrix} -p(t)^{M-1}y(t-1) \\ -p(t)^{M-1}y(t-2) \\ \vdots \\ p(t)^{M-1}u(t-n_b) \end{bmatrix} \tag{6.82}$$

After arranging terms we obtain

$$\begin{aligned} y(t) = &-\left[a_1^1 + a_1^2 p(t) + \cdots + a_1^M p(t)^{M-1}\right] y(t-1) - \cdots \\ &- \left[a_{n_a}^1 + a_{n_a}^2 p(t) + \cdots + a_{n_a}^M p(t)^{M-1}\right] y(t-n_a) \\ &+ \left[b_0^1 + b_0^2 p(t) + \cdots + b_0^M p(t)^{M-1}\right] u(t) + \cdots \\ &+ \left[b_{n_b}^1 + b_{n_b}^2 p(t) + \cdots + b_{n_b}^M p(t)^{M-1}\right] u(t-n_b) \end{aligned} \tag{6.83}$$

In compact notation, this leads to

$$A(p,q)y(t) = B(p,q)u(t) \tag{6.84}$$

with $p = p(t)$. Notice that this is exactly the original model we started with in (6.75)–(6.76).

Algorithm 6.6 Identification of LPV model parameters from input–output data

1. Specify an ARX model structure in terms of n_a and n_b.
2. Specify the basis functions $f_1(p), \ldots, f_M(p)$.
3. Define the vector $y := [y(n_a), \ldots, y(N)]^T$ and the matrices Θ, as in (6.79), and Ψ in terms of $f_i(p)$, similar as in (6.80), for $n_a \geq n_b$.
4. Expand (6.81) as in (6.83), but now in terms of $f_1(p), \ldots, f_M(p)$.

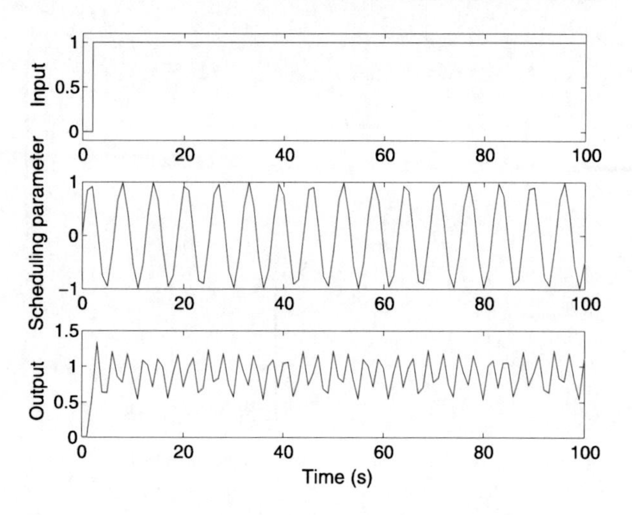

Fig. 6.9 Unit step input, scheduling parameter, and LPV step response

5. Collect products of $f_i(p), i = 1, \ldots, M$, with $y(t-l), l = 1, \ldots, n_a$ and $u(t-l)$, $l = 0, \ldots, n_b$, respectively, and define Φ.
6. Calculate from (5.10) the least-squares estimate of the unknown $M(n_a + n_b)$-dimensional parameter vector ϑ.

Let us illustrate the LPV identification procedure by an example.

Example 6.13 LPV first-order system: Let a first-order process in discrete-time with varying system parameters be described by

$$y(t) + a_1(p)y(t-1) = b_0(p)u(t) + b_1(p)u(t-1)$$

where

$$a_1(p) = a_1^1 + a_1^2 p + a_1^3 p^2 = 0.5 + 0.1p - 0.2p^2$$
$$b_0(p) = b_0^1 + b_0^2 p + b_0^3 p^2 = 0.7 + 0.1p - 0.3p^2$$
$$b_1(p) = b_1^1 + b_1^2 p + b_1^3 p^2 = 0.9 + 0.1p - 0.4p^2$$

Hence, this LPV model contains nine unknown coefficients. Assume that the scheduling parameter, time-varying set-point, is given by $p(t) = \sin(t)$. The input to this system is a shifted unit step function, namely $u(t) = 0$, $t < 2$, and $u(t) = 1$, $t \geq 2$. The input, scheduling parameter, and step response are presented in Fig. 6.9, and the resulting varying system parameters in Figs. 6.10 and 6.11. Let us select the first ten inputs and outputs (see Table 6.4) for identification, while neglecting the noninformative data at $t = 0$.

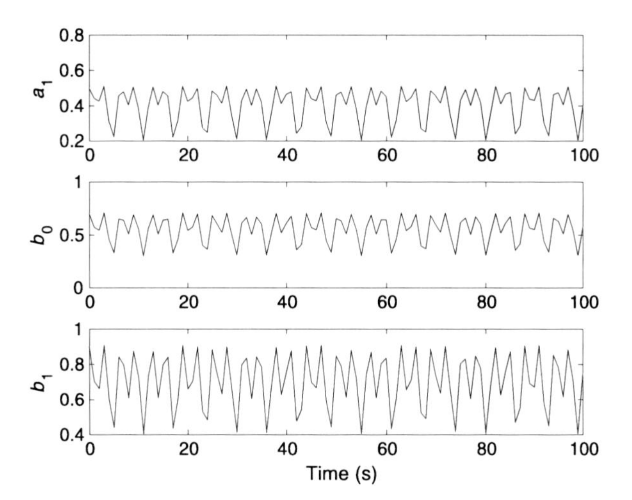

Fig. 6.10 Time-varying parameters a_1, b_0 and b_1

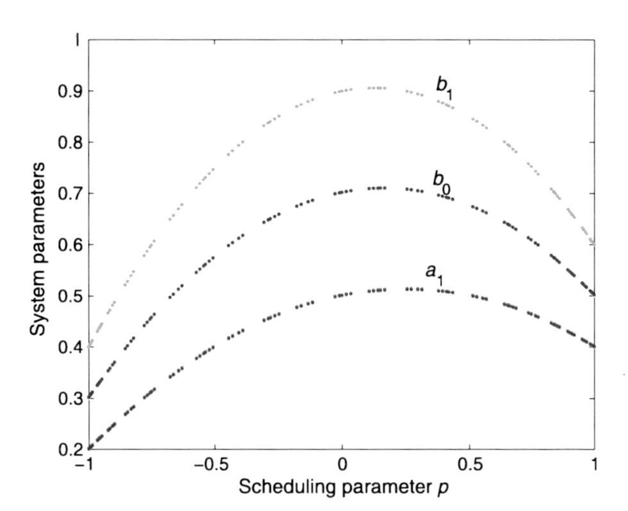

Fig. 6.11 Time-varying parameters a_1, b_0, and b_1 as functions of p

Given the input–output data, we obtain

$$
\begin{aligned}
t = 2: \quad y(2) &= \left\langle \begin{bmatrix} a_1^1 & a_1^2 & a_1^3 \\ b_0^1 & b_0^2 & b_0^3 \\ b_1^1 & b_1^2 & b_1^3 \end{bmatrix}, \begin{bmatrix} -y(1) & -y(1)\sin(2) & -y(1)\sin^2(2) \\ 1 & \sin(2) & \sin(2)^2 \\ 0 & 0 & 0 \end{bmatrix} \right\rangle \\
&= -\left[a_1^1 + a_1^2 \sin(2) + a_1^3 \sin^2(2)\right] y(1) + \cdots \\
&\quad + \left[b_0^1 + b_0^2 \sin(2) + b_0^3 \sin^2(2)\right] * 1 + \cdots \\
&\quad + \left[b_1^1 + b_1^2 \sin(2) + b_1^3 \sin^2(2)\right] * 0
\end{aligned}
$$

Table 6.4 Input and output data for the identification of LPV model

Time	1	2	3	4	5	6	7	8	9	10
$u(t)$	0	1	1	1	1	1	1	1	1	1
$y(t)$	0	0.5429	1.3373	0.6334	0.6251	1.2042	0.8520	0.7692	1.1734	0.8306

$$t=3: \quad y(3)=\left\langle \begin{bmatrix} a_1^1 & a_1^2 & a_1^3 \\ b_0^1 & b_0^2 & b_0^3 \\ b_1^1 & b_1^2 & b_1^3 \end{bmatrix}, \begin{bmatrix} -y(2) & -y(2)\sin(3) & -y(2)\sin^2(3) \\ 1 & \sin(3) & \sin^2(3) \\ 1 & \sin(3) & \sin^2(3) \end{bmatrix} \right\rangle$$

$$\begin{aligned} &= -\left[a_1^1 + a_1^2 \sin(3) + a_1^3 \sin^2(3)\right] y(2) + \cdots \\ &\quad + \left[b_0^1 + b_0^2 \sin(3) + b_0^3 \sin^2(3)\right] * 1 + \cdots \\ &\quad + \left[b_1^1 + b_1^2 \sin(3) + b_1^3 \sin^2(3)\right] * 1 \end{aligned}$$

$$t=4: \quad y(4)=\left\langle \begin{bmatrix} a_1^1 & a_1^2 & a_1^3 \\ b_0^1 & b_0^2 & b_0^3 \\ b_1^1 & b_1^2 & b_1^3 \end{bmatrix}, \begin{bmatrix} [2pt]-y(3) & -y(3)\sin(4) & -y(3)\sin^2(4) \\ 1 & \sin(4) & \sin^2(4) \\ 1 & \sin(4) & \sin^2(4) \end{bmatrix} \right\rangle$$

$$\begin{aligned} &= -\left[a_1^1 + a_1^2 \sin(4) + a_1^3 \sin^2(4)\right] y(3) + \cdots \\ &\quad + \left[b_0^1 + b_0^2 \sin(4) + b_0^3 \sin^2(4)\right] * 1 + \cdots \\ &\quad + \left[b_1^1 + b_1^2 \sin(4) + b_1^3 \sin^2(4)\right] * 1 \end{aligned}$$

$$\vdots$$

From these expressions at each time instant t we can easily build the linear regression $y = \Phi\vartheta$, where $y = [y(2), \ldots, y(10)]^T$, $\vartheta = [a_1^1, \ldots, b_1^3]^T$, and

$$\Phi = \begin{bmatrix} 0 & 0 & 0 & 1 & \sin(2) & \sin^2(2) & 0 & 0 & 0 \\ -y(2) & -y(2)\sin(3) & -y(2)\sin^2(3) & 1 & \sin(3) & \sin^2(3) & 1 & \sin(3) & \sin^2(3) \\ -y(3) & -y(3)\sin(4) & -y(3)\sin^2(4) & \vdots & \vdots & \vdots & 1 & \sin(4) & \sin^2(4) \\ \vdots & \vdots & \vdots & \vdots & \vdots & \vdots & \vdots & \vdots & \vdots \end{bmatrix}$$

Direct inversion $\Phi^{-1}y$ gives

$$\hat{\vartheta} = [0.5000\ 0.1000\ -0.2000\ 4.2743\ 0.8667\ -5.4661\ -2.6743\ -0.6667\ 4.7661]$$

with perfect estimates for the coefficients related to $a_1(p)$ and residuals that are very close to zero. A careful evaluation of Φ leads to the conclusion that the matrix is close to singular. Hence, in this case a step input is not a very good choice; a white noise input signal would have been a better choice.

6.1.8 *Orthogonal Basis Functions

Recall that the transfer function in a convolution or impulse response model is given by

$$G(q) = \sum_{k=0}^{\infty} g(k)q^{-k} \tag{6.85}$$

This expression can be interpreted as a series expansion with coefficients $g(k)$ and standard pulse basis $f_k(q) = q^{-k}$. In general terms, (6.85) can also be written as

$$G(q) = \sum_{k=0}^{\infty} L(k) f_k(q) \tag{6.86}$$

where $L(k)$ for $k = 0, 1, 2, \ldots$ is the real-valued expansion coefficient, and $f_k(q)$ is a so-called basis function. Preferably, the so-called orthogonal basis functions are chosen for efficient calculation. For example, instead of q^{-k}, we may choose

$$f_k(q, \alpha) = \frac{q^{-k}}{q - \alpha} \tag{6.87}$$

where, as a natural choice, α is the system pole closest to the unit circle. However, in practice, α has to be obtained from prior knowledge or estimated from experimental data first. Let us illustrate the idea by a simple example.

Example 6.14 First-order system: Let a stable process be described by

$$y(t) = \frac{bq^{-1}}{1 - \alpha q^{-1}}$$

Given an observed input–output data set, we can try to estimate the impulse response coefficients $g(k)$ in $G(q) = \sum_{k=0}^{\infty} L(k) f_k(q)$ for this process. Notice then that the relationship between $g(k)$ and the unknown process parameters α and b, after polynomial division of $\frac{b}{(q-\alpha)}$, is given by

$$g(k) = \alpha^k b \quad \text{for } k = 1, 2, \ldots$$

Consequently, for parameter α close to 1, many impulse response coefficients must be estimated. Alternatively, choosing $f_k(q, \alpha)$ according to (6.87) gives

$$\begin{aligned} G(q) &= \sum_{k=0}^{\infty} L(k) \frac{q^{-k}}{q - \alpha} \\ &= \frac{b}{q - \alpha} \end{aligned}$$

and thus, for α obtained a priori, only one coefficient has to be estimated from the experimental data, i.e., $L(0) = b$, because $L(k) = 0$ for $k = 1, 2, \ldots$.

Notice from this example that this identification approach using orthogonal basis functions avoids the choice of an appropriate noise model. Basically, the process dynamics are already governed by the basis functions, but in practice most often some iterations are needed to have a good estimate of the unknown parameter a.

Another type of a frequently used basis function is the Laguerre polynomial given by

$$f_k(q,\alpha) = \sqrt{1-\alpha^2}\, q \frac{(1-\alpha q)^k}{(q-\alpha)^{k+1}}, \quad |\alpha| < 1 \tag{6.88}$$

This idea has been generalized using the expansion

$$G(q) = q^{-1} \sum_{k=0}^{\infty} L(k) f_k(q) \tag{6.89}$$

Herein, $f_k(q) := (qI - A)^{-1} B G_b^k(q)$ is a so-called orthogonal basis function, with (A, B) system matrices in a linear state-space representation, and $G_b^k(q)$ is a function that is able to incorporate dynamics of any complexity.

6.1.9 *Closed-loop Identification

So far, the input–output data has been obtained from an open-loop system configuration, as in Fig. 1.5. However, identification of a system from closed-loop input–output data requires a special treatment. Consider for this the following SISO LTI control system configuration (Fig. 6.12) with $P(q)$ and $Q(q)$ rational transfer functions.

Let us first have a look at the different relationships in this configuration. For instance, the transfer function between r, d, and y is given by

$$\begin{aligned} y(t) &= P(q)u(t) + d(t) \\ &= P(q)Q(q)e(t) + d(t) \\ &= P(q)Q(q)\big(r(t) - y(t)\big) + d(t) \end{aligned}$$

Consequently,

$$y(t) = \frac{P(q)Q(q)}{1+P(q)Q(q)} r(t) + \frac{1}{1+P(q)Q(q)} d(t) \tag{6.90}$$

Commonly, it is assumed that $Q(q)$ is known. However, in many industrial situations, due to, for instance, scaling and interfacing, this knowledge may be tricky to use. Nevertheless, for now we assume that $Q(q)$ is given and thus only the parameters in $P(q)$ must be estimated. However, let us first illustrate the key problem in closed-loop identification by a simple example.

Fig. 6.12 Feedback system

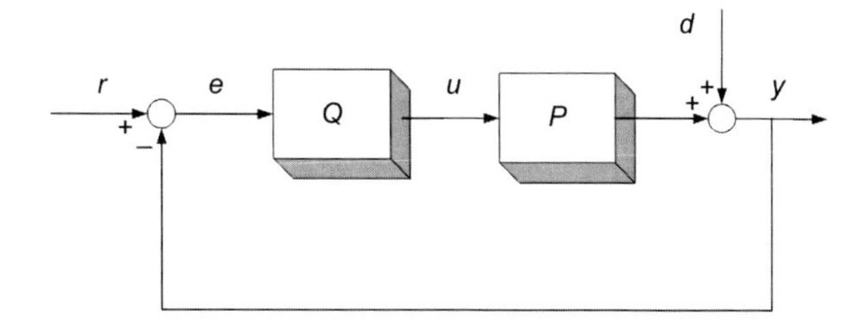

Example 6.15 P-control: Consider a plant with the transfer function

$$P(q) = \frac{bq^{-1}}{1 + fq^{-1}}$$

and assume further that the plant is controlled under proportional feedback such that

$$\begin{aligned} u(t) &= Ke(t) \\ &= K\big[r(t) - y(t)\big] \end{aligned}$$

Hence, $Q(q) = K$. Applying (6.90) gives

$$\begin{aligned} y(t) &= \frac{\frac{bq^{-1}}{1+fq^{-1}}K}{1 + \frac{bq^{-1}}{1+fq^{-1}}K} r(t) + \frac{1}{1 + \frac{bq^{-1}}{1+fq^{-1}}K} d(t) \\ &= \frac{bKq^{-1}}{1 + (bK + f)q^{-1}} r(t) + \frac{1 + fq^{-1}}{1 + (bK + f)q^{-1}} d(t) \end{aligned}$$

Thus,

$$y(t) = -(bK + f)y(t-1) + bKr(t-1) + d(t) + fd(t-1)$$

From this we can deduce the following case for $r(t) = 0$, and the filtered noise sequence $e(t) = (1 + fq^{-1})d(t)$ is white. Then,

$$y(t) = -(bK + f)y(t-1) + e(t)$$

As a consequence of this, the parameters b and f can never be estimated uniquely irrespective of knowledge of the controller gain K. However, given the gain K and using the reference signal $r(t)$ to excite the system, thus choosing $r(t) \neq 0$, the parameter b can be calculated from the estimate of bK.

Notice from the example that after substitution of $P(q)$ and $Q(q)$ into (6.90) a linear difference equation model structure results. The parameters in this model can be estimated straightforward using the methods in Sects. 6.1.2–6.1.4. Basically, for a specific application, an appropriate noise model structure must be chosen.

A fundamental problem arises when to signal-to-noise ratio is small, so that $r(t) \ll d(t)$. More specifically, let us assume that $P(q)Q(q)r(t) \ll d(t)$. Consequently, from (6.90) we have

$$y(t) \approx \frac{1}{1+P(q)Q(q)} d(t) \tag{6.91}$$

Let us assume that we have a zero-mean noise, so that $E[d(t)] = 0$. Consequently, $E[1 + P(q)Q(q)y] = E[d(t)] = 0$, and thus $P \approx \frac{-1}{Q}$. In other words, when the signal-to-noise ratio is small, closed-loop identification will not be able to find the true plant. Consequently, a persistently exciting reference signal with sufficient power must be chosen as an input to the closed-loop system.

In literature, three main groups of closed-loop identification methods have been distinguished, namely

- Direct identification
- Indirect identification
- Joint I–O identification

In particular, the direct approach can be seen as a natural approach to closed-loop data analysis. Therefore, in what follows, our focus will be on this approach. In the direct approach the input u and output y are used in the same way as in open loop, ignoring any possible feedback and not using the reference signal r. Unstable systems can be handled as well, as long as the closed loop system and the predictor are stable. The last condition implies that any of the unstable poles of $G(q)$ (see (6.12)) must be shared by $H(q)$, like in the ARX and ARMAX models.

Example 6.16 P-control: Consider again the plant with the transfer function

$$P(q) = \frac{bq^{-1}}{1+fq^{-1}}$$

with $b = 0.5$ and $f = 1.2$, so that the plant is unstable. Let the controller gain K be equal to -0.5, so that the stable transfer function from r to y is given by

$$T(q) = \frac{-0.25q^{-1}}{1+0.95q^{-1}}$$

with steady-state gain of -0.1282.

Furthermore, let $r(t)$ be a step function. The responses for the noise-free case are presented in Fig. 6.13. Clearly, for this discrete-time process with P-control, a stable closed-loop transfer function will always go together with a substantial off-set. Hence, for better closed-loop performance, a more advanced controller is needed. However, the design of such a controller is out of the scope of the book. In what follows, we will focus in particular on the identification of $T(q)$, given reference input and output data from the closed-loop system.

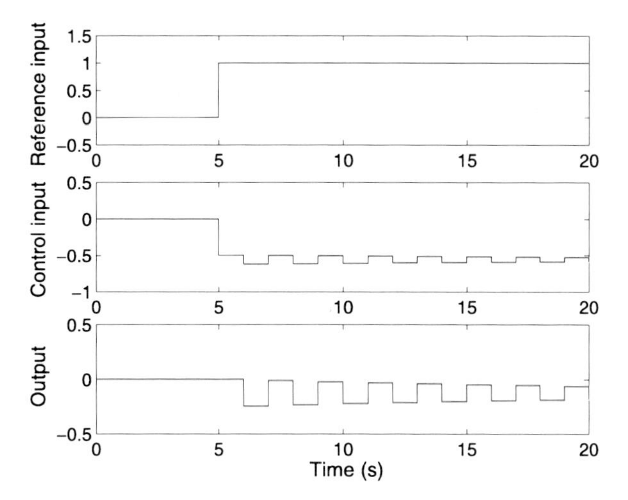

Fig. 6.13 Signals of closed-loop system

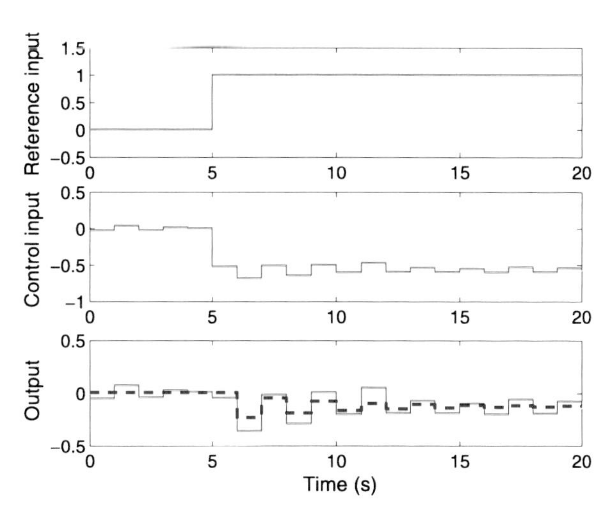

Fig. 6.14 Observed and simulated (- -) signals of closed-loop system

For the identification of this process, let us start with the following ARX model structure, relating the reference input $r(t)$ to $y(t)$:

$$y(t) = -ay(t-1) + b_r r(t-1) + e(t)$$

From the reference input–output data presented in Fig. 6.13 we obtain $\widehat{a} = 0.95$ and $\widehat{b_r} = -0.25$. Consequently, the original parameter values of $T(q)$ are recovered from the data. From these parameters of $T(q)$ we obtain the process parameters $\widehat{f} = \widehat{a} - \widehat{b}K = 1.2$ and $\widehat{b} = \frac{\widehat{b_r}}{K} = 0.5$ (see Example 6.15). If zero-mean normally distributed noise with variance of 0.0025 is added to the output data, then the following estimates of the process parameter are obtained: $\widehat{f} = 1.0169$, $\widehat{b} = 0.4736$, again via the estimates $\widehat{a}$ and $\widehat{b_r}$ in the transfer function $T(q)$ and using an ARX model structure. The closed-loop input–output data and corresponding ∞-steps ahead predictions with steady-state value of -0.1330, as a result of a step in the reference input, can be seen in Fig. 6.14. Clearly, appropriate estimation results have been obtained, and thus there is no direct need for a more advanced noise model structure.

6.2 Nonlinear Dynamic Systems

6.2.1 Simulation Models

Recall that the general description of a finite-dimensional system, see (1.2), is given by

$$\begin{aligned} \frac{\mathrm{d}x(t)}{\mathrm{d}t} &= f\big(t, x(t), u(t), w(t); \vartheta\big), \quad x(0) = x_0 \\ y(t) &= h\big(t, x(t), u(t); \vartheta\big) + v(t), \quad t \in \mathbb{R} \end{aligned} \tag{6.92}$$

where generally both function $f(\cdot)$ and $h(\cdot)$ are found from prior system knowledge. These functions may, however, be found after an approximation of an infinite-dimensional or so-called distributed parameter system. Notice that in this description, also sequences of disturbances $w(t)$ and $v(t)$ have been incorporated. Recall that these disturbances represent the errors in the modeling, due, for instance, to approximations and unmodeled effects, and measurement process, respectively. In a real implementation usually a discrete-time version of this model is used,

$$\begin{aligned} x(t+1) &= f\big(t, x(t), u(t), w(t); \vartheta\big), \quad x(0) = x_0 \\ y(t) &= h\big(t, x(t), u(t); \vartheta\big) + v(t), \quad t \in \mathbb{Z}^+ \end{aligned} \tag{6.93}$$

This representation is well suited for identification. However, it should be mentioned that neither $w(t)$ nor $v(t)$ are known in advance. These quantities can only be evaluated afterward when an estimate of the model parameter vector ϑ is available. Assuming that both $w(t)$ and $v(t)$ have zero mean, the following predictor can be derived from (6.93):

$$\begin{aligned} x(t+1) &= f\big(t, x(t), u(t), 0; \vartheta\big), \quad x(0) = x_0 \\ \widehat{y}(t) &= h\big(t, x(t), u(t); \vartheta\big), \quad t \in \mathbb{Q}^+ \end{aligned} \tag{6.94}$$

which is also called a *simulation model*. This simulation model, in which $\mathbb{Q}^+$ is the set of positive rational numbers, can thus also be used to describe the system behavior between the sampling instants to avoid large integration steps. Only at sampling instants, $t = kT_s$ with $k = 1, 2, \ldots$ and T_s the sampling interval, the full discrete-time description with disturbances is used.

Apart from simulation studies, other types of analysis most often require a linearized model, which can be found from linearizing the nonlinear system around reference trajectories $x^*(t)$ and $u^*(t)$. In Part III these reference trajectories may even be functions of the parameter estimate $\widehat{\vartheta}$, but for the moment, it suffices to disregard this dependence. For the linearization of the nonlinear system (6.94), let us introduce the differences

$$\Delta x(t) = x(t) - x^*(t)$$
$$\Delta u(t) = u(t) - u^*(t)$$
$$\Delta y(t) = y(t) - h\big(t, x^*(t), u^*(t)\big)$$

Using a Taylor series expansion of both $f(\cdot)$ and $h(\cdot)$ in (6.94) and neglecting the nonlinear (higher-order) terms in the resulting approximate model, we arrive at

$$\begin{aligned} \Delta x(t+1) &= F_x(t)\Delta x(t) + F_u(t)\Delta u(t) \\ \Delta \widehat{y}(t) &= H_x(t)\Delta x(t) + H_u(t)\Delta u(t) \end{aligned} \tag{6.95}$$

where

$$F_x(t) = \frac{\partial}{\partial x} f(t, x, u; \vartheta)\bigg|_{x^*(t), u^*(t)}, \qquad F_u(t) = \frac{\partial}{\partial u} f(t, x, u; \vartheta)\bigg|_{x^*(t), u^*(t)}$$

$$H_x(t) = \frac{\partial}{\partial x} h(t, x, u; \vartheta)\bigg|_{x^*(t), u^*(t)}, \qquad H_u(t) = \frac{\partial}{\partial u} h(t, x, u; \vartheta)\bigg|_{x^*(t), u^*(t)}$$

Notice that the simulation model (6.94) is approximated by a linear, time-varying model (6.95), which in principle is only a valid approximation around the trajectories $x^*(t)$ and $u^*(t)$.

6.2.2 *Parameter Sensitivity

Recall that for nonlinear static systems with $\vartheta \in \mathbb{R}^p$, $\widehat{y}(t) \in \mathbb{R}^s$, and $t = 1, \dots, N$, the sensitivity matrix is given by

$$X(\vartheta) = \big[\psi(1, \vartheta), \dots, \psi(M, \vartheta)\big]^T \tag{6.96}$$

with $M = N\,s$ and $\psi(t, \vartheta) := \frac{\mathrm{d}\widehat{y}(t)}{\mathrm{d}\vartheta} \in \mathbb{R}^{s \times p}$ for $t = 1, \dots, M$. For the noise-free continuous-time nonlinear dynamic case, with $x(t) \in \mathbb{R}^n$,

$$\frac{\mathrm{d}x(t)}{\mathrm{d}t} = f\big(t, x(t), u(t); \vartheta\big) \tag{6.97}$$

$$\widehat{y}(t) = h\big(t, x(t), u(t); \vartheta\big) \tag{6.98}$$

the parameter sensitivity will be calculated in two steps. First, let us define the state sensitivity matrix $S_x(t, \vartheta) := \frac{\mathrm{d}x(t)}{\mathrm{d}\vartheta} \in \mathbb{R}^{n \times p}$. Taking the derivative with respect to ϑ on both sides of (6.97) gives

$$\frac{\partial}{\partial \vartheta}\frac{\mathrm{d}x(t)}{\mathrm{d}t} = \frac{\partial f(t, x(t), u(t); \vartheta)}{\partial \vartheta} \tag{6.99}$$

If the parameters are constant, i.e., $\frac{\partial}{\partial\vartheta}\frac{\mathrm{d}x(t)}{\mathrm{d}t} = \frac{\mathrm{d}}{\mathrm{d}t}\frac{\partial x(t)}{\partial\vartheta}$, then

$$\frac{\mathrm{d}S_x(t,\vartheta)}{\mathrm{d}t} = \frac{\partial f(t,x(t),u(t);\vartheta)}{\partial x}S_x(t,\vartheta) + \frac{\partial f(t,x(t),u(t);\vartheta)}{\partial\vartheta} \tag{6.100}$$

where $\frac{\partial f(t,x(t),u(t);\vartheta)}{\partial x} \in \mathbb{R}^{n\times n}$ is the Jacobi matrix (see also $F_x(t)$ in (6.95)) and $\frac{\partial f(t,x(t),u(t);\vartheta)}{\partial\vartheta} \in \mathbb{R}^{n\times p}$. In addition to this, the initial conditions must be specified. Clearly, the initial values of $x(t)$ do not depend on the parameters and thus $S_x(0,\vartheta) = 0$. However, the initial sensitivity with respect to an initial condition is equal to 1. In a second step, the output sensitivity matrix $S_y(t,\vartheta) := \frac{\mathrm{d}\hat{y}(t)}{\mathrm{d}\vartheta} \in \mathbb{R}^{s\times p}$ is calculated from

$$\begin{aligned} S_y(t,\vartheta) &= \frac{\mathrm{d}h(t,x(t),u(t);\vartheta)}{\mathrm{d}t} \\ &= \frac{\partial h(t,x(t),u(t);\vartheta)}{\partial x}S_x(t,\vartheta) + \frac{\partial h(t,x(t),u(t);\vartheta)}{\partial\vartheta} \end{aligned} \tag{6.101}$$

Then, in a final step the sensitivity vector is defined as

$$\psi(t,\vartheta) := \mathrm{Vec}\big(S_y(t,\vartheta)^T\big) \tag{6.102}$$

where the operator Vec simply stacks the columns of $S_y(t,\vartheta)^T$ on top of each other, so that the first p elements of ψ contain the parameter sensitivities with respect to the first output.

Let us illustrate the procedure by a simple bioreactor example.

Example 6.17 Bioreactor: The substrate concentration (S in mg/l) in a fed-batch bioreactor with Monod kinetic substrate consumption can be described by

$$\frac{\mathrm{d}S}{\mathrm{d}t} = -\mu\frac{S}{K_S + S} + u, \qquad S(0) = S_0$$

where μ is the decay rate in mg/l/min, K_S is the half-saturation constant in mg/l, and u is the feed in mg/l/min. Notice that in the short-hand notation the time arguments are not explicitly shown. The Jacobi matrix is given by

$$\frac{\partial f(\cdot)}{\partial x} = -\mu\frac{K_S}{(K_S + S)^2}$$

Consequently, the parameter sensitivities are described by

$$\begin{aligned} \frac{\mathrm{d}S_x}{\mathrm{d}t} &= \frac{\mathrm{d}}{\mathrm{d}t}\Big[\tfrac{\mathrm{d}S}{\mathrm{d}\mu}\ \tfrac{\mathrm{d}S}{\mathrm{d}K_S}\Big] \\ &= \Big[-\mu\tfrac{K_S}{(K_S+S)^2}\tfrac{\mathrm{d}S}{\mathrm{d}\mu} - \tfrac{S}{K_S+S}\ \ -\mu\tfrac{K_S}{(K_S+S)^2}\tfrac{\mathrm{d}S}{\mathrm{d}K_S} + \mu\tfrac{S}{(K_S+S)^2}\Big] \end{aligned}$$

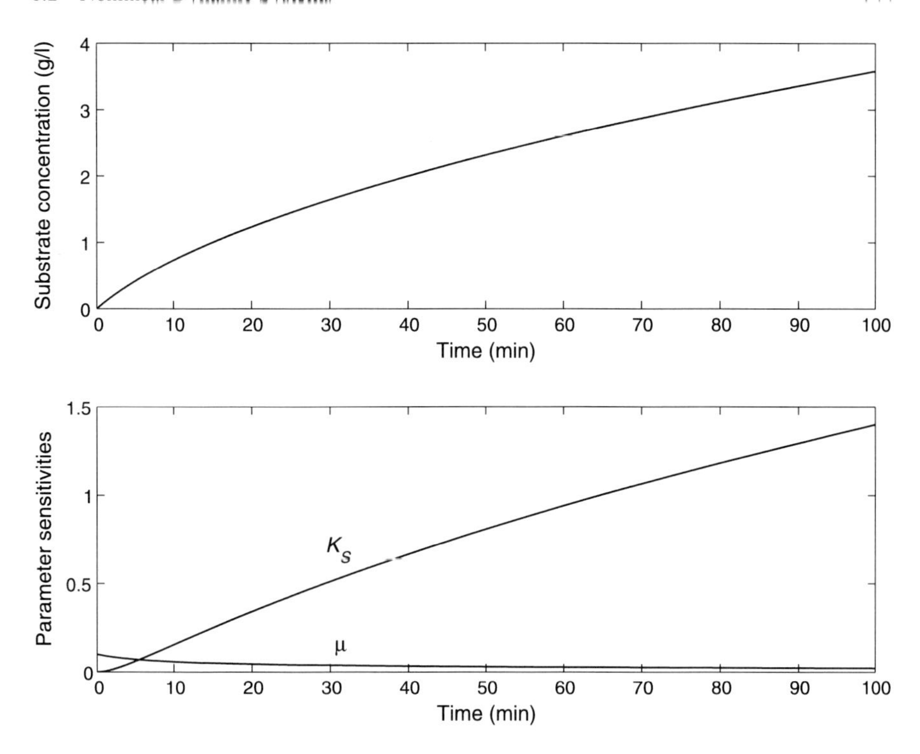

Fig. 6.15 Substrate concentration (*top figure*) and parameter sensitivities (indicated by K_S and μ, respectively, in *bottom figure*) in a fed-batch bioreactor with Monod kinetics

with $S_x(0) = [0\ 0]$. Let the feed be chosen such that $u = \mu$. Then, an analytical solution of the differential equation with $S(0) = S_0$ can be found and is given by

$$\begin{aligned} S(t) &= -K_S + \sqrt{K_S^2 + 2\mu K_S t + 2K S_0 + S_0^2} \\ &=_{[S_0=0]} - K_S + \sqrt{K_S^2 + 2\mu K_S t} \end{aligned}$$

Hence, for $S_0 = 0$, the following parameter sensitivities are found:

$$\begin{aligned} S_x(t, \vartheta) &= \left[\tfrac{\mathrm{d}S}{\mathrm{d}\mu}\ \tfrac{\mathrm{d}S}{\mathrm{d}K_S} \right] \\ &= \left[\tfrac{Kt}{\sqrt{K_S^2 + 2\mu K_S t}}\ \ -1 + \tfrac{K + \mu t}{\sqrt{K_S^2 + 2\mu K_S t}} \right] \end{aligned}$$

The trajectories of the substrate concentration, $S(t)$, and both sensitivities, $S_x(t, \vartheta)$, for $\mu = 0.1$ mg/l/min and $K_S = 1$ mg/l are shown in Fig. 6.15. Basically, the feed has been chosen such that the sensitivity of S with respect to K_S (i.e., S_{K_S}) is maximized. Notice from Fig. 6.15 that both S and S_{K_S} show a monotonically increasing behavior. Hence, for this specific choice of the feed, the sensitivity of $S(t)$ with respect to K_S is increased by increasing the substrate concentration, according to a square root law. The increasing sensitivity for K_S allows subsequently a very good estimate of K_S from fed-batch bioreactor data with $u = \mu$ and $S_0 = 0$.

6.2.3 Nonlinear Regressions

Using the linearization techniques presented in the previous section, but now with respect to ϑ, the linearized simulation model (6.95) can be written in terms of the linear regression

$$\widehat{y}(t) = \phi^T(t)\vartheta \tag{6.103}$$

which can be used to directly estimate the parameters in the linearized model. However, often this is not required; in order to estimate the physically interpretable parameters, the full nonlinear model should be used. Let us therefore introduce the nonlinear predictor

$$\widehat{y}(t) = \Pi\left(t, Z^{t-1}; \vartheta\right) \tag{6.104}$$

where Z^{t-1} denotes the set of input and output measurements available at time t. The function $\Pi(\cdot)$ can be viewed as the result of a black-box or mechanistic modeling procedure with or without filtering. In case the predictor is based on a simulation model, Z^{t-1} contains only the initial output values and the past input values. Equation (6.104) is thus a further generalization of the nonlinear predictor (5.66) derived for the static case and of the pseudo-linear regressions (6.17) and (6.20) derived for some of the linear dynamic model structures. Hence, in general, the prediction $\widehat{y}(t)$ can be constructed by evaluating the model up to time t for given inputs $u(0), \ldots, u(t-1)$, outputs $y(0), \ldots, y(t-1)$ and parameter vector ϑ. Hence, (6.93) can be written as

$$y(t) = \Pi(t, Z^{t-1}; \vartheta) + e(t) \tag{6.105}$$

where $e(t)$ can be considered now as a multi-step-ahead prediction error. This nonlinear regression model forms the starting point for parameter estimation.

6.2.4 Iterative Solution

Clearly, the unknown parameter ϑ in (6.105) cannot be found directly from the data. As in the nonlinear static case, an iterative solution is required. Starting with an initial guess $\widehat{\vartheta}^{(0)}$, the model predictions starting at $y(0)$ can be found using (6.104), which in most cases just requires a model simulation. If, however, the initial conditions are also unknown, they can be easily included in $\widehat{\vartheta}^{(0)}$, so that the initial conditions are simultaneously estimated with the unknown model parameters. On the basis of output measurements and model output predictions, the prediction errors can be evaluated to form the sum of squares. In the next iteration, new values of the estimates are required to form $\widehat{\vartheta}^{(1)}$, which can subsequently be used in the simulation model. The simulation model with parameter vector $\widehat{\vartheta}^{(1)}$ ultimately leads to a new sum of squares of the prediction errors. Notice then that this nonlinear estimation problem does not deviate too much from the one presented for the static case. The only difference is that now a nonlinear dynamic simulation, with more computational costs, instead of a nonlinear function evaluation is needed. Hence, the same kind of optimization algorithms as presented in Sect. 5.2.3 can be used here.

6.2.5 Model Reparameterization: Dynamic Case

As mentioned before, the key issue in solving nonlinear estimation problems by iterative optimization methods is that a global solution cannot be guaranteed. As in Sect. 5.2.5, we can try to reparameterize the model so that a linear regression results. As extensively shown in Chap. 5, the linear regression type of model allows a direct estimation of the unknown parameters, via matrix inversion.

The following example of model reparameterization of a dynamic system in discrete time has been inspired by Ljung [Lju87].

Example 6.18 Solar-heated house: Consider a solar-heated house with a solar panel collector constructed on the roof. The air in the solar panel collector is heated by the sun and is fanned to a heat storage. The problem then is how to find a relationship between the solar radiation I, fan velocity u, and the temperature in the heat storage y. Ljung introduced a variable $x(t)$ for the temperature of the solar panel collector at time t. In discrete time, the heating of the air in the collector $[= x(t+1) - x(t)]$ is approximately equal to heat supplied by the sun $[= d_2.I(t)]$ minus loss of heat to the environment $[= d_3.x(t)]$ minus heat transported to the storage $[= d_0.x(t).u(t)]$, so that

$$x(t+1) - x(t) = d_2.I(t) - d_3.x(t) - d_0.x(t).u(t)$$

The increase of storage temperature $[= y(t+1) - y(t)]$ is equal to the heat supplied from the collector $[= d_0.x(t).u(t)]$ minus losses to the environment $[= d_1.y(t)]$, that is,

$$y(t+1) - y(t) = d_0.x(t).u(t) - d_1.y(t)$$

Let u, I, and y be frequently measured; then the unknowns in this model are $d_0, \ldots, d_3$ and $x(t)$. However, x can be eliminated from these relationships. Then,

$$\begin{aligned} y(t) &= (1-d_1)y(t-1) + (1-d_3)\frac{y(t-1)u(t-1)}{u(t-2)} \\ &\quad + (d_3-1)(1+d_1)\frac{y(t-2)u(t-1)}{u(t-2)} + d_0 d_2 u(t-1)I(t-2) \\ &\quad - d_0 u(t-1)y(t-1) + d_0(1+d_1)u(t-1)y(t-2) \end{aligned}$$

which is clearly a nonlinear relationship in the unknown parameters $d_0, \dots, d_3$. A straightforward solution is then to apply one of the search routines. However, as mentioned before, these routines do not necessarily provide a global optimum of the nonlinear least-squares problem. In this case, Ljung suggested to reparameterize the model as

$$\begin{aligned}
&\vartheta_1 = (1 - d_1) && \phi_1(t) = y(t-1) \\
&\vartheta_2 = (1 - d_3) && \phi_2(t) = \tfrac{y(t-1)u(t-1)}{u(t-2)} \\
&\vartheta_3 = (d_3 - 1)(1 + d_1) && \phi_3(t) = \tfrac{y(t-2)u(t-1)}{u(t-2)} \\
&\vartheta_4 = d_0 d_2 && \phi_4(t) = u(t-1)I(t-2) \\
&\vartheta_5 = -d_0 && \phi_5(t) = u(t-1)y(t-1) \\
&\vartheta_6 = d_0(1 + d_1) && \phi_6(t) = u(t-1)y(t-2) \\
&\vartheta = [\vartheta_1, \vartheta_2, \dots, \vartheta_6]^T && \phi(t) = \big[\phi_1(t), \phi_2(t), \dots, \phi_6(t)\big]^T
\end{aligned}$$

which leads to the noise-free linear regression

$$y(t, \vartheta) = \phi(t)^T \vartheta$$

and thus to a direct estimation of the parameter vector ϑ. However, the algebraic relationship between these parameters has been lost, which may result in physically unrealistic estimates.

Furthermore, in addition to the possible existence of local minima, iterative optimization methods can be very time-consuming. Therefore, in practice, the number of parameters to be estimated should usually be limited to 5–7. Hence, given the input–output data set, it is important to adjust only the most sensitive parameters or parameter combinations.

Let us demonstrate this approach to the identification of the dissolved oxygen (DO) dynamics in an activated sludge plant. In this continuous-time *grey-box modeling* example, model reparameterization of the physically interpretable model structure implies a systematic reduction of the number of parameters to be estimated. In addition to this, it also leads to a reduction in the correlation between parameter estimates.

Example 6.19 Dissolved Oxygen (DO) dynamics (based on [LKvS96]): A general activated sludge plant layout is presented in Fig. 6.16, where AT indicates the aeration tank with actual DO concentration $C(t)$ and volume V.

The a priori knowledge of the DO dynamics in the completely mixed aeration tank of this pilot plant layout is represented by the following model:

$$\begin{aligned}
\frac{\mathrm{d}C(t)}{\mathrm{d}t} = &- f(C) r_{\mathrm{act}}(t) + k_L a(q_{\mathrm{air}})\big[C_s - C(t)\big] \\
&- \frac{q_{\mathrm{in}} + q_r}{V} C(t)
\end{aligned} \tag{6.106}$$

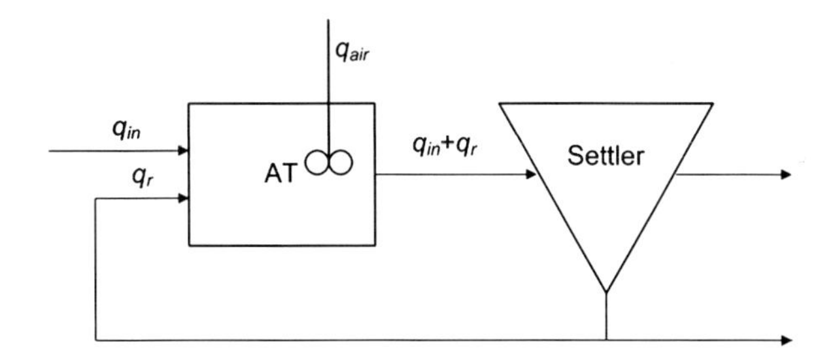

Fig. 6.16 Activated sludge plant layout

$$k_L a(q_{\text{air}}) = \alpha q_{\text{air}}(t) + \gamma \tag{6.107}$$

$$f(C) = \frac{C(t)}{K_C + C(t)} \tag{6.108}$$

where $q_{\text{in}} = q_r = 0.8$ l/min and $V = 475$ l. The first term on the right-hand side of the differential equation represents the consumption of DO for biodegradation of the substrate entering the plant, which may be limited by the DO concentration. The second term expresses the amount of DO as a result of forced aeration with air flow (q_{air}), and the last term represents the outflow of DO. The unknown parameters in this model are α, γ, K_C, and C_s.

From previous experiments it was roughly known that the saturation concentration $C_s = 9.23$ mg/l, the so-called Monod constant $K_C = 0.3$ mg/l, and the parameters in the oxygen transfer relation $\alpha = 3.34 \times 10^{-3}\ \text{l}^{-1}$ and $\gamma = 5.9 \times 10^{-2}\ \text{min}^{-1}$. The inputs in this model are $q_{\text{air}}(t)$ and $r_{\text{act}}(t)$, which have been measured directly with a respirometer. On the basis of this, experimental inputs have been designed for a period of 24 hours. The output equation is given by

$$y(t) = C(t) + e(t) \tag{6.109}$$

However, the experiment aborted, so that only data for the first 11 hours became available (see Fig. 6.17). The corresponding output $y(t)$ and disturbance $r_{\text{act}}(t)$ are also presented (see Fig. 6.18).

The model fit with optimized parameters α, γ, K_C, and C_s appeared to be unsatisfactory. Consequently, the model structure was modified. First, it was decided to extend the $k_L a$ relationship with a term proportional to the square root of q_{air}. Secondly, based on cross-correlation analysis, a dead time (Δ) was introduced for q_{air}. Thirdly, a scaling factor ($f_{\max}$) to r_{act} was introduced in order to trace a possible systematic error in this signal. Hence, the following modifications were suggested:

$$k_L a(q_{\text{air}}) = \alpha q_{\text{air}}(t - \Delta) + \beta \sqrt{q_{\text{air}}(t - \Delta)} + \gamma \tag{6.110}$$

$$f(C) = f_{\max} \frac{C(t)}{K_C + C(t)} \tag{6.111}$$

Notice that the model contains now seven unknown parameters that need to be estimated from the data. In the optimization procedure all parameters have been

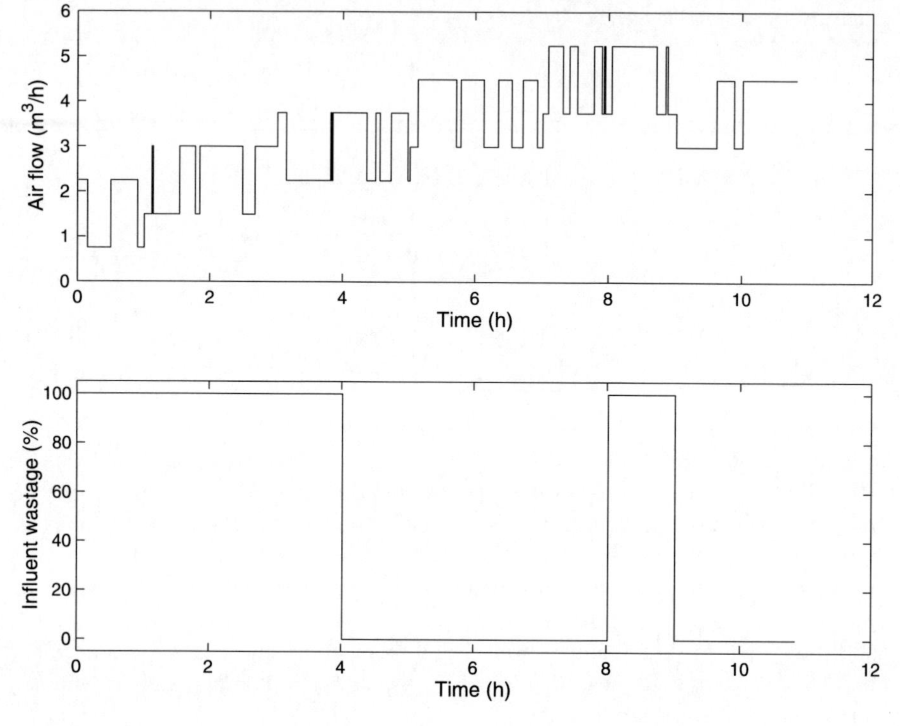

Fig. 6.17 Designed experimental air flow ($m^3\,h^{-1}$) and influent wastage (%)

scaled in advance to reduce numerical problems. The results for $t \in [0, 150]$ min with a sampling interval of one minute, which makes it a continuous-discrete time system, is presented in Table 6.5 (column 1). The last row in Table 6.5 shows the corresponding standard deviation of the prediction errors. The continuous-discrete time system description allows a detailed simulation of the DO-dynamics with, for example, time-varying integration steps between sampling time instants. Consequently, the residuals are evaluated at the sampling time instants only, and thus the objective function is, as shown before, a (weighted) finite sum of squares.

The estimated value of $f_{\max}$, close to one, confirms the correctness of the measured r_{act}. Thus, for the time being, $f_{\max}$ is fixed at one. As the dead time Δ is caused by the electro-mechanical part of the system, and thus will be largely time-invariant, it is fixed at 0.5 min. The optimal estimates for the remaining five parameters, thus with Δ and $f_{\max}$ fixed, are presented in column 2 of Table 6.5. Notice that σ_ε, the standard deviation of the prediction errors, is hardly increased by fixing $f_{\max}$ and Δ. From Fig. 6.19 we may conclude that the model output satisfactorily fits the data.

In the next two steps, successive reductions of the parameter dimensionality were made on the basis of analysis of dominant directions in the parameter space. For this purpose, consider the following covariance matrix, using (5.80), associated with the

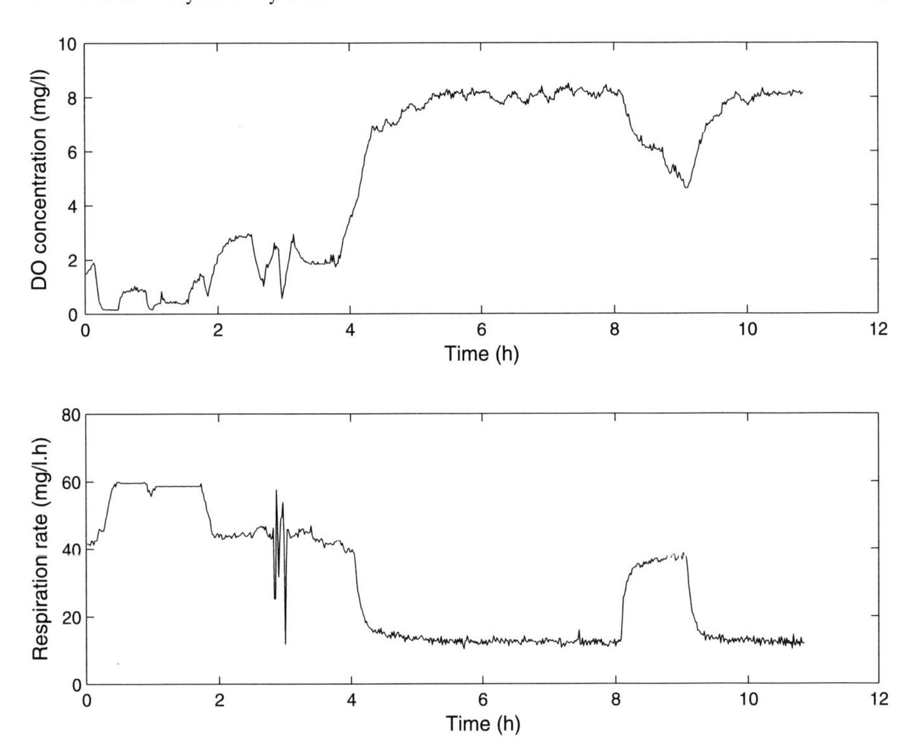

Fig. 6.18 Measured DO and r_{act} during the experiment

Table 6.5 Estimation results

Parameter	Estimated values				Exponent
	Column 1	Column 2	Column 3	Column 4	
Δ (min)	0.5				
$f_{\max}$	1.04				
α (l^{-1})	−0.82	−0.80	−1.29		10^{-3}
β ($\mathrm{l}^{-\frac{1}{2}}\,\mathrm{min}^{-\frac{1}{2}}$)	1.88	1.85	3.60	2.13	10^{-2}
γ (min^{-1})	−4.71	−4.67	−8.33	−4.51	10^{-2}
K_C ($\mathrm{mg\,l}^{-1}$)	0.54	0.54	0.29	3.16	
C_s ($\mathrm{mg\,l}^{-1}$)	17.5	17.4			
σ_ε ($\mathrm{mg\,l}^{-1}$)	6.8	6.8	9.4	9.6	10^{-2}

last estimates:

$$\mathrm{Cov}\,\widehat{\vartheta} = \begin{bmatrix} 0.0320 & -0.046 & 0.1146 & 0.0514 & 0.1543 \\ -0.046 & 0.0759 & -0.185 & -0.097 & -0.365 \\ 0.1146 & -0.185 & 0.4595 & 0.2176 & 0.8408 \\ 0.0514 & -0.097 & 0.2176 & 0.1833 & 0.6295 \\ 0.1543 & -0.365 & 0.8408 & 0.6295 & \mathbf{2.9952} \end{bmatrix}$$

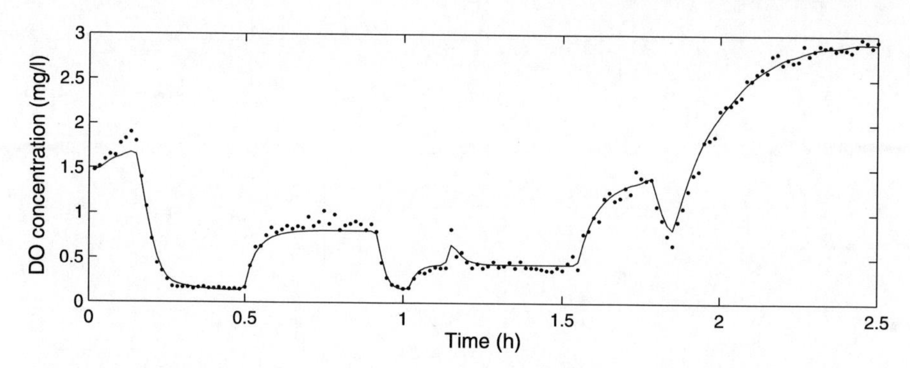

Fig. 6.19 Measured (...) and predicted (—) output for the case with five parameters

From this covariance matrix it is immediately clear that the variance of $\widehat{C_s}$, which is equal to 2.9952, is rather large. Further analysis of the matrix with eigenvectors of Cov $\widehat{\vartheta}$:

$$M = \begin{bmatrix} 0.5669 & 0.7682 & 0.0679 & 0.2839 & 0.0561 \\ 0.8144 & -0.4763 & -0.0275 & -0.3067 & -0.1225 \\ 0.1176 & -0.3931 & -0.2446 & 0.8309 & 0.2852 \\ 0.0182 & -0.1619 & 0.9565 & 0.1330 & 0.2023 \\ 0.0331 & 0.0468 & -0.1411 & -0.3424 & 0.9271 \end{bmatrix}$$

with associated eigenvalues (related to the corresponding column of M)

$$\lambda_i \in \{-0.0001,\ 0.0006,\ 0.0412,\ 0.2554,\ 3.4487\}$$

reveals that the largest eigenvalue is about 15 times the second largest, indicating an insensitive direction with large uncertainty in the parameter space (see Sect. 5.1.6 and Appendix B). The accompanying eigenvector that spans this insensitive direction (last column of M) is dominated by the fifth parameter, C_s. This result implies that the errors in the estimate of C_s have only minor influences on the sum of squares of the prediction errors. It is worth mentioning that the eigenvalues of the covariance matrix are the squared values of the singular values related to the sensitivity matrix associated with this estimation problem.

The estimated value of C_s is much larger than the physically expected value, which should be in the range of 8–10 mg/l. In view of the uncertainty in the estimate of C_s, it was decided to estimate both C_s and the parameters determining $k_L a$ from the measurements at high DO concentrations, that is, for $t \in [300, 450]$ min (see Fig. 6.18). From these data it has been found that $C_s = 9.12$ mg/l.

In the following analysis, C_s was fixed at this value, which slightly deteriorates the results for lower DO concentrations. However, it makes the model much more acceptable to engineers in the field of application. Estimation of the resulting four parameters, again for $t \in [0, 150]$ min and with Δ, $f_{\max}$, and C_s fixed, is found in column 3 of Table 6.5. Since the estimated value of α is rather small, the model can be further reduced by setting α equal to zero. The effect of this is presented in

Table 6.6 Estimation results reparameterized model

Parameter	Estimated values	
	Column 1	Column 2
α' ($\mathrm{mg\,l^{-1.5}\,min^{-\frac{1}{2}}}$)	0.166	0.1610
β' ($\mathrm{l^{-\frac{1}{2}}\,min^{-\frac{1}{2}}}$)	-1.32×10^{-2}	-6.22×10^{-3}
γ' ($\mathrm{min^{-1}}$)	5.20×10^{-2}	
δ' ($\mathrm{mg\,l^{-1}\,min^{-1}}$)	-0.429	-0.383
K_C ($\mathrm{mg\,l^{-1}}$)	0.627	0.602
σ_ε ($\mathrm{mg\,l^{-1}}$)	7.3×10^{-2}	7.4×10^{-2}

column 4 of Table 6.5, indicating a not too large loss in performance. A next step for final acceptation of the model structure is then to cross-validate this model with fresh data from a new experiment.

One problem that remains is the large correlation between the estimates of β and γ. This large correlation may well become a stumbling block when applying recursive techniques, to be treated in Part III, for online implementation. A possible way out is to reparameterize the initial model so that the parameter estimates become less correlated. Most correlation is caused by the products of β and γ with C_s in the DO model. This correlation is removed by defining a new set of parameters. Let therefore $\alpha' = \beta * C_s$, $\beta' = -\beta$, $\gamma' = -\gamma$, and $\delta' = \gamma * C_s$. Then

$$\begin{aligned}\frac{dC(t)}{dt} &= -f(C)r_{\mathrm{act}}(t) + \alpha'\sqrt{q_{\mathrm{air}}(t-\Delta)} \\ &\quad + \beta'\sqrt{q_{\mathrm{air}}(t-\Delta)}C(t) + \gamma' C(t) \\ &\quad - \frac{q_{\mathrm{in}} + q_r}{V}C(t) + \delta' \end{aligned} \tag{6.112}$$

This reparameterization offers some model whiteness for an increase in independence between the estimates. The estimated values can be viewed from Table 6.6 (column 1).

Analysis of the covariance matrix shows that γ' is insignificant. Consequently, it should be possible to fix γ' at zero without introducing a large error (see Table 6.6, column 2). Still a large correlation exists, especially between the parameter estimates of β' and K_C, but it is less profound than that between β and γ. Hence, this model structure with $\gamma' = 0$ will be used in future applications.

6.3 Historical Notes and References

Most of the basic material in this chapter can be found in the books of Norton [Nor86] and Ljung [Lju87, Lju99b], which, as mentioned before, have been a starting point for this book. Especially, the unification of identification methods for dynamic system under the umbrella of the so-called prediction-error methods has been a big step forward.

For practical use, the choice of an appropriate sampling interval is crucial, as a wrong choice may easily lead to a drastic increase of the variance of the estimates (see, for instance, [Lju87], p. 452). It has been found that very fast sampling leads to numerical problems, optimal choices of the sampling interval will lie in the range of time constants of the system, and that too fast sampling may radically increase the estimation variance. Some historical references to optimal sampling are [Sak61, DI82, BBF87, ZWR91, DW95]. In addition to the choice of the sampling rate, the choice of an appropriate presampling filter is also important. A basic and natural choice for a presampling filter is an integrator.

In this chapter only linear model structure selection was emphasized, with a focus on Akaike's criterion [Aka74]. Following Ljung [Lju87], the criterion has been formulated as the Final Prediction Error (FPE) criterion. However, many other criteria have been formulated as well, see [BA02] for an overview.

Parameter sensitivity studies are essential when identifying complex nonlinear systems. It helps to detect which parameters dominate the system's behavior. Sensitivity analysis is applied in many research areas. However, for a general theory on sensitivities, we refer to [TV72].

The algorithms presented in this chapter are at the heart of the MATLAB estimation routines, as *arx*, *armax*, *iv*, *oe*, and *pem*. For more advanced estimation routines, related to Box–Jenkins model structures, extended Instrumental Variables techniques, and subspace identification, see the MATLAB routines *bj, iv4*, and *n4sid*.

For the estimation of parameters in nonlinear models, usually iterative optimization algorithms are used, as presented in Sect. 5.2.3. However, for special classes of nonlinear systems, dedicated solutions became available, see for Hammerstein type of models (with only input nonlinearities) [Paw91, Gre00, Bai02, WZL09]. For Wiener type of models (with only output nonlinearities), we refer to [Wig93, Gre94, Gre98, Bai03b, BRJ09, GRC09], and for Hammerstein–Wiener identification, to [Bai02]. Finally, for the identification of rational systems, see also Examples 5.23–5.25 for static rational relationships, we refer to [Zhu05, DK09, KD09].

Subspace identification is essentially based on classical realization theory. In the 1990s many methods based on this principle were published, see [VD92, vOdM95, Vib95] to mention a few from the beginning. Since then many more papers have appeared, see [VH05] for a recent review. A route that was not further investigated in this book, but which may be valuable if one wants to obtain a state-space model from very limited prior knowledge, is to start from an identified input–output relationship and, using realization theory, to obtain a state-space model realization (see also Fig. 2.1). An interesting paper that investigates realization problems for system identification can be found in [VS04].

For the identification of linear parameter-varying (LPV) models, different routes have been followed, such as using linear fractional transformation (LFT), nonlinear programming, subspace identification techniques, linear regression, and orthonormal basis functions, see, for instance, [LP96, LP99, VV02, BG02, THvdH09].

The general theory of orthonormal basis function for system identification started with the work of van den Hof et al. [vdHPB95]. Other useful references in this context are [dVvdH98, AN99, Akc00, AH01, HdHvdHW04, THvdH09].

In the 1990s much emphasis was put on closed-loop identification. However, the first papers on identification in closed loop appeared in the early 1970s, see [BM74, TK75]. For overviews of closed-loop identification issues, related to the indirect, direct, and two-steps methods for linear systems, we refer to [MF95, HGDB96, FL99, GvdH01, EMT00]. An approach to nonlinear, time-varying systems is given by [DA96].

As stated in the Introduction, in this book the continuous-time model representation will only be used for demonstration. For identification and parameter estimation, the discrete-time form will mainly be explored due to the availability of sampled data and the ultimate transformation of a mathematical model into a simulation code. However, from Example 1.4 it is immediately clear that when starting from physical laws, and in particular when balance equations are used, usually differential equations or transfer functions in the continuous-time domain appear. Notice from Example 1.5 that, even for simple, linear differential equations, the input–output relation is nonlinear-in-the-parameters. A classical approach is to estimate the parameters using iterative estimation algorithms, as for simulation models, in general. However, already in the 1970s, attention had been paid to the special character of continuous-time identification, see [Phi73, SR77, Bag75, SR79]. In recent years, there has been a renewed interest in continuous-time identification methods, see, for example, [CSS08, GL09b], or in the preservation of continuous-time physical parameters in linear regression type of models, see [VKZ06, Vri08, KK09].

6.4 Problems

Problem 6.1 Consider the discrete-time system

$$G(q) = \frac{0.2q^{-2}}{1 - 0.8q^{-1}}$$

which has been found after modeling the oxygen concentration in a composting plant. (*NB*: in MATLAB z is used as forward shift operator instead of q.)

(a) Define the system in MATLAB using the function *tf* and, if necessary, $sys.inputdelay = 1$ for the specification of a unit time delay.
(b) Generate a random binary input signal (RBS) with $p = 0.2$ and $N = 50$ and determine the corresponding model output using *lsim*.
(c) Add noise ("output-error") to this output, so that one obtains a noise-corrupted output and estimate the parameter values using *oe*.
(d) Use the function *compare* to compare the identification result with the generated data.
(e) Repeat this estimation procedure, but now with the function *arx*, and again use *compare* to evaluate the result.

Problem 6.2 From an experiment in an industrial process the following data have been obtained (see Table 6.7).

Table 6.7 Experimental data

$u(t)$	−1	−1	−1	−1	1	1	1	1	1	1	1	−1	−1	−1
$y(t)$	0	−0.40	−0.69	−1.0	−1.27	−0.79	−0.35	0.0	0.4	0.73	1.03	1.30	0.82	0.38

(a) Determine the parameters a and b in the model

$$y(t) = ay(t-1) + bu(t-1)$$

on the basis of two iterations of a Markov estimation procedure and use as much as possible data. Present and interpret the results.

(b) Repeat (a), but now using an instrumental variable (IV) method.

Part III
Time-varying Systems Identification

Basically, in Parts I and II the data have been processed batch-wise, so that the resulting estimates hold for the complete time span of the data. However, in a number of real-time implementations, it is preferred to obtain estimates of the actual process parameters without processing the complete past input–output data set at each sample instant. This is especially the case in those applications with a (possible) time-varying system behavior, that is, the parameter estimates, even those in a presumed time-invariant static system description, vary with time. For these cases, in addition to the batch-wise processing of the data, *recursive identification* techniques have been introduced in the past. In the statistical literature these techniques are often identified as "sequential parameter estimation" or, when applied in signal processing, as "adaptive algorithms."

In Chap. 7 recursive estimation will be introduced and applied to static, linear, or nonlinear systems with possibly time-varying parameters. The idea is as follows. On the basis of a priori knowledge, the model parameters in the linear regression models will be considered as constant. Subsequently, the experimental data will tell how the parameter estimates vary with time. This idea can be easily extended to the case with a dynamic parameter model in the form of a linear dynamic state equation, which clearly illustrates the system theoretic concept of a model parameter as a (unobserved) state. Hence, the resemblance of the recursive least-squares parameter estimator to the well-known *Kalman filter* will be emphasized. For the nonlinear case, the concept of Extended Kalman filtering will be introduced. For practical implementation of the recursive least-squares parameter estimator/Kalman filter, modifications of the standard algorithm are needed to avoid, for instance, loss of symmetry of the covariance matrix and instabilities due to rounding errors. The numerical issues related to the Kalman filter are presented in Sect. 7.1.5. Although this section is marked as advanced material, it surely essential reading for the practitioner.

Chapter 8 focuses on the recursive parameter estimation in dynamic systems, where in general optimality of the estimation results of the linear regression models of Chap. 7 will no longer hold. Here the interchanging concept of parameter and state will be further worked out, using extended Kalman filtering and observer-based methods. And, again it will be applied to both the linear and the nonlinear cases. The theory will be illustrated by real-world examples, with most often a biological component in it, as these cases often show a time-varying behavior due to adaptation of the (micro)organisms.

Chapter 7
Time-varying Static Systems Identification

7.1 Linear Regression Models

7.1.1 Recursive Estimation

Let the inputs $u(0), u(1), \dots, u(N)$ and corresponding outputs $y(0), y(1), \dots, y(N)$ be recorded. In the previous chapters these data have been processed batch-wise, that is the input and output data were collected into $(N+1)$-dimensional vectors $u := [u(0), \dots, u(N)]^T$ and $y := [y(0), \dots, y(N)]^T$, respectively. The parameter estimates are then found by "inverting" the model with input and output data. However, for large N, this can be a heavily computational task. Furthermore, it is implicitly assumed that the parameters are constant during the experiment. Recursive estimation of the model parameters will lead to more efficient computational schemes and allows the estimation of time-varying parameters. At this point, we clearly have to distinguish between recursive and iterative data processing. Typical iterative processing schemes are presented in the Sects. 5.2.3, 6.1.3, and 6.2.4, where in each iteration step the complete data set is processed. Recursive data processing, on the other hand, only processes the data from time instant $t-1$ to t for $t = 1, \dots, N$. Let us start by illustrating the recursive estimation technique to a very simple example with just one parameter and one observation at each sampling instant. This case will subsequently be extended to the parameter vector case with p parameters and finally to the vector-output case with p parameters and n measurements at each sampling instant.

Example 7.1 Mean tracking: Consider the following regression model (see Sect. 5.1):

$$y(t) = \vartheta + e(t)$$

with unknown parameter ϑ, a noise-free constant estimate of the observations $y(t)$. The error $e(t)$ has zero-mean and variance $E[e(t)^2] = R$. Furthermore, it is assumed that $e(t)$ is uncorrelated with the estimation error at the previous time instant. Let N output measurements $y(1), y(2), \dots, y(N)$ be available. Then, an *unbiased* estimate

K.J. Keesman, *System Identification*,
Advanced Textbooks in Control and Signal Processing,
DOI 10.1007/978-0-85729-522-4_7,

of ϑ, given all N data points and, in what follows, denoted as $\widehat{\vartheta}(N)$, can be found from

$$\widehat{\vartheta}(N) = \frac{1}{N}\sum_{t=1}^{N} y(t)$$

where $\widehat{\vartheta}(N)$ is thus the mean value of $y(t)$ for $t = 1, \dots, N$. After $N + 1$ measurements the estimate becomes

$$\begin{aligned}
\widehat{\vartheta}(N+1) &= \frac{1}{N+1}\sum_{t=1}^{N+1} y(t) \\
\Longrightarrow \quad \frac{(N+1)}{N}\widehat{\vartheta}(N+1) &= \frac{1}{N}\sum_{t=1}^{N} y(t) + \frac{1}{N} y(N+1) \\
\Longrightarrow \quad \widehat{\vartheta}(N+1) &= \frac{N}{N+1}\widehat{\vartheta}(N) + \frac{1}{N+1} y(N+1) \qquad (7.1)
\end{aligned}$$

Hence, instead of repeating the calculation of the mean for each new sampling instant, we can *recursively* update the new estimate using appropriate weighting factors.

Notice that by defining $x := \vartheta$ we could also consider the linear regression model in the example as a linear output equation with constant state x (see Example 5.1). This shows that from a system-theoretical point of view a model parameter could be seen as a state; this view will be developed further in the sequel.

Let us generalize the idea on mean tracking, as presented by (7.1) in Example 7.4. Define, therefore, the estimate $\widehat{\vartheta}(t)$ as a linear combination of the preceding estimate $\widehat{\vartheta}(t-1)$ and the actual output measurement $y(t)$, that is,

$$\widehat{\vartheta}(t) := J(t)\widehat{\vartheta}(t-1) + K(t)y(t) \qquad (7.2)$$

For $\widehat{\vartheta}(t)$ to be *unbiased*, the following must hold:

$$\begin{aligned}
E\big[\widehat{\vartheta}(t)\big] &= J(t)E\big[\widehat{\vartheta}(t-1)\big] + K(t)E\big[y(t)\big] \\
&= J(t)\vartheta + K(t)\vartheta \\
&= \vartheta \qquad (7.3)
\end{aligned}$$

Hence, $J(t) + K(t) = 1$. Substituting $J(t) = 1 - K(t)$ into (7.2) then gives

$$\begin{aligned}
\widehat{\vartheta}(t) &= \big(1 - K(t)\big)\widehat{\vartheta}(t-1) + K(t)y(t) \\
&= \widehat{\vartheta}(t-1) + K(t)\big[y(t) - \widehat{\vartheta}(t-1)\big] \qquad (7.4)
\end{aligned}$$

Notice from this equation that the last estimate depends on the previous estimate, reflecting the updated prior knowledge of the parameter value, and a weighted prediction error. The prediction error $[y(t) - \widehat{\vartheta}(t-1)]$ is also called a *recursive residual* or an *innovation*, and it reflects the mismatch between predicted and measured

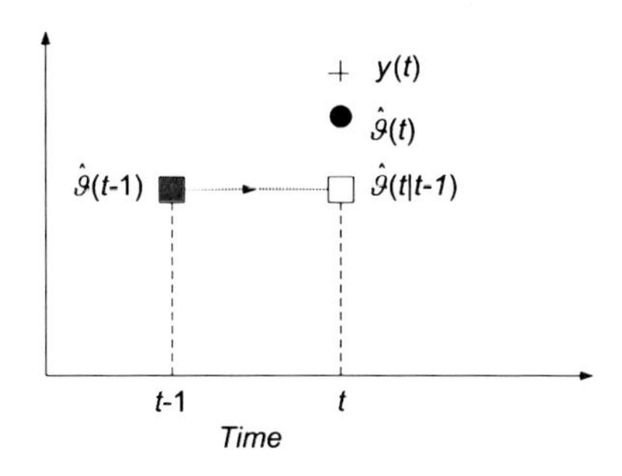

Fig. 7.1 Illustration of parameter estimate update

output. Hence, the new estimate $\widehat{\vartheta}(t)$ is a compromise between prior and posterior knowledge (see also Fig. 7.1).

Next, the question is how to choose $K(t)$, in what follows also called the gain or, more specifically, the *Kalman gain*. If, in addition to unbiasedness of the estimate, we also demand a *minimum variance* estimate, then the following can be obtained. Recall that the variance is defined as

$$P(t) := E\big[\big(\widehat{\vartheta}(t) - E\big[\widehat{\vartheta}(t)\big]\big)^2\big] \tag{7.5}$$

Substituting (7.4) with $y(t) = \vartheta + e(t)$ into (7.5) and noting that $E[y(t)] = \vartheta$ gives

$$\begin{aligned} P(t) &= E\big[\big(\widehat{\vartheta}(t-1) + K(t)\big[y(t) - \widehat{\vartheta}(t-1)\big] - \vartheta\big)^2\big] \\ &= E\big[\big(\widehat{\vartheta}(t-1) - \vartheta\big)^2 - 2K(t)\big[\widehat{\vartheta}(t-1) - y(t)\big]\big(\widehat{\vartheta}(t-1) - \vartheta\big) \\ &\quad + K(t)^2\big[y(t) - \widehat{\vartheta}(t-1)\big]^2\big] \\ &= P(t-1) - 2K(t)P(t-1) + K(t)^2 \\ &\quad \times \big\{E\big[\big(\vartheta - \widehat{\vartheta}(t-1)\big)^2\big] + 2E\big[e(t)(\vartheta - \widehat{\vartheta}(t-1)\big] + E\big[e(t)^2\big]\big\} \end{aligned} \tag{7.6}$$

Since $E[e(t)] = 0$, $E[e(t)^2] = R$ and $e(t)$ is uncorrelated with $\widehat{\vartheta}(t-1)$, the cross-term $E[e(t)(\vartheta - \widehat{\vartheta}(t-1)] = 0$, and thus (7.6) further reduces to

$$\begin{aligned} P(t) &= P(t-1) - 2K(t)P(t-1) \\ &\quad + K(t)^2 P(t-1) + K(t)^2 R \\ &= \big(1 - K(t)\big)^2 P(t-1) + K(t)^2 R \end{aligned} \tag{7.7}$$

For finding the $K(t)$ that minimizes $P(t)$ variational calculus will be used by writing $\Delta P(t)$ as a function of $\Delta K(t)$, that is,

$$\begin{aligned} \Delta P(t) &= \big(1 - \big[K(t) + \Delta K(t)\big]\big)^2 P(t-1) \\ &\quad + \big[K(t) + \Delta K(t)\big]^2 R \\ &\quad - \big(1 - K(t)\big)^2 P(t-1) - K(t)^2 R \end{aligned}$$

$$
\begin{aligned}
&= \big[1 - 2\big[K(t) + \Delta K(t)\big] + K(t)^2 \\
&\quad + 2K(t)\Delta K(t) + \Delta K(t)^2\big]P(t-1) \\
&\quad + \big[K(t)^2 + 2K(t)\Delta K(t) + \Delta K(t)^2\big]R \\
&\quad - \big(1 - K(t)\big)^2 P(t-1) - K(t)^2 R \\
&\simeq 2\Delta K(t)\big[K(t)R - \big(1 - K(t)\big)P(t-1)\big]
\end{aligned} \tag{7.8}
$$

where the approximation is a result of neglecting the second-degree terms. Recall that at a minimum point the derivative is equal to zero. Hence, for requiring a minimum variance, at least $\Delta P(t)$ must be equal to zero. Consequently,

$$
K(t)R - \big(1 - K(t)\big)P(t-1) = 0 \tag{7.9}
$$

and thus

$$
K(t) = \frac{P(t-1)}{(P(t-1) + R)} \tag{7.10}
$$

Since the neglected quadratic terms only contribute in a positive way to $\Delta P(t)$, for this choice of $K(t)$, a minimum of $P(t)$ has been achieved.

In conclusion,

$$
\widehat{\vartheta}(t) = \widehat{\vartheta}(t-1) + K(t)\big[y(t) - \widehat{\vartheta}(t-1)\big] \tag{7.11}
$$

$$
K(t) = \frac{P(t-1)}{(P(t-1) + R)} \tag{7.12}
$$

$$
P(t) = \big(1 - K(t)\big)^2 P(t-1) + K(t)^2 R \tag{7.13}
$$

for $t = 1, \ldots, N$ and $\vartheta(0)$, $P(0)$ given. Equations (7.11)–(7.13) are the scalar version of the so-called *recursive least-squares (RLS) parameter estimator* (see Appendix F for a general derivation) with the nice properties that it is linear and provides minimum variance, unbiased estimates. Notice that in the recursive estimation framework at each time instant both the estimate $\widehat{\vartheta}(t)$ and associated estimation error $P(t)$ are directly available.

From this result we can derive some particular solutions. First, setting $P(0) = 0$, that is, the initial estimate $\vartheta(0)$ is assumed to be exactly known, leads to $P(t) = 0$ and thus to $K(t) = 0$ for all t. Consequently, this choice implies that the estimator will not use any measurement information and thus $\vartheta(t) = \vartheta(0)$ for all t. Secondly, setting $R = 0$ leads to $K(t) = 1$ and $P(t) = R$ for all t, which implies that the estimate is equal to the measurement $y(t)$, and thus prior knowledge is not taken into account at all. Substituting (7.10) into (7.7) ultimately leads to

$$
P(t) = \big(1 - K(t)\big)P(t-1) \tag{7.14}
$$

a simplified expression for $P(t)$. Alternatively, substituting $P(t-1) = \frac{P(t)-K(t)^2 R}{(1-K(t))^2}$ into (7.10) gives

$$K(t) = \frac{-(R+P(t)) \pm (R-P(t))}{-2R}$$
$$= \begin{cases} \frac{P(t)}{R} \\ 1 \end{cases} \tag{7.15}$$

where only the first solution is of practical relevance. Clearly, both simplified expressions (7.14) and (7.15) cannot be used together.

Let us now evaluate the properties of the recursive estimators (7.1) and (7.11)–(7.13) for the mean tracking example.

Example 7.2 Mean tracking: For estimator (7.1) related to the mean tracking problem, we can easily derive, using (7.2), that the gain $K(t)$ is equal to $\frac{1}{t}$. Because $J(t) = \frac{t-1}{t}$ and thus $J(t) + K(t) = 1$, the recursive estimate of the mean is *unbiased*. However, for the *unbiased, minimum variance* estimator (7.11)–(7.13) the gain is $\frac{P(t-1)}{P(t-1)+R}$. Clearly, both estimators (7.1) and (7.11)–(7.13) become equivalent when $P(t-1) = \frac{R}{t-1}$, and thus $P(1) = R$, which is not a bad choice when, as an initial estimate, $\vartheta(1) = y(1)$ is chosen. Let us investigate what happens when a different choice for $P(1)$ is made.

Suppose that $P(1) = 2R$. Then from (7.1) with $K(t) = \frac{1}{t}$ and thus $K(2) = \frac{1}{2}$, we have

$$P(2) = \frac{1}{4}P(1) + \frac{1}{4}R = \frac{3}{4}R$$

However, from (7.11)–(7.13) with gain $K(2) = \frac{2}{3}$ we obtain

$$P(2) = \frac{1}{9}P(1) + \frac{4}{9}R = \frac{2}{3}R < \frac{3}{4}R$$

which clearly gives a smaller variance of the estimate. For $P(1) = \frac{1}{2}R$, using estimator (7.1), $P(2) = \frac{3}{8}R$, but from (7.11)–(7.13) with $K(2) = \frac{1}{3}$ we obtain $P(2) = \frac{1}{3}R < \frac{3}{8}R$, which again leads to a smaller variance, as predicted by the theory. Thus, estimator (7.1) gives an unbiased, but nonminimum variance, estimate.

Let us extend the *scalar* parameter case to the *parameter vector* case with p unknown parameters. If now the following univariate linear regression model is considered

$$y(t) = \phi(t)^T \vartheta + e(t) \tag{7.16}$$

where $y(t)$ is the scalar output measurement and both $\phi(t)$ and ϑ are vectors of dimension p, the recursive estimator takes the following form:

$$\widehat{\vartheta}(t) = \widehat{\vartheta}(t-1) + K(t)\left[y(t) - \phi(t)^T \widehat{\vartheta}(t-1)\right] \tag{7.17}$$

$$K(t) = P(t-1)\phi(t)\big[\phi(t)^T P(t-1)\phi(t) + R\big]^{-1} \tag{7.18}$$

$$P(t) = \big(I - K(t)\phi(t)^T\big)P(t-1)\big(I - \phi(t)K(t)^T\big) + K(t)RK(t)^T \tag{7.19}$$

for $t = 1, \dots, N$ and $\vartheta(0)$, $P(0)$ given. Notice that now $P(t)$ is a $p \times p$ covariance matrix and both $K(t)$ and $\phi(t)$ are p-dimensional column vectors. Equation (7.19) is also known as the "Joseph form" of the covariance matrix update equation and is valid for any value of $K(t)$. Alternatively, the covariance matrix update equation (7.14) can be used. This expression for the covariance matrix update is computationally cheaper but is only correct for the optimal gain. For real-time application, however, or if a nonoptimal Kalman gain is deliberately used, the simplified form (7.14) cannot be applied. In these cases, (7.19) must be used. In a later section, all this will be further extended to the vector-output case with a multivariate regression model. However, first explicit modeling of parameter variations will be considered.

7.1.2 Time-varying Parameters

In the previous section, it has been implicitly assumed that the parameters are constant, that is,

$$\vartheta(t) = \vartheta(t-1) \tag{7.20}$$

which can be seen from the presented recursive estimators when the effect of the innovations is neglected. Hence, the parameter estimates only vary due to the misfit between predicted parameter value and measurement. This very simple difference equation model of parameter invariance can be easily extended toward a Gauss–Markov stochastic difference equation. This equation will allow extra dynamics and stochastic parameter variability and thus an explicit modeling of the parameters. In the case of modeling parameter variations, the Gauss–Markov stochastic difference equation is given by

$$\vartheta(t) = \Xi\vartheta(t-1) + \Pi w(t-1) \tag{7.21}$$

where Ξ is a $p \times p$ time-invariant matrix, and Π is a $p \times q$ time-invariant input matrix. The disturbance input $w(t-1)$ at sample instant $t-1$ is a p-dimensional white noise vector with covariance matrix $Q(t-1)$. Notice at this point the resemblance with a discrete-time version of the state equation (1.3). Notice also the subtle difference in interpretation of $u(t)$ in (1.3) or a discrete-time version of it and $w(t)$ in (7.21), respectively. In the former, $u(t)$ is the deterministic input related to the dynamic state equation, while in (7.21), $w(t)$ represents the presumed stochastic time variation in the parameters.

In practice, most often the matrices Ξ and Π are not known in advance. Therefore, simplified versions of (7.21) are more frequently used, as, for instance, the

Table 7.1 Moving object data

Time t (s)	1	2	3	4	10	12	18
Distance y (ft)	9	15	19	20	45	55	78

so-called *random walk model* where $\Xi = I$ and $\Pi = I$, so that

$$\vartheta(t) = \vartheta(t-1) + w(t-1) \tag{7.22}$$

Notice that by using a parameter model like (7.22) a stochastic variation of the parameters is prespecified a priori. The parameter variation will be greatly affected by the choice of the covariance matrix $Q(t-1)$, which is commonly chosen as a diagonal matrix presuming serially independent random variables in $w(t-1)$. This adjustment of the parameter model (7.20) leads to the following linear unbiased minimum-variance estimator:

$$\widehat{\vartheta}(t) = \widehat{\vartheta}(t-1) + K(t)\big[y(t) - \phi(t)^T\widehat{\vartheta}(t-1)\big] \tag{7.23}$$

$$K(t) = \widetilde{P}(t-1)\phi(t)\big[\phi(t)^T\widetilde{P}(t-1)\phi(t) + R\big]^{-1} \tag{7.24}$$

$$\begin{aligned} P(t) &= \big(I - K(t)\phi(t)^T\big)\widetilde{P}(t-1)\big(I - \phi(t)K(t)^T\big) \\ &\quad + K(t)RK(t)^T \end{aligned} \tag{7.25}$$

for $t = 1, \ldots, N$, where $\widetilde{P}(t-1) = P(t-1) + Q(t-1)$. After replacing $P(t-1)$ by $P(t-1) + Q(t-1)$, this algorithm becomes a straightforward extension of (7.17)–(7.19). Hence, the gain and error covariance matrix is directly affected by $Q(t-1)$. For the simplest case with only one parameter, that is, $p = 1$, and $\phi(t) = 1$ for all t, choosing the variance $Q(t)$ constant and very large will give a gain that at each time instant tends to 1, and thus $\widehat{\vartheta}(t) \simeq y(t)$ with $P(t) \simeq R$. Consequently, this choice implies that no filtering of the data will take place.

Another parameter model that is frequently used is the integrated random walk model

$$\eta(t) = \eta(t-1) + w(t-1) \tag{7.26}$$

$$\gamma(t) = \gamma(t-1) + \eta(t-1) \tag{7.27}$$

with $\operatorname{Cov} w(t) = Q(t)$, a diagonal $p \times p$ matrix. In this case the parameter increments $\eta(t-1)$ are integrated stochastic variations or random walks. Consequently, both parameters η and γ are estimated, that is, $\vartheta(t) := \left[\begin{smallmatrix}\eta(t)\\ \gamma(t)\end{smallmatrix}\right]$.

Let us illustrate the recursive estimation theory, presented so far, to a linear two-parameter problem.

Example 7.3 Moving object (*constant velocity*): Let the following observations on an object moving in a straight line with constant velocity v, as presented in Table 7.1, be available (after [You84], p. 18).

From Table 7.1 we obtain that $p = 2$ and $N = 7$. Let us first plot the data (see Fig. 7.2), which shows an approximate linear relationship between time t and the

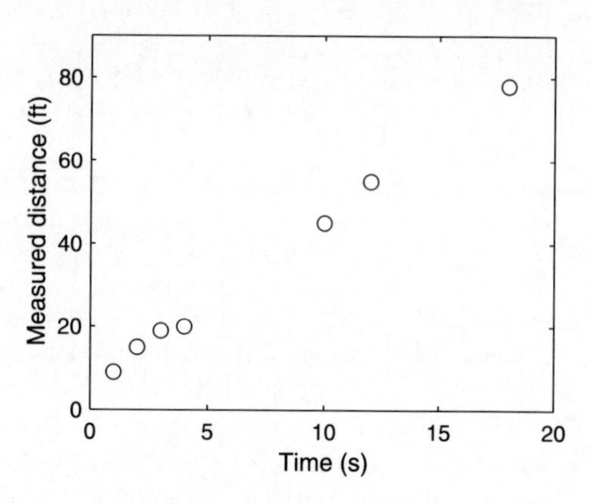

Fig. 7.2 Measured moving object data

measured distance y, as predicted by the kinematic law $s(t) = s_0 + vt$, where $s(t)$ is the noise-free distance at time instant t, with initial position s_0 (ft) and velocity v (ft/s).

Then, an appropriate model, relating the measured distance to the noise-free distance, as predicted by the kinetic law, would be

$$y(t) = s_0 + vt + e(t)$$

Define $\vartheta := \left[\begin{smallmatrix} s_0 \\ v \end{smallmatrix}\right]$ and $\phi(t) := \left[\begin{smallmatrix} 1 \\ t \end{smallmatrix}\right]$ in order to obtain a model of the form (7.16). Recursive estimation of the parameters s_0 and v on the basis of (7.17)–(7.19), with $\vartheta(0) = \left[\begin{smallmatrix} 0 \\ 0 \end{smallmatrix}\right]$, $P(0) = 1000I$, $Q = 0$, and $R = 1$ leads to the following results:

$$t = 1: \quad K(1) = \begin{bmatrix} 0.4998 \\ 0.4998 \end{bmatrix}, \qquad \vartheta(1) = \begin{bmatrix} 4.4978 \\ 4.4978 \end{bmatrix}$$

$$P(1) = \begin{bmatrix} 500.2499 & -499.7501 \\ -499.7501 & 500.2499 \end{bmatrix}$$

$$t = 2: \quad K(2) = \begin{bmatrix} -0.9921 \\ 0.995 \end{bmatrix}, \qquad \vartheta(2) = \begin{bmatrix} 3.0030 \\ 5.9970 \end{bmatrix}$$

$$P(2) = \begin{bmatrix} 4.9692 & -2.9791 \\ -2.9791 & 1.9871 \end{bmatrix}$$

$$t = 3: \quad K(3) = \begin{bmatrix} -0.6646 \\ 0.4991 \end{bmatrix}, \qquad \vartheta(3) = \begin{bmatrix} 4.4282 \\ 5.0018 \end{bmatrix}$$

$$P(3) = \begin{bmatrix} 2.3269 & -0.9972 \\ -0.9972 & 0.4988 \end{bmatrix}$$

$$t = 4: \quad K(4) = \begin{bmatrix} -0.4991 \\ 0.2997 \end{bmatrix}, \qquad \vartheta(4) = \begin{bmatrix} 6.4921 \\ 3.7025 \end{bmatrix}$$

$$P(4) = \begin{bmatrix} 1.4975 & -0.4992 \\ -0.4992 & 0.1997 \end{bmatrix}$$

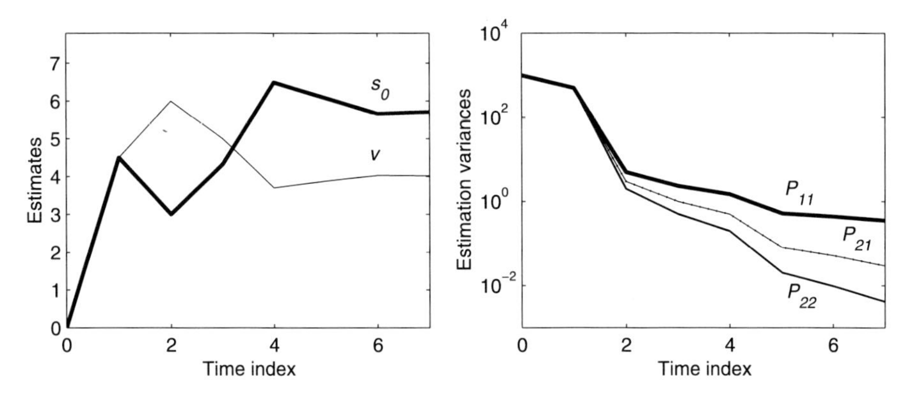

Fig. 7.3 Estimated parameter values (*left figure*) with (co)variances (*right figure*)

$$t=5: \quad K(5)=\begin{bmatrix}-0.2798\\0.1200\end{bmatrix}, \qquad \vartheta(5)=\begin{bmatrix}6.0772\\3.8804\end{bmatrix}$$

$$P(5)=\begin{bmatrix}0.5197 & -0.0800\\-0.0800 & 0.0200\end{bmatrix}$$

$$t=6: \quad K(6)=\begin{bmatrix}-0.1773\\0.0645\end{bmatrix}, \qquad \vartheta(6)=\begin{bmatrix}5.6590\\4.0325\end{bmatrix}$$

$$P(6)=\begin{bmatrix}0.4417 & -0.0516\\-0.0516 & 0.0097\end{bmatrix}$$

$$t=7: \quad K(7)=\begin{bmatrix}-0.1791\\0.0451\end{bmatrix}, \qquad \vartheta(7)=\begin{bmatrix}5.7027\\4.0215\end{bmatrix}$$

$$P(7)=\begin{bmatrix}0.3546 & -0.0296\\-0.0296 & 0.0042\end{bmatrix}$$

The estimation results are graphically presented in Fig. 7.3.

The effect of setting $Q=I$, using (7.17)–(7.19), so that the parameter estimates are allowed to vary a bit more, can be viewed from Fig. 7.4. Notice that especially the estimate of s_0 with associated estimation variance are affected.

In the following section, a state-space representation, which easily allows the incorporation of explicit parameter models and allows a natural extension to the *vector-output* case, will be presented.

7.1.3 Multioutput Case

Suppose now that n different measurements are available at each time instant. Then, at each sampling instant an algebraic equation relating the p unknown parameters

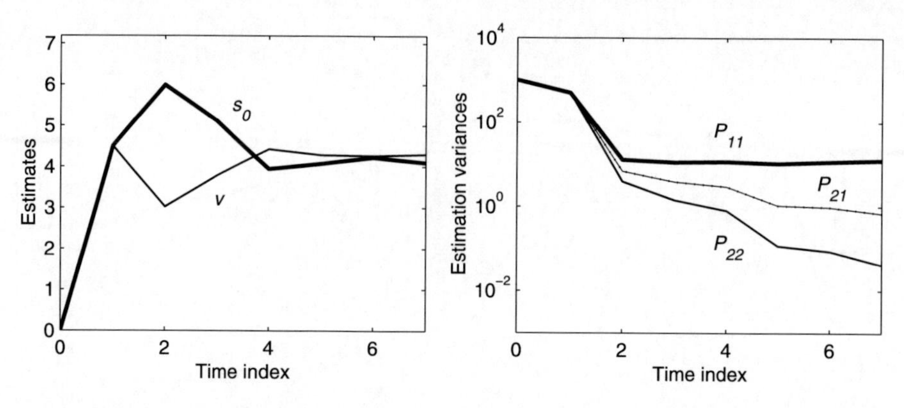

Fig. 7.4 Estimated parameter values (*left figure*) with (co)variances (*right figure*) for random walk model with $Q = I$

to the n given measurements is needed. Together with the explicit parameter model, a linear, discrete-time, time-varying state-space representation (see also Sect. 1.2.2) is obtained,

$$\begin{aligned} \vartheta(t) &= \Xi\vartheta(t-1) + \Pi w(t-1) \\ y(t) &= \Phi(t)\vartheta(t) + v(t) \end{aligned} \tag{7.28}$$

where at $t = 1, \ldots, N$, $w(t)$ and $v(t)$ are mutually uncorrelated zero-mean white noise terms with $\operatorname{Cov} w(t) = Q(t)$, $\operatorname{Cov} v(t) = R(t)$, and $E[v(t)w(t-\tau)] = 0$. Notice that by combining a dynamic parameter model and an algebraic output equation as in (7.28), two independent noise terms are introduced. Recall from Chap. 1 that these noise terms are also known as the system and sensor noises, respectively. This system representation of the vector-output case, with an $n \times p$ matrix $\Phi(t)$, leads to a full matrix version of the recursive least-squares (RLS) parameter estimator. Furthermore, for a better insight into the procedure, the estimator is presented in a prediction-correction scheme as follows. Prediction:

$$\hat{\vartheta}(t|t-1) = \Xi\hat{\vartheta}(t-1) \tag{7.29}$$

$$P(t|t-1) = \Xi P(t-1)\Xi^T + \Pi Q(t-1)\Pi^T \tag{7.30}$$

Correction:

$$K(t) = P(t|t-1)\Phi(t)^T\big[\Phi(t)P(t|t-1)\Phi^T(t) + R(t)\big]^{-1} \tag{7.31}$$

$$\hat{\vartheta}(t) = \hat{\vartheta}(t|t-1) + K(t)\big[y(t) - \Phi(t)\hat{\vartheta}(t|t-1)\big] \tag{7.32}$$

$$\begin{aligned} P(t) &= \big(I - K(t)\Phi(t)\big)P(t|t-1)\big(I - \Phi(t)^T K(t)^T\big) \\ &\quad + K(t)R(t)K(t)^T \end{aligned} \tag{7.33}$$

for $t = 1, \ldots, N$ and given $\vartheta(0)$ and $P(0)$. Notice that now $K(t)$ becomes a $p \times n$ gain matrix and that (7.11)–(7.13), (7.17)–(7.19), and (7.23)–(7.25) are special cases.

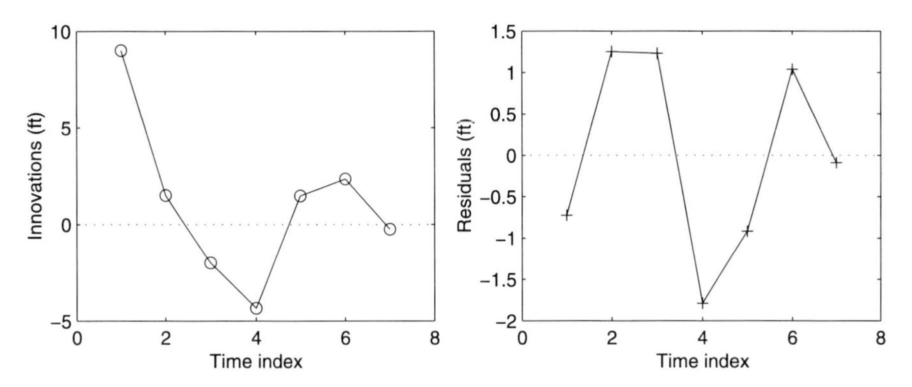

Fig. 7.5 Innovations (*left figure*) and residuals (*right figure*)

Before illustrating the recursive least-squares estimator by an example, let us first summarize the algorithm.

Algorithm 7.1 Recursive Least-squares estimation of $\vartheta(t)$ in linear time-varying static systems

1. Given $y(t)$ and $\Phi(t)$ for $t = 1, \dots, N$, specify the dynamic parameter model matrices Ξ and Π.
2. Specify the covariance matrices $Q(t)$ and $R(t)$.
3. Choose the initial parameter vector $\vartheta(0)$ and the initial error covariance matrix $P(0)$.
4. Evaluate, for $t = 1, \dots, N$, (7.29)–(7.33).
5. Check the mean and autocorrelation function of the innovations sequence ε for optimality of the solution.

Example 7.4 Moving object (constant velocity): The innovations related to the recursive estimation with $\vartheta(0) = \left[\begin{smallmatrix} 0 \\ 0 \end{smallmatrix}\right]$, $P(0) = 1000\ I$, $Q = 0$, and $R = 1$ are presented in Fig. 7.5. Clearly, due to the "wrong" initial estimate, the mean value of $\{\varepsilon\}$ is nonzero. Further, the number of data is too limited to perform a full correlation analysis. Therefore, as yet, the graphical inspection suffices. Notice especially that the effect of the initial estimate dominates the sequence of innovations. In order to appreciate the difference between the recursive residuals or innovations and the residuals obtained when using the final estimates, in Fig. 7.5 also the residuals are added.

Let us point-wise present some additional remarks on the recursive estimation problem.

Remark 7.1 Referring to the assumptions on the measurement noise $v(t)$ and initially for the linear regression case denoted by $e(t)$, made in the beginning of this section, the "whiteness" of the recursive residuals (innovations)

$$\varepsilon(t) = y(t) - \Phi(t)\widehat{\vartheta}(t|t-1) \tag{7.34}$$

is an indication of "optimal" estimation. Ideally, there should hold $\varepsilon(t) = v(t)$. Hence, the mean and autocorrelation function of the innovations sequence $\{\varepsilon\}$ should be checked.

Remark 7.2 In the time-invariant case where $\Phi(t)$, $Q(t)$, and $R(t)$ are constant matrices Φ, Q, and R, respectively, the covariance matrix $P(t)$ converges to a constant matrix P_∞ for large N, which can be found from

$$P_\infty = \left[\Phi^T R^{-1}\Phi + \left(\Phi P_\infty \Phi^T + \Pi Q \Pi^T\right)^{-1}\right]^{-1} \tag{7.35}$$

If the discrete-time linear dynamic model (7.28) is controllable and observable[1] and if $Q > 0$ and $R > 0$, then $P(t)$ converges to a unique positive definite P_∞. Let us demonstrate this by an example.

Example 7.5 Mean tracking: For the simple mean tracking problem with $\Phi(t) = 1$ for all t, $\Phi = 1$. Under the assumption of a time-varying mean with $\Xi = 1$ and $\Pi = 1$, we obtain

$$P_\infty = \frac{1}{2}\left[-Q + \sqrt{Q^2 + 4RQ}\right]$$

so that for $Q = 0$ and/or $R = 0$, $P_\infty = 0$. Otherwise, that is, for $Q > 0$ and $R > 0$, $P_\infty > 0$.

Remark 7.3 In the case of unknown initial conditions, set $\vartheta(0) = 0$ and let $P(0) \to \infty I$. In practice, setting $P(0) = 10^6 I$ is an appropriate choice. In fact, choosing $P(0)$ as a very large diagonal matrix such that $\Phi(1)(\Xi P(0)\Xi^T + \Pi Q(0)\Pi^T)\Phi^T(1) \gg R(1)$ can be interpreted as setting a very large variance on each of the initial estimates. In other words, the initial estimates in $\vartheta(0)$, and consequently the output prediction $\Phi(1)\Xi\vartheta(0)$, are assumed to be very uncertain as compared to the uncertainty in the measurement $y(1)$, specified by $R(1)$. Consequently,

$$\begin{aligned}
\Phi(1)\widehat{\vartheta}(1) &= \Phi(1)\left\{\Xi\vartheta(0) + K(1)\left[y(1) - \Phi(1)\Xi\vartheta(0)\right]\right\} \\
&= \Phi(1)\left\{\Xi\vartheta(0) + \left(\Xi P(0)\Xi^T + \Pi Q(0)\Pi^T\right)\Phi(1)^T \right. \\
&\quad \times \left[\Phi(1)\left(\Xi P(0)\Xi^T + \Pi Q(0)\Pi^T\right)\Phi^T(1) + R(1)\right]^{-1} \\
&\quad \left. \times \left[y(1) - \Phi(1)\Xi\vartheta(0)\right]\right\} \\
&\simeq \Phi(1)\Xi\vartheta(0) + \left[y(1) - \Phi(1)\Xi\vartheta(0)\right] \\
&= y(1)
\end{aligned} \tag{7.36}$$

[1] System *controllable* $\Leftrightarrow$ $\text{rank}([\Pi, \Xi\Pi, \ldots, \Xi^{p-1}\Pi]) = p$; system *observable* $\Leftrightarrow$ $\text{rank}([\Phi, \Phi\Xi, \ldots, \Phi\Xi^{p-1}]) = p$; for details, see [KS72, GGS01]

Hence, the estimate $\widehat{\vartheta}(1)$ will be fully determined by the first measurement $y(1)$, and the influence of $\vartheta(0)$ on $\widehat{\vartheta}(1)$ is negligible. Notice that setting $Q(0)$ very large will lead to similar results.

Remark 7.4 In the case of a perfect observation, set $R(t) = 0$. Then, again $\Phi(t)\widehat{\vartheta}(t) \simeq y(t)$. On the contrary, in the case of an unreliable observation, let $R(t) \to \infty$. Consequently, $K(t) \to 0$, and thus,

$$\widehat{\vartheta}(t) = \widehat{\vartheta}(t|t-1) \tag{7.37}$$

In other words, the unreliable observation $y(t)$ does not at all affect the estimate $\widehat{\vartheta}(t)$. Hence, there is no measurement update.

Remark 7.5 In general, estimator performance is more sensitive to structured modeling errors in Ξ, Π, and Φ than to uncertainty model errors in $P(0)$, Q, and R. If, however, we know in advance that the observation noise is structured, then *state augmentation* should be used to incorporate this structure in the model. Let us illustrate this by a simple example.

Example 7.6 Mean tracking: Let the constant ϑ be observed with structured observation noise $v_s(t)$, so that

$$y(t) = \vartheta + v_s(t)$$

with

$$v_s(t) = \phi v_s(t-1) + w_s(t-1)$$

Then, in state-space form,

$$\begin{bmatrix} \vartheta(t) \\ v_s(t) \end{bmatrix} = \begin{bmatrix} 1 & 0 \\ 0 & \phi \end{bmatrix} \begin{bmatrix} \vartheta(t-1) \\ v_s(t-1) \end{bmatrix} + \begin{bmatrix} 0 \\ w_s(t-1) \end{bmatrix}$$
$$y(t) = [1\ 1] \begin{bmatrix} \vartheta(t) \\ v_s(t) \end{bmatrix}$$

which is indicated as *state augmentation*. Hence, the approach is to put all the dynamics in process, sensors, and actuators in the state equation and subsequently apply the estimator (7.29)–(7.33).

Let us now further evaluate some properties of the estimator on a real-world example.

Example 7.7 Respiration rate data: Consider the measurements of the respiration rates in an activated sludge plant (see Fig. 7.6).

Let the ultimate goal be to reconstruct the noise-free respiration rates from these measurements. Therefore, we formulate the following state-space model:

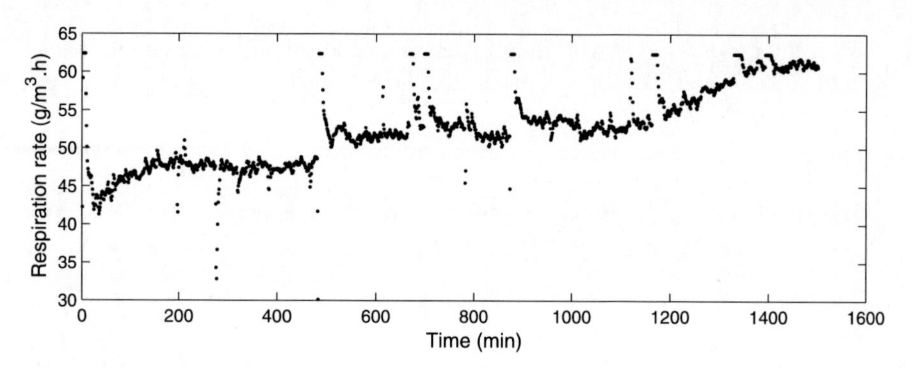

Fig. 7.6 Measured respiration rates

$$\vartheta(t) = \vartheta(t-1) + w(t-1)$$
$$y(t) = \vartheta(t) + v(t)$$

where ϑ is the noise-free respiration rate, w are the unknown variations with respect to the unknown mean value, y are the measured respiration rates, and v is the observation or sensor noise. Let us first investigate the effect of R on the performance of the estimator and thus on the estimated values of the respiration rates. The results for $\vartheta(0) = 0$ with $P_0 = 1000$, $Q(t) = 0$ for all t, and $R(t) = 1$ initially, which after 500 samples is set to 10^{-6} and after 1000 samples to $R(t) = 10^6$, are presented in Fig. 7.7.

Notice from Fig. 7.7 that initially the estimated value jumps to a value close to the first measurements and then settles. Further, the variance is drastically decreased after some measurements. Then, after 500 samples when $R(t)$ becomes very small, and thus each of the following measurements is taken very seriously, the estimates follow most of the dynamics present in the measurements. While by setting $R(t)$ very large after 1000 samples, the estimated value just remains constant, and thus it is not updated by the measurements. Finally, the effect of P_0, the initial error covariance matrix, in this case just a scalar that is set on 1000, 1, or 0.1, on the estimated respiration rates can be seen in Fig. 7.8. Clearly, setting $P(0)$ large will lead to a fast adjustment of the estimates to the measurements. Hence, as a rule of thumb mentioned before, in practice we always choose $P(0)$ large.

7.1.4 Resemblance with Kalman Filter

Consider the following linear, discrete-time state-space model (see Sect. 1.2.2) with system noise $w(t)$ and observation or sensor noise $v(t)$:

$$\begin{aligned} x(t) &= A(t)x(t-1) + B(t)u(t-1) + G(t)w(t-1) \\ y(t) &= C(t)x(t) + v(t), \quad t \in \mathbb{Z}^+ \end{aligned} \tag{7.38}$$

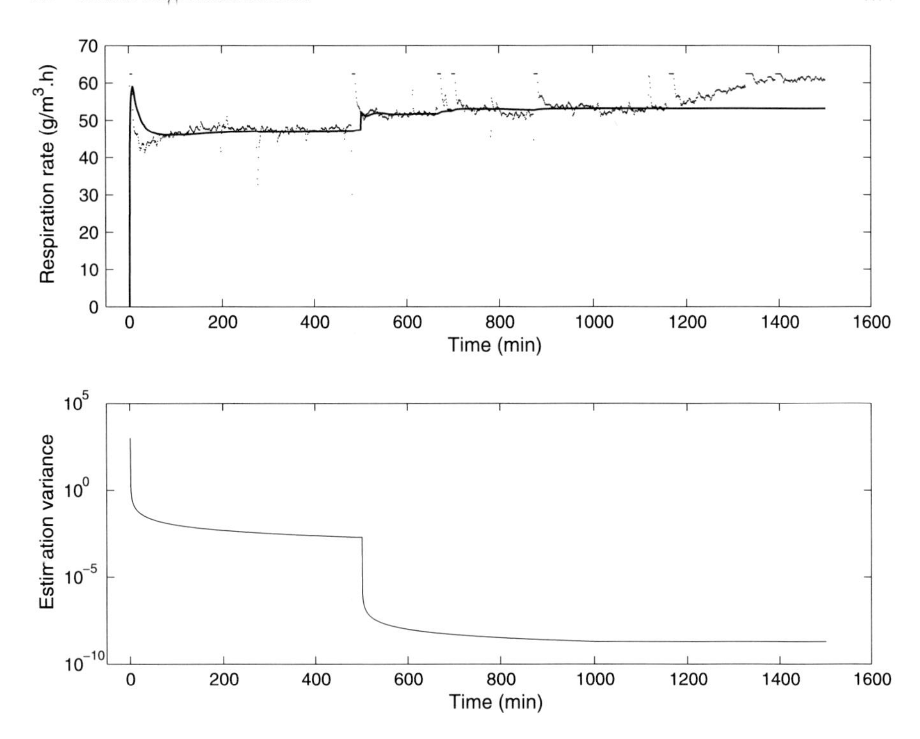

Fig. 7.7 Measured respiration rates (*dots*) with their estimates (*solid line*) (*top figure*) and variances (*bottom figure*)

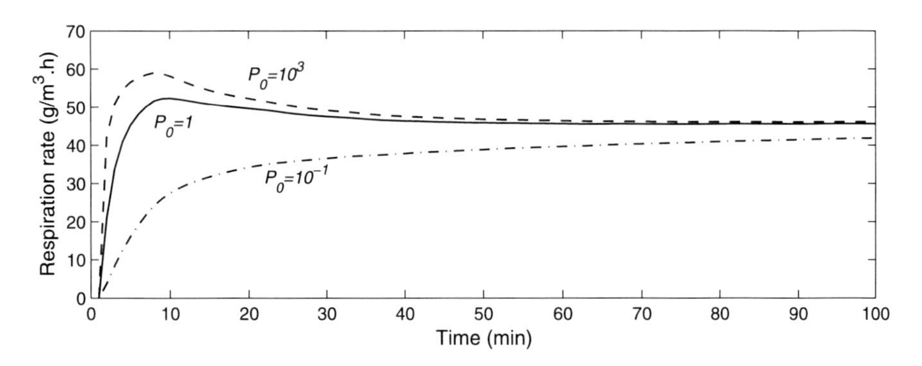

Fig. 7.8 Effect of P_0 on estimates

where all the vectors and matrices have appropriate dimensions. Further, assume again that $w(t)$ and $v(t)$ are zero-mean, statistically independent, white noise terms with $\text{Cov}\, w(t) = Q(t)$ and $\text{Cov}\, v(t) = R(t)$. Under these assumptions, the well-known *Kalman filter* related to (7.38) reads as follows.

Prediction:

$$\widehat{x}(t|t-1) = A(t)\widehat{x}(t-1) + B(t)u(t-1) \tag{7.39}$$

$$P(t|t-1) = A(t)P(t-1)A(t)^T + G(t)Q(t-1)G(t)^T \tag{7.40}$$

Correction:

$$K(t) = P(t|t-1)C(t)^T\big[C(t)P(t|t-1)C^T(t) + R(t)\big]^{-1} \tag{7.41}$$

$$\widehat{x}(t) = \widehat{x}(t|t-1) + K(t)\big[y(t) - C(t)\widehat{x}(t|t-1)\big] \tag{7.42}$$

$$\begin{aligned} P(t) = {} & \big(I - K(t)C(t)\big)P(t|t-1)\big(I - C(t)^T K(t)^T\big) \\ & + K(t)R(t)K(t)^T \end{aligned} \tag{7.43}$$

Hence, the recursive estimator described by (7.29)–(7.33) shows a great resemblance with these filter equations. More specifically, by setting $x = \vartheta$, $A(t) = \varXi$, $B(t) = 0$, $G(t) = \varPi$, and $C(t) = \varPhi(t)$ the resemblance becomes more clear. However, the essential difference is that the Kalman filter has initially been derived for state estimation, while the recursive estimator in the previous sections is dedicated to parameter estimation. Notice then that, for linear (time-invariant) regression-type models as (7.16), the observation matrix in a state-space representation becomes time-varying, that is, $C(t) = \varPhi(t)$, unlike the Kalman filter for state estimation in linear time-invariant systems. From this point of view, the concept of time-varying parameters as (unobserved) state variables becomes very transparent! Consequently, the mean tracking problem could also have been seen as a state estimation problem for a linear, static model.

7.1.5 *Numerical Issues

For implementation of the Kalman filter and the recursive least-squares parameter estimator in practical situations, some modifications of the equations are needed. First, by definition the covariance matrix $P(t)$ is symmetric. However, due to the asymmetric form of the Kalman gain expression in (7.42), $P(t)$ may become asymmetric. From this perspective the application of the error covariance matrix expression as in (7.33), i.e., in Joseph's form, is much more preferred than the simplified version $P(t) = [I - K(t)\varPhi(t)]P(t|t-1)$, which is the more general matrix counterpart of (7.14). A simple remedy could be to mirror the upper triangular part at each time instant so that $P(t) = P(t)^T$. A more advanced technique uses the modified Choleski or *UD-decomposition* of P,

$$P = LDL^T \tag{7.44}$$

where L is a lower triangular matrix with only ones along the principal diagonal, and D is a diagonal matrix. In some numerical schemes the decomposition $P = LL^T$, where the nonunique lower triangular matrix L can be interpreted as the square root of the error covariance matrix, is used to maintain symmetry of $P(t)$ at each time instant. Secondly, the condition number of $P(t)$, that is, the quotient of maximum and minimum eigenvalues, may become very large when accurate measurements are used. This often leads to negative eigenvalues of $P(t)$, which in turn usually causes instability of the filter algorithm. The use of factorization methods often prevents the occurrence of large condition numbers and thus of negative eigenvalues. For this purpose, the most well-known factorization methods are eigenvalue decomposition, singular value decomposition, and, again, UD decomposition. In particular, the last decomposition method has been widely used in Kalman filtering problems, leading to the so-called *square root filter*, which is more robust than the original Kalman filter. Using square roots reduces the range of number magnitudes, and thus the computation becomes less sensitive to rounding errors. In the square root filtering algorithm (see [May79]), using the state-space representation of (7.38) with constant covariance matrices Q and R, the prediction and correction of L, together with the expression for the Kalman gain, are given by

$$L(t|t-1) = \left[A(t)L(t-1) \vdots G(t)Q^{\frac{1}{2}}\right]\mathscr{U}(t) \tag{7.45}$$

$$\begin{aligned} K(t) &= L(t|t-1)L(t|t-1)^T C(t)^T \\ &\quad \times \left[C(t)L(t|t-1)L(t|t-1)^T C(t)^T + R\right]^{-1} \end{aligned} \tag{7.46}$$

$$\begin{aligned} L(t) &= L(t|t-1) \\ &\quad \times \left[I - L(t|t-1)^T C(t)^T \mathscr{V}(t)^{-T}\left(\mathscr{V}(t) + R^{\frac{1}{2}}\right)^{-1} \right. \\ &\quad \left. \times\, C(t)L(t|t-1)\right] \end{aligned} \tag{7.47}$$

where $\mathscr{U}(t)$ is an orthogonal matrix such that the last m rows of $A(t)L(t-1)$ become zero, using, for example, the Modified Gram-Schmidt procedure, and furthermore $\mathscr{V}(t)\mathscr{V}(t)^T = C(t)L(t|t-1)L(t|t-1)^T C(t)^T + R$.

Algorithm 7.2 Square root filtering for the estimation of $x(t)$ in (7.38)

1. Given input–output data $u(t)$ and $y(t)$ for $t = 1, \dots, N$ and the state-space matrices $A(t)$, $B(t)$, $C(t)$, and $G(t)$, specify the covariance matrices $Q(t)$ and $R(t)$.
2. Choose the initial state vector $x(0)$ and the initial square root of the error covariance matrix $L(0)$.
3. Evaluate, for $t = 1, \dots, N$, (7.39), (7.45)–(7.47), (7.42).

Let us illustrate the effect of accurate measurements on the estimation result by the following example (after [Gel74]).

Example 7.8 Square root filter: Consider the recursive estimation of two unknowns from a single measurement. Let $P(t) = I$, $C = [1\ 0]$, and $R = \varepsilon^2$, where $\varepsilon \ll 1$. To simulate computer word length roundoff, it is assumed that $1 + \varepsilon \neq 1$, but $1 + \varepsilon^2 \simeq 1$.

Then, the exact value of $P(t+1|t)$ is found from $P(t+1|t) = \begin{bmatrix} \frac{\varepsilon^2}{1+\varepsilon^2} & 0 \\ 0 & 1 \end{bmatrix}$, whereas the value calculated in the computer using the standard Kalman filter algorithm gives $P(t+1|t) = \begin{bmatrix} 0 & 0 \\ 0 & 1 \end{bmatrix}$. Using the square root filter algorithm gives $P(t+1|t) = \begin{bmatrix} \varepsilon^2 & 0 \\ 0 & 1 \end{bmatrix}$. Since $K(t+1) = P(t+1|t)CR^{-1}$, it follows that

$$K(t+1) =_{[\text{exact}]} \begin{bmatrix} \frac{1}{1+\varepsilon^2} \\ 0 \end{bmatrix}$$

$$K(t+1) =_{[\text{conventional}]} \begin{bmatrix} 0 \\ 0 \end{bmatrix}$$

$$K(t+1) =_{[\text{squareroot}]} \begin{bmatrix} 1 \\ 0 \end{bmatrix}$$

Obviously, the conventional Kalman filter algorithm may lead to divergence problems.

Clearly, the price for a more accurate and robust result is a significant increase of the number of calculations. Although square root algorithms are more robust than the standard Kalman filter, they are, in general, not more efficient, and therefore the algorithm presented above cannot be directly used for large-scale models.

The so-called reduced-rank square root (RRSQRT) filter is a special formulation of the Kalman filter or, more specifically, of the square root filter for assimilation of data in large-scale models. In most large-scale applications the time update of the error covariance matrix ($P(t)$) is the most problematic part. The number of operations needed for a time update of $P(t)$ is of order $O(n^2)$. In the RRSQRT filter the covariance matrix is expressed in a small number of modes, stored in a lower-rank square root matrix. The algorithm includes a reduction step that reduces the number of modes if it becomes too large in order to ensure that the problem is feasible. When different scales in the model are considered, some sort of normalization of the square root matrix is required in the reduction step. The approximated error covariance matrix is found by a truncated eigenvalue decomposition. The optimal rank q approximation of a positive semi-definite symmetric matrix is given by a projection onto the q leading eigenvectors. The smaller rank can be exploited to reduce both the computational burden of the Kalman filter and the memory requirements.

In particular, for constant Q and R, the following steps in the algorithm can be distinguished.

Prediction:

$$\widehat{x}(t|t-1) = A(t)\widehat{x}(t-1) + B(t)u(t-1) \tag{7.48}$$

$$L(t|t-1) = \left[A(t)L(t-1) \vdots G(t)Q^{\frac{1}{2}}\right] \tag{7.49}$$

Reduction:

$$L(t|t-1)^T L(t|t-1) = U(t)D(t)U(t)^T \tag{7.50}$$

$$L^*(t|t-1) = \left[L(t|t-1)U(t)\right]_{1:n,1:q} \tag{7.51}$$

Correction:

$$H(t) = L^*(t|t-1)^T C(t)^T \tag{7.52}$$

$$\beta(t) = \left[H(t)^T H(t) + R\right]^{-1} \tag{7.53}$$

$$K(t) = L^*(t|t-1)H(t)\beta(t) \tag{7.54}$$

$$\widehat{x}(t) = \widehat{x}(t|t-1) + K(t)\left[y(t) - C(t)\widehat{x}(t|t-1)\right] \tag{7.55}$$

$$L(t) = L^*(t|t-1) - K(t)H(t)^T\left[1 + \left(\beta(t)R\right)^{\frac{1}{2}}\right]^{-1} \tag{7.56}$$

Algorithm 7.3 Reduced-rank square root (RRSQRT) filtering for the estimation of $x(t)$ in (7.38)

1. Given input–output data $u(t)$ and $y(t)$ for $t = 1, \ldots, N$ and the state-space matrices $A(t)$, $B(t)$, $C(t)$, and $G(t)$, specify the *constant* covariance matrices Q and R.
2. Choose the initial state vector $x(0)$ and the initial square root of the error covariance matrix $L(0)$.
3. Evaluate, for $t = 1, \ldots, N$, (7.48)–(7.56).

In the next section, the recursive estimation theory will be applied to nonlinear static systems.

7.2 Nonlinear Static Systems

7.2.1 State-space Representation

In what follows, only simple dynamic parameter models, that is, either the constant or the random walk parameter model will be considered. Then, a nonlinear regression model with possibly time-varying parameters can be cast in the state-space framework as follows:

$$\begin{aligned} \vartheta(t) &= \vartheta(t-1) + w(t-1) \\ y(t) &= h\big(\phi(t), \vartheta(t)\big) + v(t) \end{aligned} \tag{7.57}$$

where $h(\phi(t), \vartheta(t))$ is a vector function relating the explanatory variables in $\phi(t)$ to the output vector $y(t)$, and ϑ contains all unknown parameters that have to be estimated from the available data. Let us illustrate this to a moving vehicle example.

Example 7.9 Moving vehicle: Consider a moving vehicle which is equipped with a differential global positioning system (DGPS) receiver and a radar velocity sensor. According to the kinetic law, the position at time instant t in both the x- and y-directions can be described by the linear algebraic equation

$$s(t) = s_0 + vt + \frac{1}{2}at^2$$

where $s(t)$ is the position (m), v is the velocity (m/s), and a is the acceleration (m/s^2). The radar velocity is a composition of both velocities in the x- and y-directions, that is, $v_{\text{radar}} = \sqrt{v_x^2 + v_y^2}$. Assuming zero acceleration and setting $s_0 = 0$, so that velocities in both directions, v_x and v_y, are the only unknowns. A discrete-time state-space representation of this system is given by

$$\begin{bmatrix} v_x(t) \\ v_y(t) \end{bmatrix} = \begin{bmatrix} v_x(t-1) \\ v_y(t-1) \end{bmatrix} + w(t-1)$$

$$\begin{bmatrix} y_x(t) \\ y_y(t) \\ y_v(t) \end{bmatrix} = \begin{bmatrix} v_x(t)t \\ v_y(t)t \\ \sqrt{v_x(t)^2 + v_y(t)^2} \end{bmatrix} + v(t)$$

where the system and measurement noise consists of two, respectively three, statistically independent white noise terms. Notice that only one nonlinear term appears due to the measured radar velocity.

A common approach to nonlinear estimation problems is to linearize the set of equations. In recursive estimation schemes one usually linearizes around the currently available estimate, so that (7.57) is approximated by

$$\begin{aligned} \Delta\vartheta(t) &= \Delta\vartheta(t-1) + w(t-1) \\ \Delta y(t) &= H(t)\Delta\vartheta(t) + v(t) \end{aligned} \tag{7.58}$$

where $w(t)$ and $v(t)$ are now noise terms related to perturbations in the trajectories of $\vartheta(t)$ and $y(t)$. The Jacobi matrix $H(t) = H(\phi, \widehat{\vartheta})$, where its elements h_{ij} are defined in the following manner:

$$h_{ij} := \left[\frac{\partial h_i(\phi(t), \widehat{\vartheta}(t-1))}{\partial \widehat{\vartheta}_j(t-1)}\right] \quad \text{for } i, j = 1, 2, \ldots \tag{7.59}$$

Hence, the Jacobi matrix contains all the partial differential coefficients of the vector function $h(\phi, \vartheta)$ with respect to all p elements in the last estimated parameter vector $\widehat{\vartheta}(t-1)$.

Example 7.10 Moving vehicle: The linearized set of state-space equations related to the moving vehicle problem simply becomes

$$\begin{bmatrix} \Delta v_x(t) \\ \Delta v_y(t) \end{bmatrix} = \begin{bmatrix} \Delta v_x(t-1) \\ \Delta v_y(t-1) \end{bmatrix} + w(t-1)$$

$$\begin{bmatrix} \Delta y_x(t) \\ \Delta y_y(t) \\ \Delta y_v(t) \end{bmatrix} = \begin{bmatrix} t(t-1) & 0 \\ 0 & t(t-1) \\ \frac{v_x(t-1)}{\sqrt{v_x(t-1)^2+v_y(t-1)^2}} & \frac{v_y(t-1)}{\sqrt{v_x(t-1)^2+v_y(t-1)^2}} \end{bmatrix} \times \begin{bmatrix} \Delta v_x(t) \\ \Delta v_y(t) \end{bmatrix} + v(t)$$

which is a linear, discrete-time, time-varying state-space representation in perturbation variables.

In the following, the *Extended Kalman Filter* (EKF) algorithm, based on the linearized state-space representation, for the static, nonlinear case with unknown parameter vector ϑ will be presented.

7.2.2 Extended Kalman Filter

Since the EKF is based on the Kalman filter, in principle, a similar prediction-correction structure of the algorithm as in (7.39)–(7.43) will be used to present the EKF algorithm.

Prediction:

$$\widehat{\vartheta}(t|t-1) = \widehat{\vartheta}(t-1) \tag{7.60}$$

$$P(t|t-1) = P(t-1) + Q(t-1) \tag{7.61}$$

Correction:

$$K(t) = P(t|t-1)H(t)^T\big[H(t)P(t|t-1)H^T(t) + R(t)\big]^{-1} \tag{7.62}$$

$$\widehat{\vartheta}(t) = \widehat{\vartheta}(t|t-1) + K(t)\big[y(t) - h\big(\phi(t), \widehat{\vartheta}(t-1)\big)\big] \tag{7.63}$$

$$P(t) = \big(I - K(t)H(t)\big)P(t|t-1)\big(I - H(t)^T K(t)^T\big) + K(t)R(t)K(t)^T \tag{7.64}$$

Notice that the calculation of the innovations, and thus the update of $\widehat{\vartheta}$, is fully based on the nonlinear relationship using the currently available estimate, that is, $\varepsilon(t) = y(t) - h(\phi(t), \widehat{\vartheta}(t-1))$. Hence, the linearization step is only needed for the calculation of $K(t)$ and the update of $P(t)$, which is in fact a first-order variance propagation step. Consequently, for this type of application related to nonlinear static systems, for the linearization, it just suffices to compute the Jacobi matrix $H(t)$ at each time instant.

To summarize, the Extended Kalman Filter is given by the next algorithm.

Algorithm 7.4 Extended Kalman filtering for the estimation of $\vartheta(t)$ in a static nonlinear system

Table 7.2 Moving vehicle data

Time t (s)	0	0.2	0.4	0.6	0.8	1.0	1.2	1.4	1.6	1.8	2.0
x (m)	–0.01	0.10	0.15	0.23	0.32	0.37	0.48	0.57	0.59	0.75	0.83
y (m)	0.01	0.07	0.07	0.11	0.16	0.21	0.21	0.23	0.35	0.36	0.37
Radar velocity	0.41	0.41	0.44	0.44	0.40	0.49	0.41	0.44	0.43	0.41	0.49

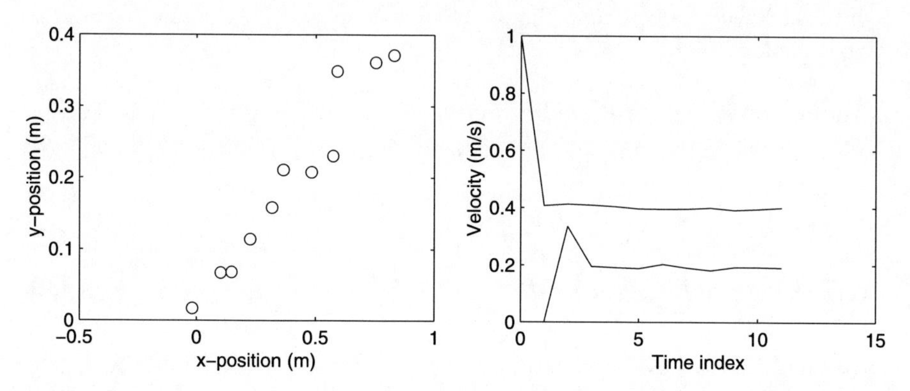

Fig. 7.9 Measured positions (*left figure*) and velocity estimates (*right figure*, with $v_x(0) = 1$ and $v_y(0) = 0$)

1. Given input–output data $u(t)$ and $y(t)$ for $t = 1, \dots, N$ and the nonlinear state-space representation (7.57), specify the covariance matrices $Q(t)$ and $R(t)$.
2. Choose the initial parameter vector $\vartheta(0)$ and the initial error covariance matrix $P(0)$.
3. Evaluate, for $t = 1, \dots, N$, (7.60)–(7.64).

Example 7.11 Moving vehicle: Let the following data (see Table 7.2) be available for estimation of both v_x and v_y.

The position in x- and y-coordinates is also presented in Fig. 7.9, which indicates that the vehicle is moving along a straight line. Hence, the assumption that the acceleration is zero appears to be valid.

The estimated velocities under the assumption that $\vartheta(0) = \left[\begin{smallmatrix} v_x(0) \\ v_y(0) \end{smallmatrix}\right] = \left[\begin{smallmatrix} 1 \\ 0 \end{smallmatrix}\right]$ with $P_0 = 1000I$, $Q(t) = 0$, and $R(t) = 0.1I$ for all t are presented in Fig. 7.9.

In a second experiment, where the acceleration in x-direction is equal to 1 m/s^2 while keeping the acceleration in y-direction zero, the data presented in Table 7.3 have been generated.

In order to allow more time variation in the velocity estimates, the covariance matrix related to the systems noise $Q(t)$ is set to $0.1\,I$ for all t. The noise-corrupted position (y) and estimates are presented in Fig. 7.10.

Clearly, the estimated velocity in x-direction shows a trajectory that tends to a constant increase of the velocity with time, which is obviously related to the acceleration in this direction. In a next step, the parameter vector could therefore be

Table 7.3 Moving vehicle data with acceleration

Time t (s)	0	0.2	0.4	0.6	0.8	1.0	1.2	1.4	1.6	1.8	2.0
x (m)	0.05	0.07	0.25	0.42	0.68	0.93	1.20	1.49	1.95	2.33	2.81
y (m)	0.01	0.07	0.07	0.11	0.16	0.21	0.21	0.23	0.35	0.36	0.37
Radar velocity	0.41	0.41	0.44	0.44	0.40	0.49	0.41	0.44	0.43	0.41	0.49

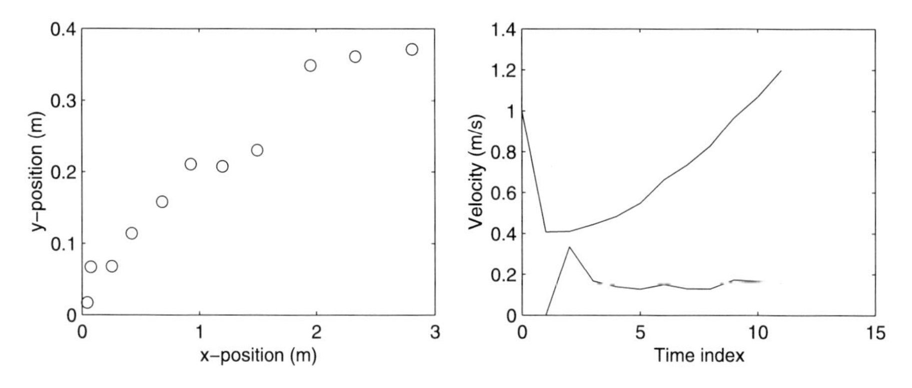

Fig. 7.10 Measured positions (*left figure*) and velocity estimates (*right figure*, with $v_x(0) = 1$ and $v_y(0) = 0$) for $a_x = 1$ m/s^2

extended to include an unknown acceleration, where an initial guess can be obtained from the slope of v_x in Fig. 7.10. It is well known that in the EKF algorithm the parameter estimation error covariance matrix is usually rather poorly estimated. Therefore, the estimates of $P(t)$, especially in highly nonlinear cases, should be handled carefully.

7.3 Historical Notes and References

For an overview of recursive least-squares (RLS) estimation techniques and their implementations, we refer to [Gel74, LS83, You84, Tur85, Ver89]. For recursive estimation of statistical parameters, as introduced in the first sections of Chap. 7, see the historical papers of [Sak65, Whi70, Maj73].

Shortly after the introduction of the Kalman filter [Kal60, KB61], Mayne [May63] and Lee [Lee64] were among the first who pointed out the link between parameter and state estimation, so that the resemblance between the Kalman filter and the recursive least-squares parameter estimator became very clear. In this interpretation, parameters are seen as time-varying unobserved states, see also [Kau69, Che70]. However, apart from the unbiased, minimum variance concept, as used in this book for the derivation of the Kalman filter, many other derivations have been presented in literature as well. For instance, in the derivation, a Bayesian framework, orthogonal projection, and dynamic programming concepts have also been used, see Sorenson [Sor85] for an overview on this.

For time-varying parameter tracking, as an alternative to dynamic parameter modeling, a forgetting factor [SS83, You84, BBB85, BC94] or covariance resetting [SGM88] in the recursive algorithm has been used.

Especially for real-time implementation and for large-scale systems, the numerical implementation of the estimator becomes important; for details, see [Gel74, Bie77, May79, GVL89]. In particular, the square root filter [Car73, Pet75, MV91b, Car90] has been introduced, while nowadays in data assimilation studies the reduced-rank square root filter is popular, see [VH97, BEW02, TM03, CKBR08].

For nonlinear estimation problems, Jazwinski [Jaz70] introduced the so-called Extended Kalman Filter, usually abbreviated to EKF. However, in case the model is highly nonlinear, the extended Kalman filter may not always give reliable results, basically because the mean and covariance are propagated through linearization of the underlying nonlinear model. As an alternative to the EKF, the unscented Kalman filter (UKF) [JU97] has been introduced. Instead of linearization, the UKF uses a deterministic sampling technique to pick sample points around the mean. Via simple simulation-based propagation, these sample points are used to recover the mean and covariance of the estimate. In addition to this, in general, application of (7.60)–(7.64) will not guarantee stability of the algorithm. Therefore, a somewhat modified EKF, based on regularization theory [BRD97, RU99], has been suggested in literature.

7.4 Problems

Problem 7.1 Consider again the discrete-time system of Problem 6.1,

$$G(q) = \frac{0.2q^{-2}}{1 - 0.8q^{-1}}$$

and repeat Problem 6.1a–c to generate synthetic noisy data.

(a) Recursively estimate the parameter values using the MATLAB function *roe*. Set adm = 'kf' and adg = a∗eye(2) with a = 0, 0.1, 10. Evaluate the recursive estimation results.
(b) Repeat this procedure, but now by varying the initial covariance matrix of the estimates, that is choose P0 = b∗eye(2) with b = 1, 100, 1e6. Evaluate again the recursive estimation results.

Problem 7.2 Let us compare the results from a recursive parameter estimation and a state estimation for the moving object example, i.e., an object moving in a straight line with constant velocity (Example 5.2). Notice that this process can be described by kinetic and dynamic models. Thus,

$$s(t) = s_0 + vt \tag{7.65}$$

or

$$ds(t)/dt = v \tag{7.66}$$

(a) Define a discrete-time state-space model, including the noise terms with their stochastic characterization, for the case that both *parameters* s_0 and v have to be estimated recursively from the measured output data. HINT: start with the (kinematic model) equation (7.65) as output equation and add two discrete-time state (difference) equations to describe the expected changes of the parameters s_0 and v. Assume that s_0 is constant and v is a random walk process due to unmodeled accelerations.
(b) Recursively estimate, using a standard Kalman filter, the position of the object under the assumption that the parameters are completely unknown, for example, $s_0 = v = 0$ with $P_0 = 10^3 I$. Hence, first estimate the trajectories of s0 and v and subsequently use (7.65) to calculate the position at each time instant. Evaluate the result.
(c) Use the results from (a) as prior knowledge to obtain new recursive estimates of s_0 and v. Evaluate the result.
(d) Vary the diagonal elements in the covariance matrices P_0, R, and Q to obtain some feeling of their effect on the final estimation result. Evaluate the result.
(e) Using the previous results give an appropriate prediction (including the prediction uncertainty) of the position at time instant $t = 25$ s.
(f) Instead of using first a parameter estimation step, one could also directly estimate the position (state) on the basis of model (7.66) and a standard Kalman filter. Formulate a discrete-time state-space model for this state estimation problem under the assumption that, for example, $v = 4$ m/s.
(g) Repeat the steps in (b)–(e) for this new model.
(h) Compare the results from both approaches.

Chapter 8
Time-varying Dynamic Systems Identification

8.1 Linear Dynamic Systems

8.1.1 Recursive Least-squares Estimation

In Chap. 7, a time-varying static system representation has been introduced for recursive estimation of parameters. In this representation, state dynamics were not considered. Thus, in addition to the algebraic output equation, it contains only differential or difference equations related to the possible dynamics in the parameter estimates. In this chapter, the idea of recursive estimation of the model parameters is further developed for the estimation of unknowns in a dynamic system. Let us start with an example that illustrates how to estimate inputs and parameters in a continuous-time linear dynamic system. In this particular example, the process dynamics are described by *piece-wise linear* differential equations with *piece-wise constant* inputs. Consequently, given the explicit solution of the differential equations (see footnote 1 in Chap. 1), a time-varying static system representation results, and thus the algorithms of Chap. 7 can be applied directly.

Example 8.1 NH_4/NO_3 dynamics in pilot plant Bennekom (based on [LKvS99]): The layout of the pilot-activated sludge plant (ASP) is presented in Fig. 8.1.

In the alternating (anoxic/aerobic) reactor the air flow is manipulated by a dissolved oxygen controller (DO-ctrl) that receives its alternating set-point from a higher-level nitrogen controller (N-ctrl). Furthermore, the amount of activated sludge is regulated by a sludge controller (X-ctrl). The NH_4/NO_3 dynamics in alternating ASPs on the time scale of hours can be explained by only three processes: reactor's influent load, nitrification, and denitrification. Hence, the combined NH_4/NO_3 balances in alternating aerated reactors can be modeled as

$$\begin{bmatrix} \frac{dNH_4}{dt} \\ \frac{dNO_3}{dt} \end{bmatrix} = \frac{-q^{in}}{V} \begin{bmatrix} NH_4 \\ NO_3 \end{bmatrix} + \begin{bmatrix} -r_{NH} \\ r_{NH} + r_{NO} \end{bmatrix} u + \begin{bmatrix} \frac{q^{in}}{V} NH_4^{in} \\ -r_{NO} \end{bmatrix} \tag{8.1}$$

K.J. Keesman, *System Identification*,
Advanced Textbooks in Control and Signal Processing,
DOI 10.1007/978-0-85729-522-4_8,

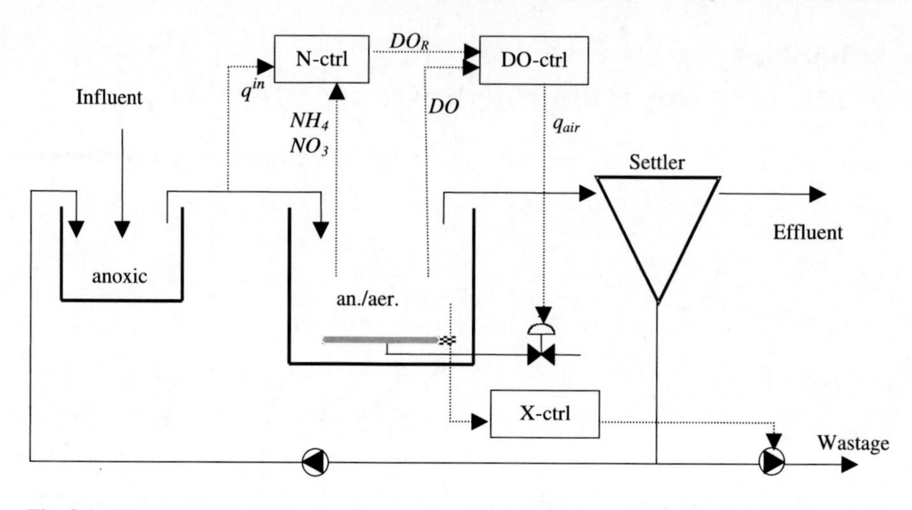

Fig. 8.1 Pilot activated sludge plant layout

$$r_{\mathrm{NH}} = \begin{cases} r_{\mathrm{NH,max}} & \text{if } \mathrm{NH}_4 > 0 \\ \frac{q^{\mathrm{in}}}{V}\mathrm{NH}_4^{\mathrm{in}} & \text{if } \mathrm{NH}_4 = 0 \end{cases} \tag{8.2}$$

$$r_{\mathrm{NO}} = \begin{cases} r_{\mathrm{NO,max}} & \text{if } \mathrm{NO}_3 > 0 \\ 0 & \text{if } \mathrm{NO}_3 = 0 \end{cases} \tag{8.3}$$

$$\begin{bmatrix} y_1(t) \\ y_2(t) \end{bmatrix} = \begin{bmatrix} \mathrm{NH}_4(t-\tau) \\ \mathrm{NO}_3(t-\tau) \end{bmatrix} \tag{8.4}$$

where q^{in} is the influent flow, V is the reactor volume, $r_{\cdot,\mathrm{max}}$ is the maximum consumption rate of NH_4 or NO_3, respectively. Furthermore, τ is the measurement time delay, and $u \in \{0, 1\}$, thus, u is "off" or "on," i.e., in Fig. 8.1, $\mathrm{DO}_R = 0$ (anoxic) or $\mathrm{DO}_R = 3$ mg/l (aerobic, no DO limitation). In particular we define

$$\vartheta := \left[\mathrm{NH}_4^{\mathrm{in}}\ r_{\mathrm{NH,max}}\ r_{\mathrm{NO,max}}\right]^T$$

Using the random walk parameter model, (7.22), using (8.1)–(8.4) and after eliminating the state variables NH_4 and NO_3 to arrive at an equivalent discrete-time system, the following state-space model is obtained:

$$\vartheta(t+1) = \vartheta(t) + w(t)$$

$$y(t+1) - \mathrm{e}^{\frac{-q^{\mathrm{in}}(t-\tau)}{V}T} y(t) = X(t)\vartheta(t) + v(t)$$

where $\vartheta \in \mathbb{R}^p$, $y \in \mathbb{R}^n$, $w \in \mathbb{R}^p$, $v \in \mathbb{R}^n$; in this application, $p = 3$ and $n = 2$. Furthermore,

Table 8.1 Jacobi matrix elements in different operating modes

	$\{y(t+1), y(t)\} > 0$	$\{y_1(t+1), y_1(t)\} = 0$	$\{y_2(t+1), y_2(t)\} = 0$
$X_{11}(t)$	$\frac{q^{\text{in}}(t-\tau)}{V}$	0	$\frac{q^{\text{in}}(t-\tau)}{V}$
$X_{12}(t)$	$-u(t-\tau)$	0	$-u(t-\tau)$
$X_{13}(t)$	0	0	0
$X_{21}(t)$	0	$\frac{q^{\text{in}}(t-\tau)}{V}$	0
$X_{22}(t)$	$u(t-\tau)$	0	0
$X_{23}(t)$	$u(t-\tau)-1$	$u(t-\tau)-1$	0

$$X(t) = \frac{\partial \widehat{y}(t+1|t)}{\partial \vartheta}$$
$$= \frac{1 - \mathrm{e}^{\frac{-q^{\text{in}}(t-\tau)}{V} T_s}}{\mathrm{e}^{\frac{-q^{\text{in}}(t-\tau)}{V}}} \times \begin{bmatrix} X_{11}(t) & X_{12}(t) & X_{13}(t) \\ X_{21}(t) & X_{22}(t) & X_{23}(t) \end{bmatrix}$$

where T_s is the sampling interval (5 min). The elements $X_{11}(t), \ldots, X_{23}(t)$ are defined in Table 8.1.

Consequently, the piece-wise linear dynamic system with time-varying parameters has been cast into the framework of a time-varying static system. In particular, (7.28) with $\Xi = I$, $\Pi = I$, $\Phi = X$, and output $y(t+1) - \mathrm{e}^{\frac{-q^{\text{in}}(t-\tau)}{V} T_s} y(t)$. Hence, under the assumption that $w(t)$ and $v(t)$ are white, the recursive estimator of (7.29)–(7.33) can be used to estimate the unknowns in ϑ.

Experimental data has been collected from the alternating aerated pilot scale ASP with continuously mixed aeration tank ($V = 475$ l, Mixed Liquor Suspended Solids ("biomass") MLSS = 3.5 g/l, pH = 7) continuously fed with presettled municipal waste water. The inputs, DO and q^{in}, and measured output data, NH_4 and NO_3, are shown in Figs. 8.2 and 8.3.

The tuning matrices of the recursive least-squares estimator $P(0)$, Q, and R are set to

$$P(0) = 10^6 I$$
$$Q = \begin{bmatrix} 10 & 0 & 0 \\ 0 & 5 \times 10^{-7} & 0 \\ 0 & 0 & 1 \times 10^{-5} \end{bmatrix}$$
$$R = 0.1 I$$

where $P(0)$ is chosen large enough (see Remark 7.3), Q has been derived from the diagonal of the final covariance matrix $P(N)$, related to the case of $Q = 0$, and is chosen as 0.1 diag($P(N)$), and R is related to the accuracy of the measurement devices. The estimated parameter trajectories are presented in Fig. 8.4.

Contrary to expectation, the influent related parameter NH_4^{in} does not show much diurnal variation, probably due to attenuation of diurnal influent cycles in the over-estimated presettler (see Fig. 8.1). As expected, the estimated $r_{\text{NH,max}}$ in Fig. 8.4

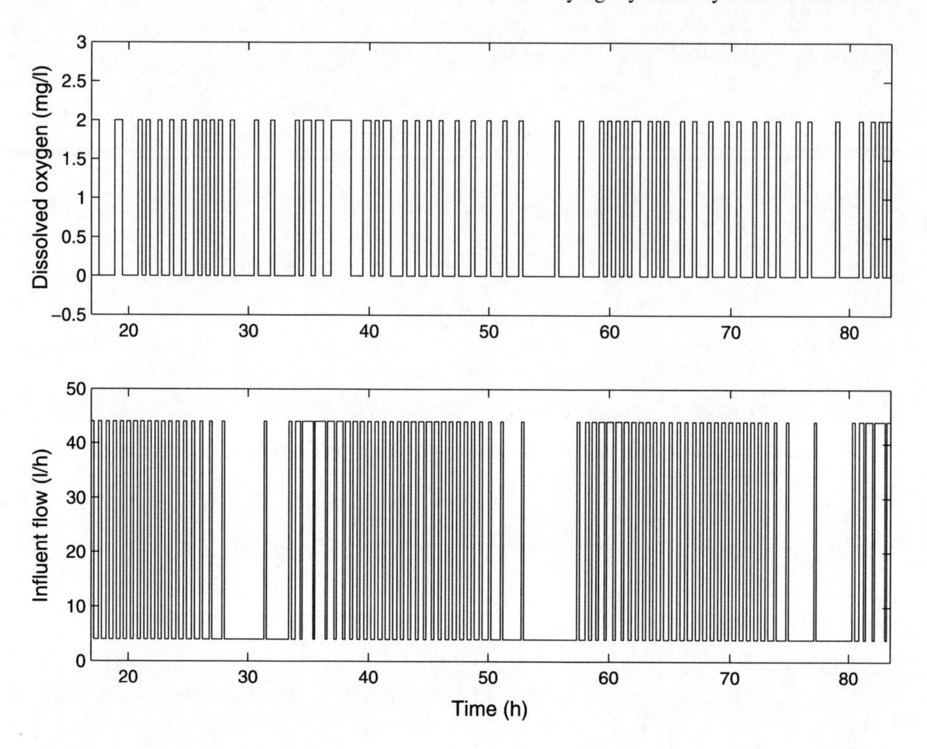

Fig. 8.2 Input signals, dissolved oxygen concentration (*top figure*), and influent flow (*bottom figure*) for the pilot-activated sludge plant

just represented without the subscript "max", shows little variation on a short time scale, but a clear change is observed on a larger time scale. A clear diurnal variation is observed in the estimates of $r_{NO,max}$; it consistently reaches its minimum at about noon, just after a period with low loads. On the basis of these results, several hypotheses can be stated, but this is beyond the scope of this book. It should, however, be noted that for practical implementation, special actions with respect to prediction errors and the matrix R are required in unusual situations. These unusual situations occur, for instance, in case of auto-calibration of the sensors (see fat over-bars at the top of the subplots in Fig. 8.3), outliers, or when NH_4 or NO_3 is depleted, while due to some off-set, the sensor indicates a nonzero value.

To summarize, Example 8.1 illustrates how to recursively estimate both inputs and parameters in a continuous-time linear dynamic system. Since the process dynamics are described by *piece-wise linear* differential equations with *piece-wise constant* inputs, explicit solutions of the differential equations (see footnote 1 in Chap. 1) were found. In this case, it further appears that, after some rewriting, the explicit solution is linear in the parameter vector, $\vartheta = [\mathrm{NH}_4^{\mathrm{in}}\ r_{\mathrm{NH,max}}\ r_{\mathrm{NO,max}}]^T$. Consequently, a multioutput, time-varying static system representation, as in Sect. 7.1.3, results, and thus algorithm (7.29)–(7.33) from Chap. 7 can be applied directly. However, if we intend to recursively estimate, for instance, the influent flow q^{in} (input)

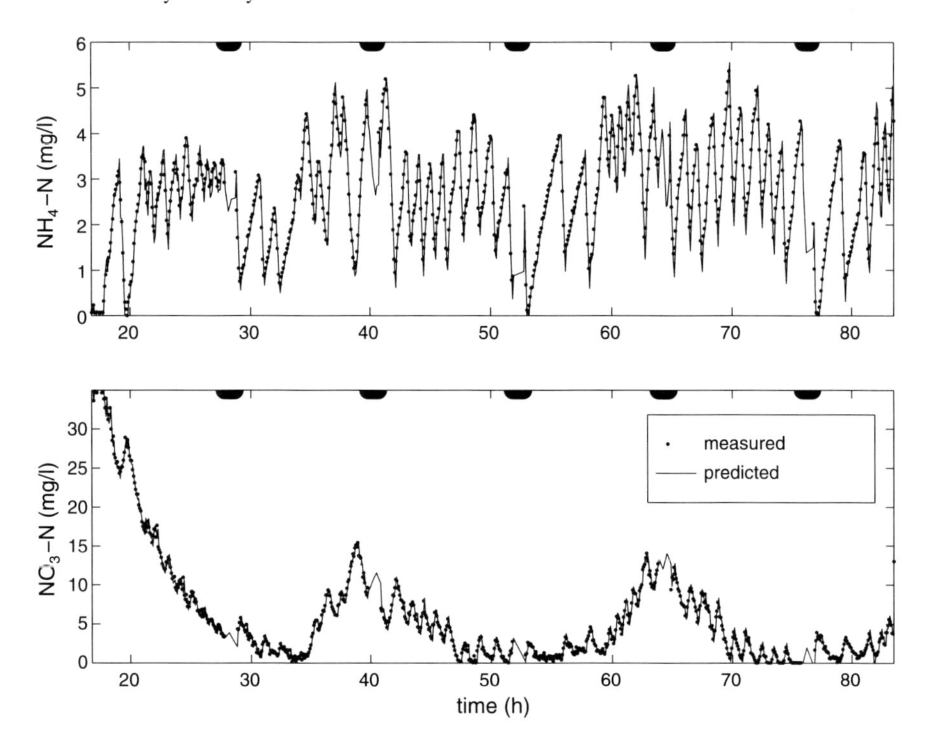

Fig. 8.3 Measured (*dots*) and predicted model output (*solid line*) signals, NH_4–N (*top figure*) and NO_3–N (*bottom figure*), for the pilot-activated sludge plant

or volume V (parameter) as well, a nonlinear regression between $y(t)$ and $\vartheta(t)$ will result. Notice that this input and parameter are directly related to the states and thus appear in the exponent of the resulting exponential function. Hence, for this case, the EKF algorithm ((7.60)–(7.64)) can be used.

Thus, for the recursive estimation of parameters, and possibly inputs as well, of continuous-time linear dynamic systems, for which explicit solutions exist, the algorithms from Chap. 7 can be used.

8.1.2 Recursive Prediction Error Estimation

In addition to the recursive least-squares estimation algorithms presented sofar, which are basically related to the equation-error identification problem, several modified recursive schemes related to the output-error identification problem with its colored noise have been proposed as well. Typical examples of these schemes, which will not be worked out here, are the extended least-squares and the instrumental variables algorithms (see Sect. 6.1.3).

Recall that, in particular in Sect. 7.1, the emphasis has been on unbiased, minimum variance estimates. As an alternative to this, when a model is, for instance,

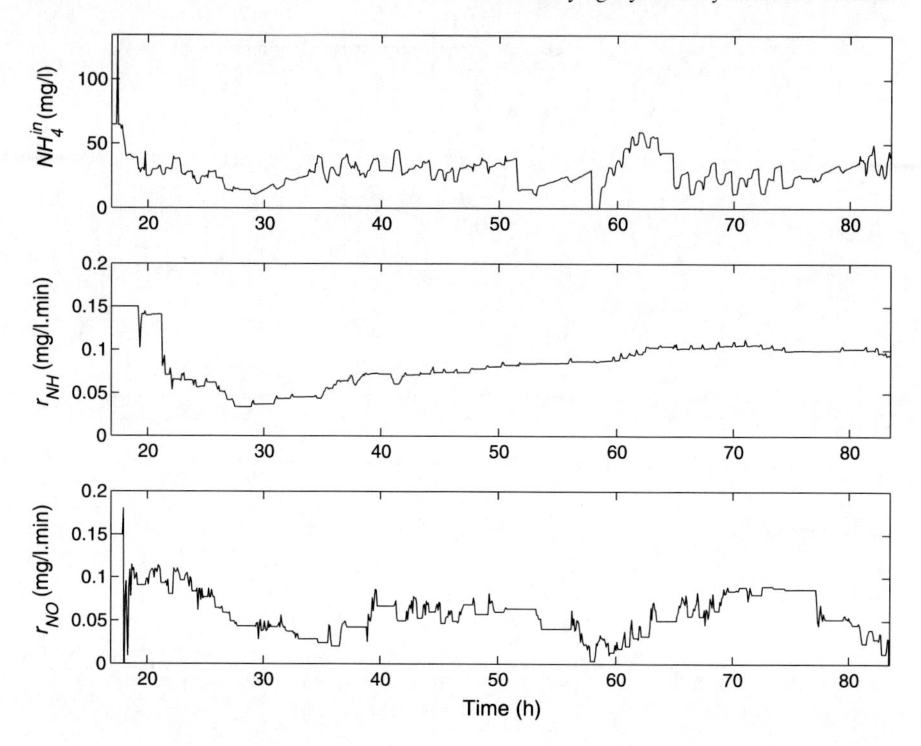

Fig. 8.4 Estimated parameter, NH_4^{in} (*top figure*), r_{NH} (*middle figure*), and r_{NO} (*bottom figure*) trajectories

developed for prediction, the algorithm should be chosen such that some scalar function of the prediction errors is minimized. Algorithms that focus on this particular model application are called *Prediction-Error algorithms* that in fact provide a general framework for identification (see Sect. 6.1.4).

For the development of a recursive prediction-error (RPE) algorithm, first an expression for $\psi(t, \vartheta)$, the gradient of the prediction, must be found. Recall the definition of the gradient, that is,

$$\psi(t, \vartheta) := \frac{d\widehat{y}(t, \vartheta)}{d\vartheta} = -\frac{d\varepsilon(t, \vartheta)}{d\vartheta} \tag{8.5}$$

Example 8.2 Output error model: Recall from (6.19) and (6.33) that for an output error model structure with $A(q) = C(q) = D(q) = 1$, the one-step-ahead prediction $\widehat{y}(t|t-1)$, and further denoted by $\widehat{y}(t, \vartheta)$ to express its dependency on ϑ, is given by

$$\widehat{y}(t, \vartheta) = \frac{B(q)}{F(q)} u(t) = \xi(t, \vartheta)$$

Then,

$$\frac{\partial \widehat{y}(t,\vartheta)}{\partial b_k} = \frac{1}{F(q)}u(t-k) \tag{8.6}$$

$$\begin{aligned}\frac{\partial \widehat{y}(t,\vartheta)}{\partial f_k} &= -\frac{B(q)}{F(q)F(q)}u(t-k) \\ &= -\frac{1}{F(q)}\xi(t-k,\vartheta)\end{aligned} \tag{8.7}$$

$$\begin{aligned}\Longrightarrow \quad \frac{\partial \widehat{y}(t,\vartheta)}{\partial \vartheta} &= \frac{1}{F(q)}\big[u(t-1),\dots,u(t-n_b), \\ &\quad -\xi(t-1,\vartheta),\dots,-\xi(t-n_f,\vartheta)\big]^T \\ &= \psi(t,\vartheta)\end{aligned} \tag{8.8}$$

Consequently, $\psi(t,\vartheta) = \frac{1}{F(q)}\phi(t,\vartheta)$ with regression vector $\phi(t,\vartheta) = [u(t-1), \dots, u(t-n_b), -\xi(t-1,\vartheta), \dots, -\xi(t-n_f,\vartheta)]^T$, so that the gradient of the prediction is a filtered regression vector.

Notice that in the time recursions ϑ is not known. What is available is the approximation $\widehat{\vartheta}(t-1)$. Consequently, the idea is to substitute ϑ by $\widehat{\vartheta}(t-1)$ in the variables $\widehat{y}(t,\vartheta)$ and $\psi(t,\vartheta)$, which are further denoted by $\widehat{y}(t)$ and $\psi(t)$. For the calculation of $\widehat{y}(t)$ and $\psi(t)$ we introduce the 'state' vector, which is related to the generalized model structure (6.11) with $n_k = 0$, and which is given by

$$\begin{aligned}\phi(t,\vartheta) = \big[&-y(t-1),\dots,-y(t-n_a),u(t-1),\dots,u(t-n_b), \\ &-\xi(t-1,\vartheta),\dots,-\xi(t-n_f,\vartheta),\varepsilon(t-1,\vartheta),\dots,\varepsilon(t-n_c,\vartheta), \\ &-v(t-1,\vartheta),\dots,-v(t-n_d,\vartheta)\big]^T\end{aligned} \tag{8.9}$$

with $\xi(t,\vartheta) = \frac{B(q)}{F(q)}u(t)$, $\varepsilon(t,\vartheta) = y(t) - \widehat{y}(t) = \frac{D(q)}{C(q)}[A(q)y(t) - \frac{B(q)}{F(q)}u(t)]$ and $v(t,\vartheta) = A(q)y(t) - \xi(t,\vartheta)$. The corresponding parameter vector of the generalized model structure (6.11) $\vartheta \in \mathbb{R}^p$ is given by

$$\vartheta = [a_1 \dots a_{n_a}\; b_1 \dots b_{n_b}\; f_1 \dots f_{n_f}\; c_1 \dots c_{n_c}\; d_1 \dots d_{n_d}]^T \tag{8.10}$$

After some algebraic manipulation and assuming a linear time-invariant finite-dimensional model, as (6.11), the output prediction is found from

$$\begin{aligned}\phi(t+1) &= \mathscr{F}\big(\widehat{\vartheta}(t)\big)\phi(t) + \mathscr{G}\big(\widehat{\vartheta}(t)\big)\begin{bmatrix} y(t) \\ u(t)\end{bmatrix} \\ \widehat{y}(t) &= \mathscr{H}\big(\widehat{\vartheta}(t-1)\big)\phi(t)\end{aligned} \tag{8.11}$$

where $\mathscr{F}$, $\mathscr{G}$ and $\mathscr{H}$ are properly chosen. Let us illustrate this model description to a simple output error model, so that proper choices for $\mathscr{F}$, $\mathscr{G}$ and $\mathscr{H}$ become clear straightforwardly. For simplicity of notation, no reference is made to the estimates in a time recursion.

Example 8.3 Output error model: Consider the following output error model with $n_b = 2$ and $n_f = 2$:

$$\xi(t,\vartheta) = b_1 u(t-1) + b_2 u(t-2) - f_1 \xi(t-1,\vartheta) - f_2 \xi(t-2,\vartheta)$$

Then, (8.11) with $\phi(t,\vartheta) = [u(t-1), u(t-2), -\xi(t-1,\vartheta), -\xi(t-2,\vartheta)]^T$, becomes

$$\begin{bmatrix} u(t) \\ u(t-1) \\ -\xi(t,\vartheta) \\ -\xi(t-1,\vartheta) \end{bmatrix} = \begin{bmatrix} 0 & 0 & 0 & 0 \\ 1 & 0 & 0 & 0 \\ -b_1 & -b_2 & -f_1 & -f_2 \\ 0 & 0 & 1 & 0 \end{bmatrix} \begin{bmatrix} u(t-1) \\ u(t-2) \\ -\xi(t-1,\vartheta) \\ -\xi(t-2,\vartheta) \end{bmatrix} + \begin{bmatrix} 0 & 1 \\ 0 & 0 \\ 0 & 0 \\ 0 & 0 \end{bmatrix} \begin{bmatrix} y(t) \\ u(t) \end{bmatrix}$$

$$\widehat{y}(t,\vartheta) = \xi(t,\vartheta) = [b_1\ b_2\ f_1\ f_2] \begin{bmatrix} u(t-1) \\ u(t-2) \\ -\xi(t-1,\vartheta) \\ -\xi(t-2,\vartheta) \end{bmatrix}$$

Consequently,

$$\mathscr{F} = \begin{bmatrix} 0 & 0 & 0 & 0 \\ 1 & 0 & 0 & 0 \\ -b_1 & -b_2 & -f_1 & -f_2 \\ 0 & 0 & 1 & 0 \end{bmatrix}$$

$$\mathscr{G} = \begin{bmatrix} 0 & 1 \\ 0 & 0 \\ 0 & 0 \\ 0 & 0 \end{bmatrix}$$

$$\mathscr{H} = [b_1\ b_2\ f_1\ f_2]$$

After differentiating (8.11) with respect to $\vartheta_1, \ldots, \vartheta_p$ and introducing

$$\chi(t) = \left[\phi(t)^T\ \frac{\partial}{\partial\vartheta_1}\phi(t)^T\ \ldots\ \frac{\partial}{\partial\vartheta_p}\phi(t)^T\right]^T \tag{8.12}$$

the following approximation is obtained

$$\begin{aligned} \chi(t+1) &= \mathscr{A}\big(\widehat{\vartheta}(t)\big)\chi(t) + \mathscr{B}\big(\widehat{\vartheta}(t)\big)\begin{bmatrix} y(t) \\ u(t) \end{bmatrix} \\ \begin{bmatrix} \widehat{y}(t) \\ \psi(t) \end{bmatrix} &= \mathscr{C}\big(\widehat{\vartheta}(t-1)\big)\chi(t) \end{aligned} \tag{8.13}$$

Let us illustrate this extended model description, thus including the gradient of the prediction, $\psi(t) = \frac{\partial \widehat{y}(t)}{\partial \vartheta}$, as in (8.13), to a first-order output error model and again no reference is made to the estimates in a time recursion.

Example 8.4 Output error model: Consider the following output-error model with $n_b = 1$ and $n_f = 1$:

$$\widehat{y}(t, \vartheta) = \xi(t, \vartheta) = b_1 u(t-1) - f_1 \xi(t-1, \vartheta)$$

so that $\phi(t, \vartheta) = [u(t-1), -\xi(t-1, \vartheta)$. The gradients of the prediction are found from

$$\begin{aligned} \frac{\partial \widehat{y}(t, \vartheta)}{\partial b_1} &= \psi_1(t, \vartheta) = u(t-1) - f_1 \frac{\partial \xi(t-1, \vartheta)}{\partial b_1} \\ \frac{\partial \widehat{y}(t, \vartheta)}{\partial f_1} &= \psi_2(t, \vartheta) = -\xi(t-1, \vartheta) - f_1 \frac{\partial \xi(t-1, \vartheta)}{\partial f_1} \end{aligned}$$

See also (8.6) and (8.7). Consequently, for

$$\begin{aligned} \chi(t, \vartheta) = \big[&u(t-1), -\xi(t-1, \vartheta), u_{b_1}(t-1), -\xi_{b_1}(t-1, \vartheta), u_{f_1}(t-1), \\ &-\xi_{f_1}(t-1, \vartheta)\big]^T \end{aligned}$$

using the short-hand notation $u_x := \frac{\partial u}{\partial x}$ and $\xi_x := \frac{\partial \xi}{\partial x}$, the dynamic systems (8.13) becomes

$$\begin{aligned} \begin{bmatrix} u(t) \\ -\xi(t, \vartheta) \\ u_{b_1}(t) \\ -\xi_{b_1}(t, \vartheta) \\ u_{f_1}(t) \\ -\xi_{f_1}(t, \vartheta) \end{bmatrix} &= \begin{bmatrix} 0 & 0 & 0 & 0 & 0 & 0 \\ -b_1 & -f_1 & 0 & 0 & 0 & 0 \\ 0 & 0 & 0 & 0 & 0 & 0 \\ -1 & 0 & 0 & -f_1 & 0 & 0 \\ 0 & 0 & 0 & 0 & 0 & 0 \\ 0 & -1 & 0 & 0 & 0 & -f_1 \end{bmatrix} \begin{bmatrix} u(t-1) \\ -\xi(t-1, \vartheta) \\ u_{b_1}(t-1) \\ -\xi_{b_1}(t-1, \vartheta) \\ u_{f_1}(t-1) \\ -\xi_{f_1}(t-1, \vartheta) \end{bmatrix} \\ &+ \begin{bmatrix} 0 & 1 \\ 0 & 0 \\ 0 & 0 \\ 0 & 0 \\ 0 & 0 \\ 0 & 0 \end{bmatrix} \begin{bmatrix} y(t) \\ u(t) \end{bmatrix} \end{aligned}$$

$$\begin{bmatrix} \widehat{y}(t,\vartheta) \\ \psi_1(t,\vartheta) \\ \psi_2(t,\vartheta) \end{bmatrix} = \begin{bmatrix} b_1 & f_1 & 0 & 0 & 0 & 0 \\ 1 & 0 & 0 & f_1 & 0 & 0 \\ 0 & 1 & 0 & 0 & 0 & f_1 \end{bmatrix} \begin{bmatrix} u(t-1) \\ -\xi(t-1,\vartheta) \\ u_{b_1}(t-1) \\ -\xi_{b_1}(t-1,\vartheta) \\ u_{f_1}(t-1) \\ -\xi_{f_1}(t-1,\vartheta) \end{bmatrix}$$

from which the matrices $\mathscr{A}$, $\mathscr{B}$ and $\mathscr{C}$ (see (8.13)) can be directly deduced.

Hence, the required approximations $\widehat{y}(t)$ and $\psi(t)$ are found from the dynamic system (filter) given by (8.13), as illustrated by the example.

A general recursive algorithm is given by

$$\widehat{\vartheta}(t) = \widehat{\vartheta}(t-1) + \gamma(t)R^{-1}(t)\psi\big(t,\widehat{\vartheta}(t-1)\big)\varepsilon\big(t,\widehat{\vartheta}(t-1)\big) \tag{8.14}$$

Consequently, using (8.14), for a specific choice of $R(t)$, for example,

$$R(t) = \gamma(t)\sum_{k=1}^{t}\beta(t,k)\psi(k)\psi(k)^T$$

based on the Gauss–Newton method with gain $\gamma(t)$ and (least-squares) weighting sequence $\beta(t,k)$ both defined below, the following so-called *recursive Gauss–Newton prediction-error algorithm* is obtained:

$$\varepsilon(t) = y(t) - \widehat{y}(t) \tag{8.15}$$

$$R(t) = R(t-1) + \gamma(t)\big[\psi(t)\psi(t)^T - R(t-1)\big] \tag{8.16}$$

$$\widehat{\vartheta}(t) = \widehat{\vartheta}(t-1) + \gamma(t)R^{-1}(t)\psi(t)\varepsilon(t) \tag{8.17}$$

For details of the derivation of (8.15)–(8.17), we refer to [Lju99b], Sect. 11.2. What is important for now is to notice that, unlike the previously presented recursive least-squares algorithms, in the derivation of (8.15)–(8.17) no statistical information in terms of means and covariances is taken into account; it directly starts from explicit search schemes. Especially, for model structures that cannot be written as linear regressions, the general algorithm (8.15)–(8.17) provides a good alternative, because the recursive least-squares algorithms of Chap. 7 are only optimal for linear regression models. However, the tuning parameters $\beta(t,k)$ or $\gamma(t)$ must be properly chosen in order to obtain a good behavior of the estimator. This behavior is usually expressed in terms of a trade-off between *tracking ability* and *noise sensitivity*. Unfortunately, no unique tuning rules that take into account this trade-off are available. Ljung [Lju99b] summarizes the relationships between the forgetting profile $\beta(t,k)$ with forgetting factors $\lambda(t)$ and the gain $\gamma(t)$ as

$$\beta(t,k) = \prod_{j=k+1}^{t}\lambda(j) = \frac{\gamma(k)}{\gamma(t)}\prod_{j=k+1}^{t}\big(1-\gamma(j)\big) \tag{8.18}$$

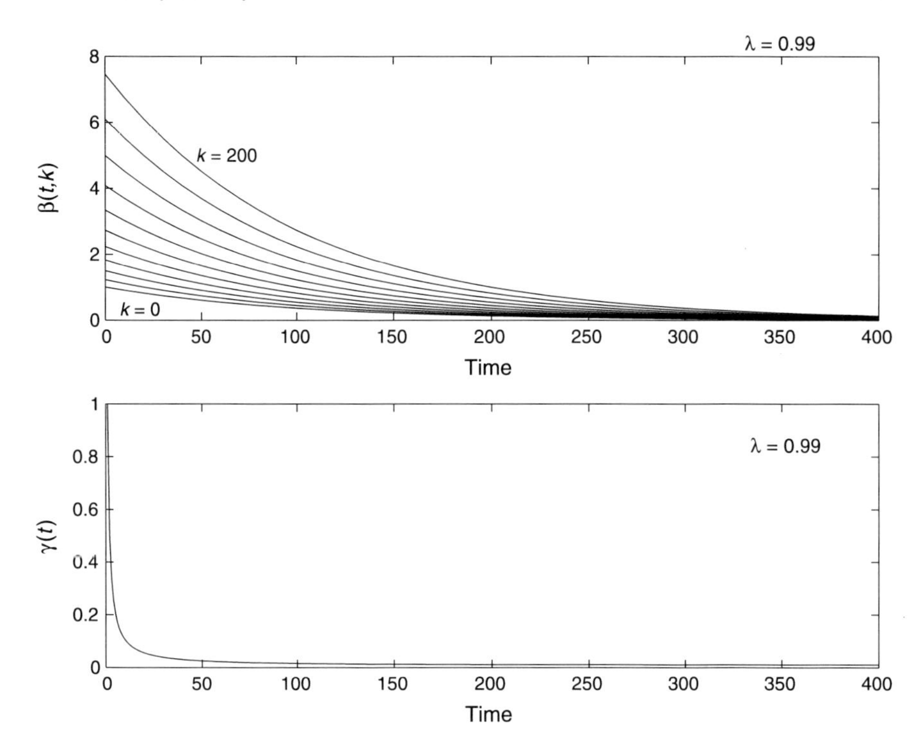

Fig. 8.5 Forgetting profiles $\beta(t, k)$ (*top figure*) and gain $\gamma(t)$ (*bottom figure*)

$$\lambda(t) = \frac{\gamma(t-1)}{\gamma(t)}\big(1 - \gamma(t)\big) \tag{8.19}$$

$$\gamma(t) = \frac{1}{1 + \frac{\lambda(t)}{\gamma(t-1)}} \tag{8.20}$$

In the following example, the profiles are evaluated for specific choices of λ, $\gamma(0)$ and k.

Example 8.5 RPE-algorithm: Let $\lambda(t) \equiv \lambda = 0.99$, $\gamma(0) = 1$, and $k = 0 : 20 : 200$. Then, the following profiles for $t \in [0, 400]$, as presented in Fig. 8.5, are found.

Let us round off this section by summarizing the RPE-algorithm.

Algorithm 8.1 Recursive Prediction-Error estimation of $\vartheta(t)$ in linear time-varying dynamic systems

1. Choose the initial parameter vector $\vartheta(0)$ and the forgetting factor $\beta(t)$ or gain $\gamma(t)$.
2. Evaluate, for $t = 1, \ldots, N$, the prediction $\widehat{y}(t)$ and the gradient of the prediction $\psi(t)$, and subsequently calculate $\varepsilon(t)$, $R(t)$, and $\widehat{\vartheta}$ from (8.15)–(8.17).

In the next section, we will focus on the third type of estimation problem already mentioned in Sect. 1.1.2, i.e., smoothing.

8.1.3 Smoothing

Recall that smoothing is the estimation of $x(t)$, $0 \leq t \leq T$, from $y(t)$, $0 \leq t \leq T$. If, for example, the final goal is to reconstruct possibly time-varying parameter estimates in a model of a dynamic system from a (short) data set, smoothing is a good alternative to filtering, because at any time instant it builds into the estimates the information contained in the present, future, and past measurements. Notice from the material presented so far that filtering always introduces some time lag; the new estimate depends on the current output and also on the previous estimate. Especially in short data sets with limited prior parameter knowledge, this phenomenon becomes visible quite clearly. Therefore, algorithms have been proposed that do not only take into account past data, but also future data, when available, in the data set. Hence, instead of the estimation of $\vartheta(t|t)$, i.e., the estimate at time instant t given information up to t, the focus is now on the estimation of $\vartheta(t|N)$ and which is known as smoothing. Given the explicit time-varying parameter model with output equation (7.28) with $\text{Cov}\, w(t) = Q(t)$, $\text{Cov}\, v(t) = R(t)$, and $\Pi = I$, a *fixed-interval optimal smoothing algorithm* for the time-varying parameters can be formulated as

$$\widehat{\vartheta}(t+1|N) = \widehat{\vartheta}(t+1|t) - P(t+1|t)\lambda(t), \quad \text{for } t = 0, 1, \ldots, N-1 \tag{8.21}$$

$$\begin{aligned}\lambda(t) = {} & \left(I - \Phi(t+1)^T R(t+1)^{-1} \Phi(t+1) P(t+1|t+1)\right) \\ & \left(\Xi^T \lambda(t+1) - \Phi(t+1)^T R(t+1)^{-1}\right) \\ & \left(y(t+1) - \Phi(t+1)\widehat{\vartheta}(t+1|t)\right), \\ & \text{for } t = N-1, N-2, \ldots, 0\end{aligned} \tag{8.22}$$

where $\widehat{\vartheta}(t+1|t)$ and $P(t+1|t)$ are found from the nonsmoothing forward recursion (7.29)–(7.30), and $\lambda(t)$ is the so-called Lagrange multiplier related to the explicit time-varying parameter model. In fact, the smoothing algorithm minimizes

$$\begin{aligned}V_N = \sum_{t=1}^{N} & \left[\left(y(t) - \Phi(t)\widehat{\vartheta}(t|N)^T\right) R(t)^{-1} \left(y(t) - \Phi(t)\widehat{\vartheta}(t|N)\right)\right] \\ & + \widehat{w}(t-1)^T Q(t-1)^{-1} \widehat{w}(t-1) \\ & + \left(\widehat{\vartheta}(0|N) - \widehat{\vartheta}(0|0)\right)^T P(0|0)^T \left(\widehat{\vartheta}(0|N) - \widehat{\vartheta}(0|0)\right)\end{aligned} \tag{8.23}$$

under the equality constraints

$$\vartheta(t) = \Xi\vartheta(t-1) + w(t-1) \tag{8.24}$$

Table 8.2 Moving object data

Time t (s)	1	2	3	4	10	12	18
Distance y (ft)	9	15	19	20	45	55	78

In (8.24), the first term on the right-hand side reflects the costs of prediction errors, the second one represents the costs of parameter variations, and the last term is related to the final cost of the parameter deviations. Alternative forms, in which $\widehat{\vartheta}(t+1|N)$ is expressed in terms of $\widehat{\vartheta}(t+1|t+1)$ instead of $\widehat{\vartheta}(t+1|t)$, can be formulated as well, but this is not further shown here.

Algorithm 8.2 Fixed-interval optimal smoothing of $\vartheta(t)$ in linear time-varying dynamic systems

1. Given $y(t)$ and $\Phi(t)$ for $t = 1, \ldots, N$, specify the dynamic parameter model matrix Ξ.
2. Specify the covariance matrices $Q(t)$ and $R(t)$.
3. Choose the initial parameter vector $\vartheta(0)$ and the initial error covariance matrix $P(0)$.
4. Evaluate, for $t = 1, \ldots, N$, (7.29)–(7.30) and subsequently, for $t = N-1, \ldots, 0$, (8.21)–(8.22).

Example 8.6 Moving object (constant velocity): Let us focus again on the moving object data of [You84], p. 18, see also Table 8.2.

A simple linear dynamic model describing the behavior of a moving object along a straight line is

$$\frac{\mathrm{d}s(t)}{\mathrm{d}t} = \upsilon(t) + \omega(t)$$
$$y(t) = s(t) + \nu(t)$$

where in this example ω and ν are used instead of w and v to avoid confusion with the velocity υ. If s_0 is fixed at $s_0 = 5.7$ ft, υ is assumed to be constant, and only a smoothed estimate of this constant velocity υ is required, an appropriate model is

$$\upsilon(t+1) = \upsilon(t)$$
$$y(t) - s_0 = \upsilon(t)t + e(t)$$

Hence, under the assumption that υ is constant, $\Xi = 1$, $\Pi = 0$, and $Q(t) = 0\ \forall t$. Let further $R(t) = 1\ \forall t$ and $P(0|0) = 10^8$. Then, the following smoothed estimates with corresponding estimation variances are found (see Figs. 8.6–8.7). Notice from these figures that, as expected, the trajectory of the smoothed estimates does not show any effect of the unknown initial estimate. Furthermore, smoothing shows that the estimation variances are significantly smaller than in the case of filtering.

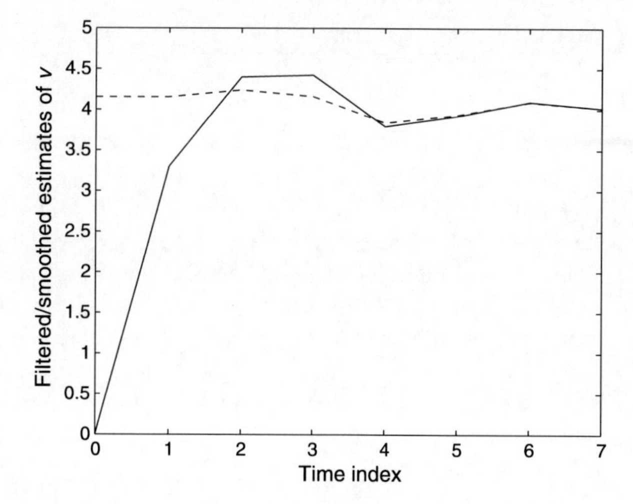

Fig. 8.6 Filtered (*solid line*) and smoothed (*dashed line*) estimates

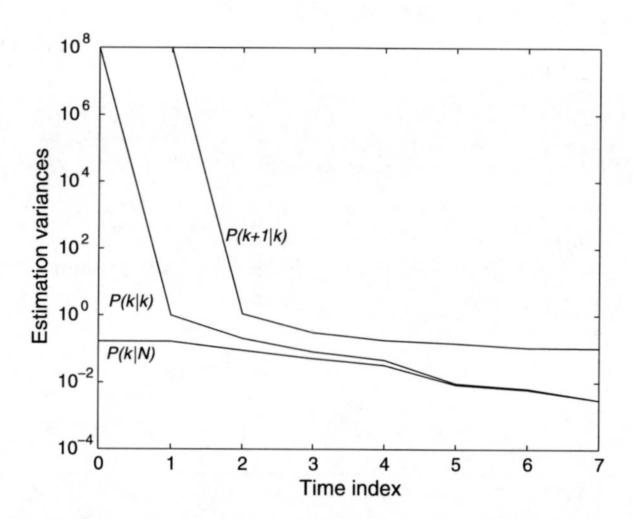

Fig. 8.7 Estimation variances related to one-step-ahead predictions $P(k+1|k)$, filtered estimates $P(k|k)$, and smoothed estimates $P(k|N)$

Recall that Chap. 7 focusses on static systems or (non)linear regression type of models. In this chapter so far (piece-wise) linear dynamic systems with time-varying parameters have been considered. Because of the linear system dynamics, explicit solutions of the state differential equations exist. In Example 8.1, these explicit solutions have been substituted in the output equation (see footnote 1 in Chap. 1), so that the algorithms of Chap. 7 were finally used. However, in the next section, algorithms and examples will be presented for the general nonlinear dynamic case, where the differential or difference equations related to the process states and output equations remain present. Thus, in the following, the question is: "given a nonlinear dynamic system description, how can we recursively estimate the unknown parameters?"

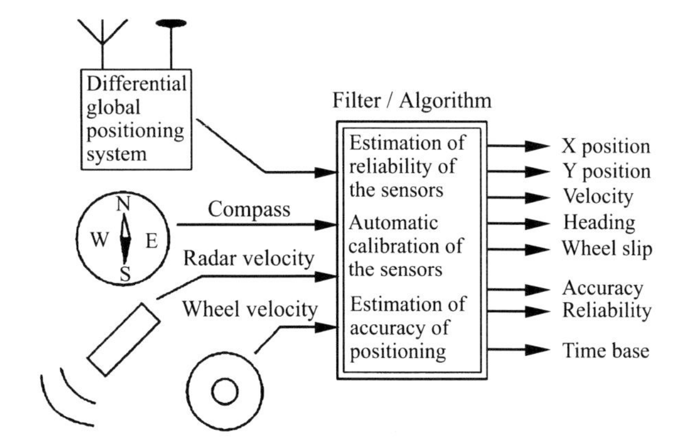

Fig. 8.8 Components and prespecified outputs of the positioning system

8.2 Nonlinear Dynamic Systems

8.2.1 Extended Kalman Filtering

Basically, all the ingredients for solving a recursive estimation problem of a nonlinear dynamic system have been presented in Chap. 7. Let us start with an example of a moving vehicle, which shows linear dynamics and a nonlinear relationship between the states and the measured outputs.

Example 8.7 Moving vehicle—real-world case (based on [vBGKS98]): Consider a moving vehicle which, unlike the previous case, is equipped with a differential global positioning system (DGPS) receiver, a radar velocity sensor, a wheel velocity sensor, and an electronic compass. The structure of the positioning system is schematically presented in Fig. 8.8. The sensors for position, velocity, and heading were connected to a PC-based data acquisition system equipped with analogue-to-digital conversion, counter inputs and an RS-232 port. The data-logging rate for the DGPS receiver was restricted to the maximum update rate of 4 Hz. Velocity and heading measurements were collected at a sampling frequency of 40 Hz.

Assume that, for the online estimation of the x- and y-positions, the system behavior can be described by the difference equations

$$
\begin{aligned}
s(t) &= s(t - T_s) + v(t - T_s)T_s + \frac{1}{2}a(t - T_s)T_s^2 \\
v(t) &= v(t - 1) + a(t - T_s)T_s \\
a(t) &= a(t - T_s) + \omega(t)
\end{aligned}
$$

where $s(t)$ is the position at time instant t (m), $v(t)$ is the velocity (m/s), $a(t)$ is the acceleration (m/s^2), and T_s is the sampling interval (s). As in Example 8.6, we will use $\omega(t)$ and $\nu(t)$ to represent the system and sensor noise. In what follows, we distinguish between x- and y-directions, so that the six-dimensional state vector becomes $x(t) = [s_x\ s_y\ v_x\ v_y\ a_x\ a_y]$. Given the set of difference equations, the state matrix related to the position, velocity, and acceleration in both x- and y-directions is given by

$$A = \begin{bmatrix} 1 & T_s & \frac{1}{2}T_s^2 & 0 & 0 & 0 \\ 0 & 1 & T_s & 0 & 0 & 0 \\ 0 & 0 & 1 & 0 & 0 & 0 \\ 0 & 0 & 0 & 1 & T_s & \frac{1}{2}T_s^2 \\ 0 & 0 & 0 & 1 & T_s & 0 \\ 0 & 0 & 0 & 0 & 0 & 1 \end{bmatrix}$$

Assuming equal system noise properties in the accelerations in x- and y-directions, the noise matrix is defined as $G := [0\ 0\ 1\ 0\ 0\ 1]$. Furthermore, the nonlinear vector function $h(x(t), \vartheta(t))$ relates the state variables and parameters to the output vector $y(t)$, containing measurements from the DGPS receiver, radar velocity sensor, wheel velocity sensor, and electronic compass, and is defined as

$$h\big(x(t), \vartheta(t)\big) := \begin{bmatrix} s_x(t) \\ s_y(t) \\ \sqrt{v_x(t)^2 + v_y(t)^2} \\ \sqrt{v_x(t)^2 + v_y(t)^2} \\ \arctan(\frac{v_x(t)}{v_y(t)}) \end{bmatrix}$$

Notice that the state equations are linear and the output equation is nonlinear. For an EKF implementation, the vector function $h(\cdot, \cdot)$ must be linearized, usually at time instant $t-1$, leading to the observation matrix

$$H(t) = \begin{bmatrix} 1 & 0 & 0 & 0 & 0 & 0 \\ 0 & 0 & 0 & 1 & 0 & 0 \\ 0 & \frac{v_x}{\sqrt{v_x^2+v_y^2}} & 0 & 0 & \frac{v_y}{\sqrt{v_x^2+v_y^2}} & 0 \\ 0 & \frac{v_x}{\sqrt{v_x^2+v_y^2}} & 0 & 0 & \frac{v_y}{\sqrt{v_x^2+v_y^2}} & 0 \\ 0 & \frac{v_y}{v_x^2+v_y^2} & 0 & 0 & -\frac{v_x}{\sqrt{v_x^2+v_y^2}} & 0 \end{bmatrix}$$

The results of a real-world experiment with, for all t,

$$Q(t) = 0.1$$

$$R(t) = \begin{bmatrix} 1.39 & 0 & 0 & 0 & 0 \\ 0 & 1.39 & 0 & 0 & 0 \\ 0 & 0 & 0.001 & 0 & 0 \\ 0 & 0 & 0 & 0.002 & 0 \\ 0 & 0 & 0 & 0 & 0.14 \end{bmatrix}$$

can be seen in Fig. 8.9. Notice that at the northeast corner, when the vehicle crossed the tree line, the DGPS lost satellite fixed and jumped to positions 20–30 m away

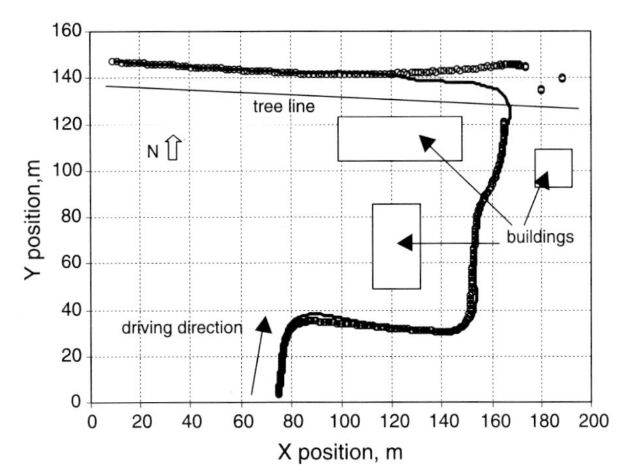

Fig. 8.9 Filtered positions (*solid line*) and DGPS measurements (*circles*) during a satellite no fix period in the northeast corner

from the real position. The positioning system then relied on dead reckoning, using only the radar velocity sensor, wheel velocity sensor, and electronic compass. The DGPS receiver needed about 80 m or 12 s, to recover from the satellite loss.

So far in Part III, the emphasis was on recursive parameter estimation, and the resemblance with state estimation was underlined. However, in those applications where mainly indirect or rather uncertain measurements are available, state estimation using a mathematical model of the system is indispensable. In general, these models contain uncertain or even unknown parameters. Hence, in addition to state estimation, in fact some or all of the parameters must be estimated as well from experimental data. For these cases, one commonly applies a simultaneous state/parameter estimation approach. This approach uses so-called *state augmentation* by regarding the parameters as states (see also Example 7.6). Consequently, given (7.28a) and (7.38a), the following state equation with $x'(t) := \left[\begin{smallmatrix} x(t) \\ \vartheta(t) \end{smallmatrix}\right]$ is obtained:

$$x'(t) = \begin{bmatrix} A(t) & 0 \\ 0 & \Xi \end{bmatrix} x'(t-1) + \begin{bmatrix} B(t) \\ 0 \end{bmatrix} u(t-1) + \begin{bmatrix} G(t) & 0 \\ 0 & \Pi \end{bmatrix} \omega'(t-1) \tag{8.25}$$

Notice that in Example 8.7 state augmentation was already implicitly used. That is, in addition to the four states s_x, s_y, v_x, and v_y, the acceleration in both x- and y-directions, which can be considered as parameters of the model, is simultaneously estimated from the data.

Let us now demonstrate the simultaneous estimation of states and parameters, using state augmentation and extended Kalman filtering, in a real-world application with nonlinear dynamics. However, for this example, we need an essential modification of the basic EKF for static, nonlinear systems, as in (7.60)–(7.64).

Let us first introduce a continuous-discrete time system description,

$$\begin{aligned}\frac{\mathrm{d}x(t)}{\mathrm{d}t} &= f\big(t, x(t), u(t)\big) + w(t), \quad x(0) = x_0 \\ y(t_k) &= h\big(t_k, x(t_k), u(t_k)\big) + v(t_k), \quad k = 0, 1, \ldots, N\end{aligned} \tag{8.26}$$

where $x(t)$ may be an augmented state vector, so that we do not distinguish between states and parameters. Furthermore, because of the continuous-discrete time description, we introduce the new notation t_k, indicating the kth sampling time instant. As before, $w(t)$ is the system noise, representing modeling error and unknown inputs, and $v(t_k)$ is the measurement noise at time instant t_k. Define the Jacobi matrix $F = (f_{ij})$ with elements

$$f_{ij} = \left[\frac{\partial f_i(t, \widehat{x}(t), u(t))}{\partial \widehat{x}_j}\right]_{t=t_{k-1}} \tag{8.27}$$

and from this the transition matrix $A(t_k) = A(\widehat{x}(t_{k-1})) := e^{F\Delta t}$ with $\Delta t = t_k - t_{k-1}$ for equidistant measurements. Notice that, in general, $e^{F\Delta t}$ is the exponential of a matrix (see Appendix A). Furthermore, define the matrix $H(t_k)$ with elements

$$h_{ij} = \frac{\partial h_i(t, \widehat{x}(t_{k-1}), u(t_{k-1}))}{\partial \widehat{x}_j} \tag{8.28}$$

Then, the Extended Kalman filter equations for the continuous-discrete time dynamic system, (8.26), are given as follows.

Prediction:

$$\widehat{x}(t_k|t_{k-1}) = \widehat{x}(t_{k-1}) + \int_{t_{k-1}}^{t_k} f\big(\tau, x(\tau), u(\tau)\big)\,\mathrm{d}\tau \tag{8.29}$$

$$P(t_k|t_{k-1}) = A(t_k)P(t_{k-1})A(t_k)^T + Q(t_k) \tag{8.30}$$

Correction:

$$K(t_k) = P(t_k|t_{k-1})H(t_k)^T\big[H(t_k)P(t_k|t_{k-1})H^T(t_k) + R(t_k)\big]^{-1} \tag{8.31}$$

$$\widehat{x}(t_k) = \widehat{x}(t_k|t_{k-1}) + K(t_k)\big[y(t_k) - h\big(t_k, x(t_k|t_{k-1}), u(t_k)\big)\big] \tag{8.32}$$

$$\begin{aligned}P(t_k) = {}& \big(I - K(t_k)H(t_k)\big)P(t_k|t_{k-1})\big(I - H(t_k)^T K(t_k)^T\big) \\ & + K(t_k)R(t_k)K(t_k)^T\end{aligned} \tag{8.33}$$

where $P(t_k)$ is the covariance matrix of the estimates at time instant t_k, $Q(t_k)$ is the covariance matrix associated with the system noise ($w(t_k)$), and $R(t_k)$ is the covariance matrix of the measurement noise ($v(t_k)$) at t_k. As before, the argument "$t_k|t_{k-1}$" denotes prediction from time instant t_{k-1} to t_k. Notice then from the definition of F that $A(t_k)$ is evaluated at each new sampling instant.

Algorithm 8.3 Extended Kalman filtering for the estimation of both $x(t)$ and $\vartheta(t)$ in a continuous-discrete time nonlinear system

1. Given input–output data $u(t_k)$ and $y(t_k)$ for $t_k = 1, \ldots, N$ and the nonlinear (state-augmented) state-space representation (8.26), specify the covariance matrices $Q(t_k)$ and $R(t_k)$.
2. Choose the initial (augmented) state vector $x(0)$ and the initial error covariance matrix $P(0)$.
3. Evaluate, for $t_k = 1, \ldots, N$, (8.29)–(8.33).

Example 8.8 Dissolved Oxygen (DO) dynamics (based on [LKvS96]): Recall from Example 6.19 that for a specific application, the DO dynamics in an aeration tank are at last described by (6.106), (6.108), and (6.110). Recall that this is a continuous-discrete time system, as the output was sampled every minute. After reparameterization of (6.106), the following DO balance was obtained:

$$\begin{aligned}\frac{\mathrm{d}C(t)}{\mathrm{d}t} &= -f(C)r_{\mathrm{act}}(t) + \alpha'\sqrt{q_{\mathrm{air}}(t-\Delta)} \\ &\quad + \beta'\sqrt{q_{\mathrm{air}}(t-\Delta)}C(t) + \gamma' C(t) - \frac{q_{\mathrm{in}} + q_r}{V}C(t) + \delta'\end{aligned}$$

Let us now try to estimate the continuous-time parameters K_C, α', β', and δ' and the DO concentration $C(t)$ recursively from the first 11 hours of experimental data in Fig. 6.18a. Clearly, simultaneous estimation of parameters and states will, in general, lead to nonlinear estimation problems. Assuming a simple random walk model for the parameters, the system matrix of the linearized system with $x = [K_c\ \alpha'\ \beta'\ \delta'\ C]$, and linearized around the last estimates (not explicitly shown here) and inputs, becomes

$$A(t) = \begin{bmatrix} 0 & 0 & 0 & 0 & 0 \\ 0 & 0 & 0 & 0 & 0 \\ 0 & 0 & 0 & 0 & 0 \\ 0 & 0 & 0 & 0 & 0 \\ \frac{C}{(K_C+C)^2}r_{\mathrm{act}} & \sqrt{q_{\mathrm{air}}} & C\sqrt{q_{\mathrm{air}}} & 1 & \frac{K_C}{(K_C+C)^2}r_{\mathrm{act}} + \beta'\sqrt{q_{\mathrm{air}}} - \frac{q_{in}+q_r}{V} \end{bmatrix}$$

For specific choices of Q and R such that a good trade-off between tracking ability and noise reduction is obtained, the following trajectories of the continuous-time parameter estimates, as presented in Fig. 8.10, result. These trajectories were obtained by using a continuous-discrete time EKF implementation, see (8.29)–(8.33). The estimates of $C(t)$ are not shown here, as they smoothly follow the measured DO concentrations.

8.2.2 *Observer-based Methods*

Consider a simplified version of (1.2) without the noise terms w and v and obtained after state augmentation, so that the model parameter vector ϑ is assimilated in the

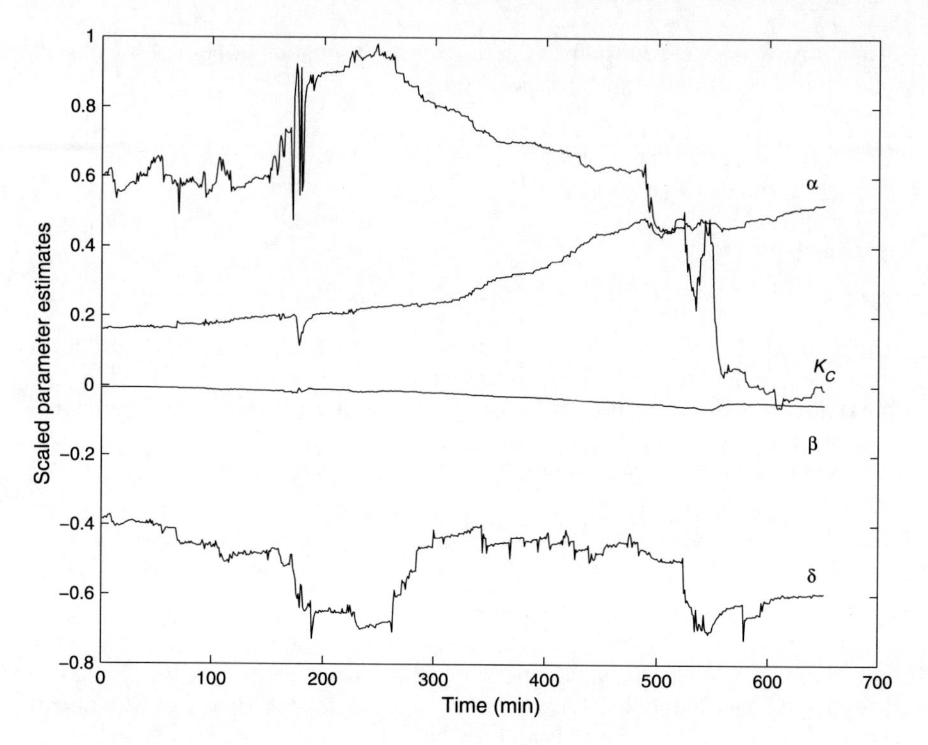

Fig. 8.10 Recursive parameter estimates of re-parameterized DO model

state vector x,

$$\begin{aligned} \frac{\mathrm{d}x(t)}{\mathrm{d}t} &= f\big(t, x(t), u(t)\big) \\ y(t) &= h\big(t, x(t), u(t)\big) \end{aligned} \tag{8.34}$$

Hence, our starting point is a noise-free system representation, where x has to be estimated from the available data. Notice then that the focus is on state estimation, which originates from mathematical systems theory, rather than parameter estimation. It may, however, be clear from the previous sections that parameters can be considered as unobserved states. Thus, for parameter estimation problems, state estimation techniques may be deployed as well. Suppose now that an estimate $\widehat{x}(t)$ of $x(t)$ in (8.34) is given. We will then use a linearized version of (8.34) to see how the estimate propagates. Linearization around the given estimate gives

$$\begin{aligned} \frac{\mathrm{d}x(t)}{\mathrm{d}t} &\approx f\big(t, \widehat{x}(t), u(t)\big) + \left.\frac{\partial f(t, x(t), u(t))}{\partial x}\right|_{\widehat{x}(t)} \big[x(t) - \widehat{x}(t)\big] \\ y(t) &\approx h\big(t, \widehat{x}(t), u(t)\big) + \left.\frac{\partial h(t, x(t), u(t))}{\partial x}\right|_{\widehat{x}(t)} \big[x(t) - \widehat{x}(t)\big] \end{aligned} \tag{8.35}$$

Define

$$F_x := \left.\frac{\partial f(t, x(t), u(t))}{\partial x}\right|_{\widehat{x}(t)}, \qquad H_x := \left.\frac{\partial h(t, x(t), u(t))}{\partial x}\right|_{\widehat{x}(t)}$$

$$D_x := h\big(t, \widehat{x}(t), u(t)\big) - \left.\frac{\partial h(t, x(t), u(t))}{\partial x}\right|_{\widehat{x}(t)} \widehat{x}(t)$$

$$E_x := f\big(t, \widehat{x}(t), u(t)\big) - \left.\frac{\partial f(t, x(t), u(t))}{\partial x}\right|_{\widehat{x}(t)} \widehat{x}(t)$$

where D_x, E_x, F_x, H_x are time-varying matrices that depend on $\widehat{x}(t)$ and $u(t)$. The linearized system, after rearranging (8.35) and using the definitions of $D_x, \ldots, H_x$, becomes

$$\begin{aligned} \frac{\mathrm{d}x(t)}{\mathrm{d}t} &= F_x x(t) + E_x \\ y(t) &= H_x x(t) + D_x \end{aligned} \tag{8.36}$$

Following classical observer theory (see, for instance, [KS72]), a *linear* observer related to (8.36) is given by

$$\frac{\mathrm{d}\widehat{x}(t)}{\mathrm{d}t} = F_x \widehat{x}(t) + E_x + K\big[y(t) - H_x \widehat{x}(t) - D_x\big] \tag{8.37}$$

After substituting $D_x, \ldots, H_x$ in (8.37), we obtain, in terms of the nonlinear functions $f(t, \widehat{x}(t), u(t))$ and $h(t, \widehat{x}(t), u(t))$,

$$\frac{\mathrm{d}\widehat{x}(t)}{\mathrm{d}t} = f\big(t, \widehat{x}(t), u(t)\big) + K\big[y(t) - h\big(t, \widehat{x}(t), u(t)\big)\big] \tag{8.38}$$

There are basically two common ways of designing the observer gain K. One way is to take fixed matrices for D_x, E_x, F_x, H_x and use these to design a single fixed K based on linear observer theory. Alternatively, we can design a gain K that depends on $\widehat{x}(t)$ at every instant of time, as, for example, in the EKF algorithm in Sect. 8.2.1. Notice that the structure of the nonlinear observer (8.38) also fits into the recursive prediction error schemes of Sect. 8.1.2, where K can be chosen on the basis of a selected search method.

8.3 Historical Notes and References

The first recursive estimation algorithms were driven by one-step ahead prediction errors. However, to generalize the idea to multistep ahead prediction errors, recursive prediction-error (RPE) algorithms have been introduced by [MW79, Lju81, MB86]. Extensions of the RPE algorithm to nonlinear systems [CB89, LB93] and to a continuous-discrete time version [SB04] followed.

It is interesting to notice that already in the mid 1970s, it has been recognized that the recursive parameter estimates contain information that could be used in the model structure selection procedure. In particular, trends and jumps in the reconstructed parameter trajectories indicate model deficiencies, see [BY76, SB94, LB07] for real-world applications of this.

However, we could go a step further by inferring the nonlinear model structure from the data. So far, it has always been assumed that some prior knowledge in terms of (non)linear differential equations was available. This is most often the case if we start with a physical model and put this into a semi-physical modeling approach, as outlined in this book. In the mid 1990s, several papers appeared on what is called *state-dependent parameter* modeling (see [YB94, KJ97, You98]). In these papers, transfer function models have always been taken as a starting point. The key idea behind this approach is that time-varying parameter estimates, preferably as a result of smoothing, can be modeled in terms of the known states/outputs of the system. Hence, nonlinear data-based mechanistic models result if the resulting nonlinear model allows a mechanistic interpretation. In the earliest papers on state-dependent parameter modeling, correlation techniques were used to find relationships between the time-varying parameters and the states/outputs. However, by plotting parameter values against an appropriate choice of states/inputs/outputs, or a combination of these, in general, nonlinear relationships will be revealed, see [YG06]. These (non)linear relationships can subsequently be substituted into the transfer function model, thus modifying the original input and output variables. Thus, basically the data instead of prior physical knowledge is taken as a starting point and hopefully a physically interpretable model structure results. This approach shows some resemblance with the linear parameter-varying modeling approach. However, in the latter case, unlike the data-based mechanistic modeling approach, the nonlinear parameter relationship is specified a priori, see Example 6.13.

The first smoothing algorithms were published in the early 1970s, see [BS72, Blu72, Nor75, Nor76], where Norton was among the first who used smoothing algorithms to estimate time-varying parameters in linear models. Since then, many papers have appeared on this subject of smoothing.

The specific problem of joint parameter and state estimation has been recognized in the late 1970s [JY79, SAML80]. Using local linearization techniques and state augmentation, the problem was cast in the EKF framework. However, state augmentation usually leads to a large dimension of the augmented state vector. Hence, there is a need for model reduction while maintaining the physical insights, see [Kee02] for useful decomposition methods. As an alternative to this, Goodwin and Sin [GS84] suggested an alternated parameter and state estimation scheme. Recently, Keesman, and Maksimov [KM08a, KM08b] presented an algorithm that solves the simultaneous state and parameter estimation problem and that is stable with respect to bounded informational noises and computational errors. The algorithm is based on the principle of auxiliary models with adaptive controls.

In addition to EKF and UKF (see Sect. 7.3), a third Kalman filter type of algorithm, suited for solving nonlinear estimation problems, is known as the Ensemble Kalman Filter (EnKF) [Eve94]. As with the UKF, the EnKF also uses sampling techniques, in particular Monte Carlo sampling. It is very popular in data-assimilation studies of dynamic systems and, in particular, in weather forecast applications. This idea of using sampling techniques in estimation problems can be found in many books and articles on Bayesian estimation using Monte Carlo methods, see, for example, [Liu94, GRS96, BR97, CSI00, DdFG01, LCB⁺07]. The application of sequential Monte Carlo methods, as the Monte Carlo Markov Chain (MCMC) algorithm, in estimation of dynamic systems is also known as "particle filtering." The simulation-based Bayesian estimation methods, as a result of the increasing computing power, will increase in popularity, see [Nin09].

For further reading on observer-based methods for recursive parameter estimation in nonlinear dynamic systems, as an alternative to the EKF, we refer to [She95, KH95, OFOFDA96, PDAFD00].

8.4 Problem

Problem 8.1 This exercise, presented here as a project problem related to the identification and prediction of a continuous-time dynamic system, will lead you through a couple of different estimation methods introduced in this book. Fill in your answers at the appropriate places, presented by $\langle\cdot\rangle$. The full "real-world" data set[1] can be found in Appendix G, Table 6.1. The exercise focuses, in particular, on dissolved oxygen (DO) prediction uncertainty evaluation, which has also been treated in a couple of papers, see [KvS89, Kee89, Kee90].

Problem formulation:

Given $N = 196$ hourly measurements of the dissolved oxygen concentration in $\mathrm{g/m^3}$, the saturated DO concentration (C_S) in $\mathrm{g/m^3}$ and the radiation (I) in $\mathrm{W/m^2}$, from the lake "De Poel en 't Zwet" (The Netherlands) over the period 111.875–120 days (see Fig. 8.11), predict the DO concentration at time instant 120.25 d, i.e., at 06:00 a.m. of the next day.

In the following we distinguish between a nonparametric, using only the available data, and a parametric approach which incorporates prior knowledge in the form of a commonly used mass balance equation of the DO concentration (C), i.e.,

$$\frac{\mathrm{d}C(t)}{\mathrm{d}t} = k_r\big(C_s(t) - C(t)\big) + \alpha I(t) - R \tag{8.39}$$

where the first term on the right-hand side describes the reaeration process, the second term describes the effects of photosynthesis, and R ($\mathrm{g/m^3\,h}$) represents the respiration rate due to decay of organic matter. Furthermore, k_r is the reaeration coefficient ($1/\mathrm{h}$), and α the photosynthesis rate coefficient ($\mathrm{g/m\,h\,W}$).

[1] The data from "De Poel en 't Zwet," a lake situated in the western part of the Netherlands, for the period 21–30 April 1983, were collected by students of the University of Twente.

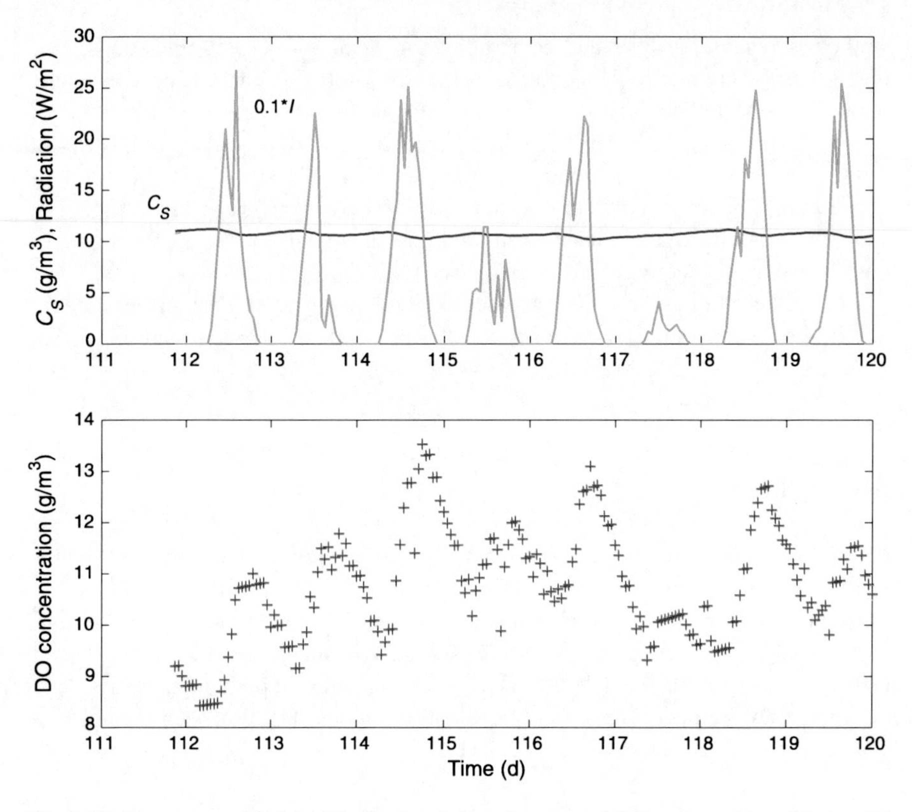

Fig. 8.11 Measurements in lake "De Poel en 't Zwet", saturated DO concentrations (*dark solid line*), radiation (*light solid line*) (*top figure*), and DO concentrations (*plus signs*) (*bottom figure*)

Let us start with a *nonparametric prediction approach*. Then, a rough prediction of the DO concentration at 120.25 d, on the basis of data only, can be given by the mean value plus standard deviation. Hence, $C(120.25) = \langle 1 \rangle$ g/m^3 with a standard deviation of $\langle 2 \rangle$ g/m^3, a rather uncertain estimate! Alternatively, the unknown-but-bounded estimate can be given. This estimate is given by

$$C(120.25) \in \left[\min C(t), \max C(t)\right] \quad \text{for } 111.875 \leq t \leq 120$$

that is, the interval $[\langle 3a \rangle, \langle 3b \rangle]$ g/m^3, a very wide range! For obtaining more accurate results, in what follows a stochastic uncertainty modeling approach is used.

A more advanced prediction is found when the trend is taken into account. Hence, the following predictor is formulated:

$$C(120.25) = C(t_0) + a(t - t_0)$$

with $t_0 = 111.875$ d. Notice that we had to find first estimates of the unknowns $C(t_0)$ and a. Hereto the following regression (or static linear) model in vector-matrix notation is formulated:

$$y = \begin{bmatrix} 1 & t_1 - t_0 \\ 1 & t_2 - t_0 \\ \vdots & \vdots \\ 1 & t_N - t_0 \end{bmatrix} \begin{bmatrix} C(t_0) \\ a \end{bmatrix} + v$$

where $y = [C(t_1), C(t_2), \ldots, C(t_N)]^T$ and $t_1 = 111.917$, $t_2 = 111.958, \ldots, t_N =$ 120 d. From the available data we find the following estimate of $\vartheta^T = [C(t_0)\ a]^T$, i.e., ⟨4⟩ with covariance matrix of the estimation errors $\text{Cov}\,\widehat{\vartheta} =$ ⟨5⟩. Consequently, the prediction is given by $C(120.25) =$ ⟨6⟩ g/m^3 with prediction uncertainty of ⟨7⟩ g/m^3.

In conclusion, these short-term predictions, implicitly using linear static models, seem to be not very reliable, although they can be good estimates for the mean value on day 120. Clearly the dynamics present in the available data are missing. Extension toward sinusoidal models, as, for instance, in Fourier analysis (see Appendix C for details on the Fourier transform), could be a possibility.

However, in the following, we will take a different point of view by incorporating prior knowledge when making predictions and which is known as the *parametric* or *model-based prediction approach*.

Let us first derive the discrete-time equivalence of (8.39) with sampling interval T_s, i.e., using the general solution to a set of linear differential equations as in (1.3) (see also footnote 1 in Chap. 1), with $u(t)$ piece-wise constant and t, T_s in hours,

$$x(t) = e^{AT_s} x(t - T_s) + \left(e^{AT_s} - I\right) A^{-1} B u(t - T_s)$$

Notice that this solution is a generalization of the linear differential equation solution presented in Example 1.4, but now for the special case that $u(t)$ is constant on the interval $[t - T_s, t]$.

Hence,

$$C(t) = e^{-k_r T_s} C(t - T_s) + \frac{e^{-k_r T_s} - 1}{-k_r} \left(k_r C_s(t - T_s) + \alpha I(t - T_s) - R\right)$$

Recall that $e^{-k_r T_s} = 1 - k_r T_s + \frac{1}{2}(k_r T_s)^2 - \frac{1}{6}(k_r T_s)^3 + \cdots$, so that with $k_r T_s$ small, the following approximation is valid: $e^{-k_r T_s} \approx 1 - k_r T_s$, and thus the DO model (8.39) can be written as

$$C(t) = (1 - k_r T_s) C(t - T_s) + k_r T_s C_s(t - T_s) + \alpha T_s I(t - T_s) - R T_s \tag{8.40}$$

the so-called Euler approximation of (8.39). Since the data is hourly sampled, the sampling interval $T_s = 1$ h!

Assume that $k_r = 0.1$ 1/h, $\alpha = 0.002$ g/m h W, and $R = 0.1$ g/m^3 h. Consequently, $k_r T_s = 0.1$ is small, so that higher-order terms in the approximation of

$e^{-k_r T_s}$ can be neglected, and thus the Euler approximation is valid. Furthermore, assume that C_s on the interval [120, 120.25] d is equal to 10.7 g/m^3 and the radiation is zero (see Fig. 8.11). Hence, at the next sampling instant (1:00 a.m.) and using (8.40), we obtain, with $T_s = 1$ h at day 120.042,

$$C(t+1) = (1-k_r)C(t) + k_r C_s(t) + \alpha I(t) - R$$
$$= \langle 8 \rangle$$

in g/m^3. At the next time instant we obtain $\langle 9 \rangle$ g/m^3 etc., so that at day 120.25, $C(t) = \langle 10 \rangle$ g/m^3. The prediction uncertainty at 1:00 a.m., given an initial uncertainty in terms of the covariance matrix $P(0)$ in the DO concentration of 0.1 g/m^3 at day 120.000, and in this case simply the variance of the prediction, is found from $[1\ k_r]^T\ P(0)\ [1\ k_r] = \langle 11 \rangle$ (see Chap. 7). Notice that the contribution of the measurement uncertainty at 1:00 a.m. is not taken into account. Hence, the noise-free model output and not the sensor output is predicted! Consequently, at day 120.25 we obtain a variance of $\langle 12 \rangle$ (g/m^3)^2, that is, a standard deviation of $\langle 13 \rangle$ g/m^3.

If, on the contrary, the parameters k_r, α, and R are unknown, we should first estimate these from the available data. Hereto the following regression model in vector-matrix notation with parameter vector $\vartheta := [k_r\ \alpha\ R]^T$ is defined:

$$y = \Phi\vartheta + e$$

where

$$y = \begin{bmatrix} C(1) - C(0) \\ C(2) - C(1) \\ \vdots \\ C(N) - C(N-1) \end{bmatrix}$$

and

$$\Phi = \begin{bmatrix} C_s(0) - C(0) & I(0) & -1 \\ C_s(1) - C(1) & I(1) & -1 \\ \vdots & & \\ C_s(N-1) - C(N-1) & I(N-1) & -1 \end{bmatrix}$$

Notice that the parameter estimation problem of the discretized dynamic model is formulated as a linear regression problem. Then standard least-squares estimation leads to

$$\widehat{\vartheta} = \begin{bmatrix} \widehat{k_r} \\ \widehat{a} \\ \widehat{R} \end{bmatrix} = \langle 14 \rangle, \qquad \text{Cov}\,\widehat{\vartheta} = \langle 15 \rangle$$

Consequently, as before, the model output at 1:00 a.m., that is, at day 120.042, is calculated as follows using the estimated values of k_r and R: $C(t=1) =$

$(1-k_r)C(0)+k_r C_s(0)+R = \langle 16 \rangle$ g/m^3. Finally, at 6:00 a.m., thus for day 120.25, we find $C(t) = \langle 17 \rangle$ g/m^3. Notice that now smaller values are found because the aeration coefficient is approximately 20% smaller than before and the respiration is some 20% higher. For the propagation of the parameter estimation uncertainty to the uncertainty in the prediction, the discretized DO model is rewritten in terms of the estimated parameter vector $\widehat{\vartheta}$, that is,

$$C(t) = \left[C_s(t-1) - C(t-1) \; I(t-1) \; -1\right]\widehat{\vartheta} + C(t-1)$$

Hence, at 1:00 a.m. the variance of the prediction $P(1)$ is given by

$$\begin{aligned} P(1) &= \Phi(0)\,\mathrm{Cov}\,\widehat{\vartheta}\,\Phi(0)^T + P(0) \\ &= \left[C_s(0) - C(0) \; 0 \; -1\right]\mathrm{Cov}\,\widehat{\vartheta} \begin{bmatrix} C_s(0) - C(0) \\ 0 \\ -1 \end{bmatrix} + P(0) \\ &= \langle 18 \rangle \end{aligned}$$

where the uncertainty effect in $C(0)$ on the parameter uncertainty propagation, represented by the term $\Phi(0)\,\mathrm{Cov}\,\widehat{\vartheta}\,\Phi(0)^T$, is neglected; only the direct effect is taken into account via the covariance matrix $P(0) = 0.1$. Hence, the standard deviation in this one-step-ahead prediction is equal to $\langle 19 \rangle$ g/m^3. Notice that for the next step, this procedure can be repeated, but now the predicted value of the DO concentration is needed, giving $P(2) = \langle 20 \rangle$. Finally, at day 120.25 we find $P(t) = \langle 21 \rangle$, and thus the standard deviation is equal to $\langle 22 \rangle$ g/m^3, which is mainly affected by the initial DO concentration uncertainty at $t = 0$, thus at day 120.000.

Alternatively, a Monte Carlo approach can be performed, but this is out of the scope of this exercise. In general, a Monte Carlo approach is preferred when the system is complex and analytical error propagation rules cannot be easily found. In this case the analytical approach is chosen, because more insight in the error propagation process is obtained. If, instead of the continuous-time, DO model is used, nonlinear estimation and error propagation problems will appear (see [vSK91]).

Part IV
Model Validation

In the previous parts, from data-based identification to time-invariant/time-varying system identification, many methods have been introduced to find an appropriate model structure, with or without using prior knowledge, from experimental data. From Fig. 1.7 it can be seen that the final step in a single system identification loop is model validation. In this step the user has to decide whether the identified model is appropriate or not. This part of the book will therefore focus on methods that support the user in making the right decisions about the validity of the mathematical model of the system. To be a little bit more precise, and in line with the Popperian philosophy, validation does not usually guarantee validity, but just tries to test adequacy or fails to establish invalidity.

The use of prior knowledge, model experience, and experimental data in the model validation step is emphasized and basically illustrated by a couple of examples. After introducing the methods for model validation, a real-world application, related to perishable food storage, will be extensively introduced and discussed in terms of model validation.

Chapter 9
Model Validation Techniques

After having identified a model, in a model validation step the identified model is usually evaluated with respect to

(i) prior knowledge,
(ii) model behavior in numerical experiments,
(iii) experimental data.

Notice that these items are also input to the system identification procedure (see Fig. 1.7), where the second item is related to the final modeling objective. In what follows, each of these aspects in a model validation step will be considered and illustrated in some more detail.

9.1 Prior Knowledge

A first test whether the identified model is appropriate is by evaluating the estimated parameter values. In particular, the knowledge of a priori parameter bounds is very useful. For example, in a physical modeling approach we expect positive parameter values, and thus negative estimates found after a formal calibration step indicate model inappropriateness. Consider the following biochemical example as an illustration of this.

Example 9.1 Substrate consumption: Frequently, the substrate consumption in a reactor is expressed in terms of Michaelis–Menten kinetics. Hence, the following discrete-time model with unit time step is a good starting point for describing the substrate concentration in a batch reactor:

$$S(t) = S(t-1) - \mu \frac{S(t-1)}{K_S + S(t-1)}, \quad S(0) = S_0 \tag{9.1}$$

with $S(t)$ the substrate concentration at time instant t, $\mu > 0$ the maximum degradation rate of a substrate, and $K_S > 0$ the corresponding half saturation constant.

K.J. Keesman, *System Identification*,
Advanced Textbooks in Control and Signal Processing,
DOI 10.1007/978-0-85729-522-4_9,

Table 9.1 Substrate data

t (min)	7	17	19
$y(t)$ (g/m^3)	16.9	4.48	2.42

Notice that this model is nonlinear in the parameters μ and K_S. Let measurements of $S(t)$ be denoted by $y(t)$; Table 9.1 shows the next three measurements.

In fact, in this example the measurements have been generated under the assumption that $S_0 = 30\ \text{g/m}^3$, $\mu = 2\ \text{min}^{-1}$, and $K_S = 2\ \text{g/m}^3$ and using some additive noise. The application of a nonlinear least-squares estimation procedure, as implemented by the MATLAB function *lsqnonlin*, with starting values $\mu = 2$ and $K_S = 0.001$ results in $[\widehat{\mu}\ \widehat{K}_S] = [2.26158\ -0.15417]$. Hence, the estimate $\widehat{K}_S$ violates the positivity constraint on the possible values of K_S.

Hence, this example clearly shows that, if no physical knowledge in terms of bounds is used and the estimation is merely treated as a fitting problem, we could easily end up in, as we will see later, a local minimum, because of a possible singularity in the equation. Recall that, as illustrated in Example 5.25 for a similar model, model reparameterization of (9.1) into a linear regression would avoid this problem of a local minimum.

Introduction of parameter bounds will keep the parameter estimates in the right physical range, but this will at the same time limit the model output to fit the data. However, positive estimates alone do not suffice, because it is also important to evaluate the corresponding estimation variances. For instance, a positive parameter estimate with a coefficient of variation (that is, the ratio of standard deviation to mean value) of one is not very reliable. In this case, we could consider to fix or to remove the corresponding term from the model. However, before removing terms, it is always good practice to evaluate the output sensitivity with respect to this parameter for checking its practical identifiability. If, for instance, the practical identifiability is low, a better experiment design should be considered, if possible.

9.2 Experience with Model

9.2.1 Model Reduction

For further use of the model of an LTI system, it is always sensible to investigate the possibilities for model reduction by pole-zero cancelation, as in the next example.

Example 9.2 Pole-zero cancelation: Consider a discrete-time state-space model with

$$A = \begin{bmatrix} -1 & 0 \\ 0 & 0 \end{bmatrix}, \qquad B = \begin{bmatrix} 1 \\ 0 \end{bmatrix}, \qquad C = \begin{bmatrix} 1 & 0 \end{bmatrix}, \qquad D = 0$$

The corresponding transfer function in z-domain (see Appendix C), using the expression $G(z) = C(zI - A)^{-1}B + D$ (see also (E.4)), is given by

$$G(z) = \frac{z}{z(z+1)} = \frac{1}{(z+1)}$$

Consequently, after eliminating the pole and zero at $z = 0$, a less complex model results with the same input–output behavior as the original system. Consequently, most likely fewer parameters need to be estimated.

It is common practice to cancel poles and zeros that are close to each other. Consequently, the input–output properties of the original and reduced model after pole-zero cancelation will not be exactly the same, but this deviation can be specified beforehand.

9.2.2 Simulation

In addition to the evaluation of the parameter estimates and, for LTI systems, checking possible pole-zero cancelation, the output of the model should also be evaluated. Typically, the model output is evaluated in simulation mode. Let us first demonstrate this by the substrate consumption example.

Example 9.3 Substrate consumption: Calculation of the corresponding model output using difference equation (9.1), which describes the substrate concentration in the batch reactor, with the estimates

$$\left[\widehat{\mu}\ \widehat{K}_S\right] = [2.26158\ -0.15417]$$

leads to the response as shown in Fig. 9.1. In this figure, the original noise-free model output (thin line) and the measurements ($\oplus$) are also shown.

This graphical result directly shows why the particular parameter combination leads to a local minimum for a negative parameter value of K_S. It, furthermore, illustrates the sudden unrealistic increase in substrate concentration when the concentration comes close to zero.

Simulation results may also indicate over- and under-modeling, as in the next example.

Example 9.4 Integrator: A pure integrator in discrete-time is given by

$$x(t) = x(t-1) + u(t-1)$$
$$y(t) = x(t)$$

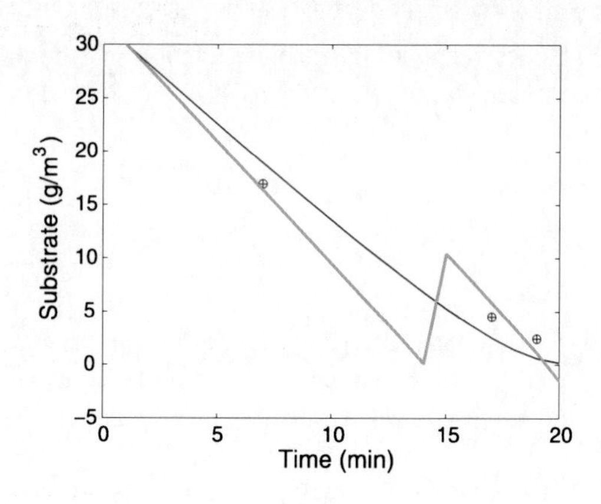

Fig. 9.1 Model responses (noise-free: *thin line*, estimated: *bold line*) and measurements ($\oplus$) of the substrate concentration in a batch reactor

with discrete-time transfer function

$$y(t) = \frac{q^{-1}}{1 - q^{-1}} u(t)$$

For the noise-free case, several ARX models, as ARX(2, 2, 1) and ARX(3, 3, 1), give a perfect fit. If we add a noise term $e(t)$ to the output, such that $e(t) = 0.2y(t)w(t)$ with $w(t) \in N(0, 1)$ for all t, we obtain for an ARX(5, 5, 1) model structure the following estimation result:

$$\begin{array}{llllll} B(q) = & 0.9293q^{-1} & + 0.7220q^{-2} & + 0.6299q^{-3} & + 0.3144q^{-4} & + 0.2508q^{-5} \\ & \pm 0.1561 & \pm 0.2076 & \pm 0.2220 & \pm 0.2337 & \pm 0.2073 \end{array}$$

$$\begin{array}{llllll} A(q) = 1 & - 0.2862q^{-1} & - 0.0534q^{-2} & - 0.3654q^{-3} & - 0.1243q^{-4} & - 0.1418q^{-5} \\ & \pm 0.1075 & \pm 0.1116 & \pm 0.1092 & \pm 0.1197 & \pm 0.1161 \end{array}$$

Consequently, the last two terms of each polynomial and the third term of $A(q)$ are unreliable and can possibly be neglected. In Fig. 9.2 the simulation results of the pure integrator and of the ARX(5, 5, 1) model are presented. Clearly, the ARX(5, 5, 1) model fits the noise to a large extent. This phenomenon is even more visible when we evaluate the high-frequency region in the Bode plots (see Appendix D) of the pure integrator and ARX(5, 5, 1) model, as in Fig. 9.3. Hence, both the parameter estimation and simulation results indicate over-modeling. Notice that, especially in the low frequency region of the Bode plots, a large misfit appears. Hence, in time domain we expect a large difference when applying a low frequency signal, as a unit step, to the model. Figure 9.4 shows the unit step responses of both models, and this figure confirms what we expected from the analysis of the Bode plots.

On the contrary, let us neglect the effects of the input signal and consider the very simple model

$$x(t) = x(t - 1) \tag{9.2}$$

$$y(t) = x(t) \tag{9.3}$$

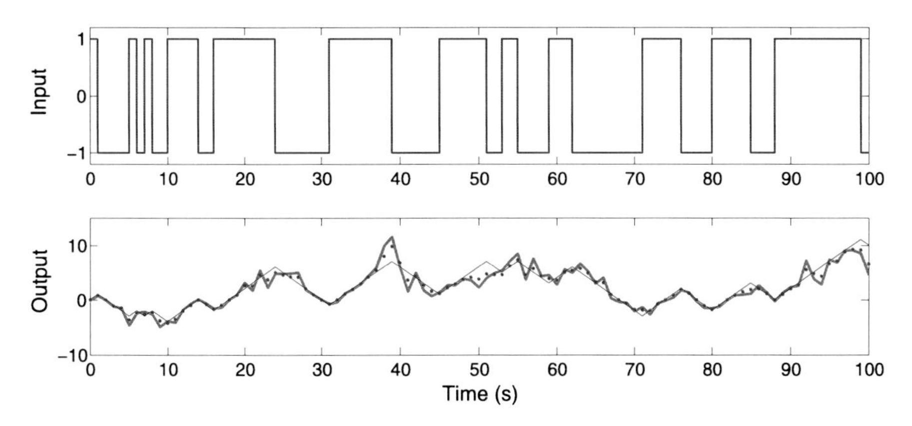

Fig. 9.2 Input signal (*top panel*), noisy output ($\cdot$) and simulation results of a pure integrator (*thin line*) and ARX(5, 5, 1) model (*bold line*) (*bottom panel*)

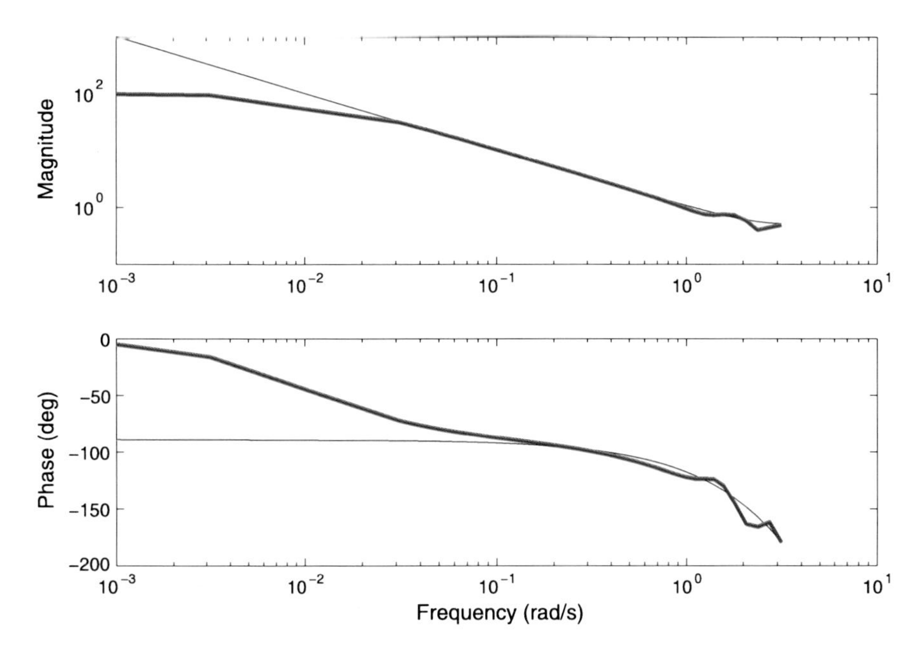

Fig. 9.3 Bode plots of pure integrator (*thin line*) and ARX(5, 5, 1) model (*bold line*)

Given this model and the data in Fig. 9.2, we find that $\widehat{x}(t) = 2.1703 \pm 3.1623$ for all t. Consequently, the predicted model output is constant and rather uncertain. Notice from Fig. 9.2 that this result is not fully supported by the data, unless we attribute the resulting structured noise to sensor noise. Recall that we start from the introduced structured noise into this example, and thus the last conclusion is, in fact, partly correct. Hence, applying (9.2) will typically lead to under-modeling.

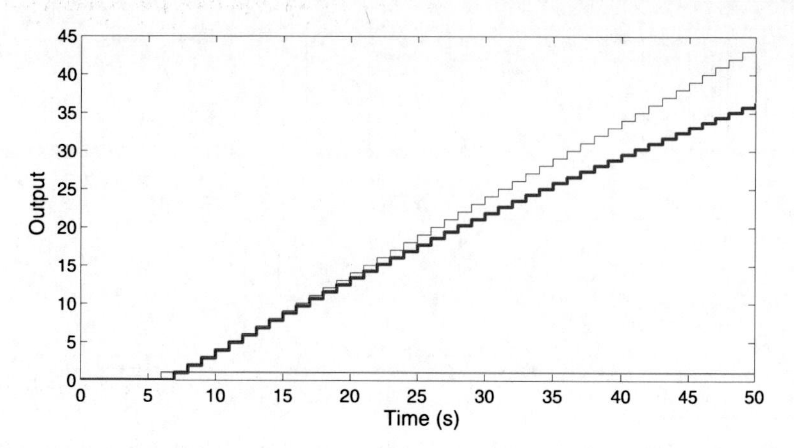

Fig. 9.4 Step responses of pure integrator (*thin line*) and ARX(5, 5, 1) model (*bold line*)

Table 9.2 Process data

t	1	2	3	4
$y(t)$	5.2	5.3	5.1	5.0

9.2.3 Prediction

Sometimes model simulation studies are not able to elucidate the deficiencies in the model structure that could be made visible by model predictions. In the next example, we demonstrate the use of model predictions in the model validation step.

Example 9.5 Constant process with noise: Presume that from prior expert knowledge we know that the process is more or less in steady state. Let, furthermore, the following measurements, as in Table 9.2, be given.

The polynomial model

$$y(t) = \vartheta_0 + \vartheta_1 t + \vartheta_2 t^2 + \vartheta_3 t^3$$

with $\vartheta_0 = 4.4$, $\vartheta_1 = 1.2833$, $\vartheta_2 = -0.55$, and $\vartheta_3 = 0.0667$ results in a perfect fit on the interval [1, 4], see Fig. 9.5. Thus, the model simulation results do not indicate any model deficiency. However, model predictions outside the range and for increasing t tend to go to infinity. Clearly, from the prior process knowledge this is not expected, and thus the polynomial model does not pass the validation test. The sine function

$$y(t) = \vartheta_0\left(\sin\left(\frac{2\pi t}{0.25} + \vartheta_1\right) + \vartheta_2\right)$$

with $\vartheta_0 = 0.7418$, $\vartheta_1 = 1.4464$, and $\vartheta_2 = 5.9499$ shows a good fit with respect to the data, too. However, neither the data nor the prior knowledge do support a model output with this frequency. As in the integrator example (Example 9.4), we

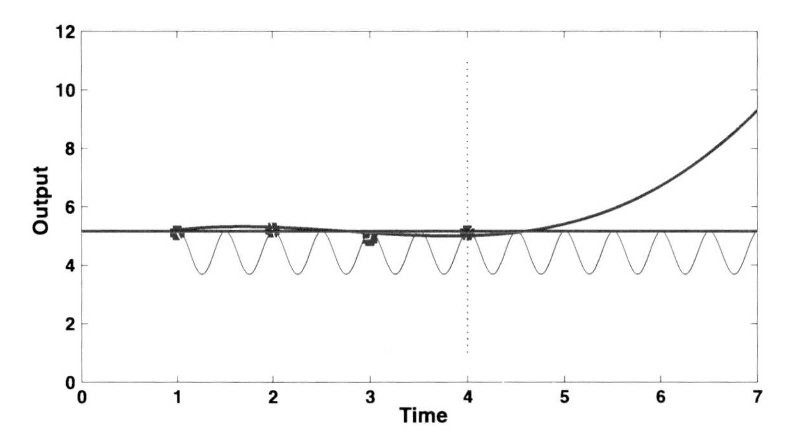

Fig. 9.5 Measurements and model predictions

see in both cases the effect of over-modeling. Given the prior process knowledge, the model $y(t) = c$ with $\widehat{c} = 5.15$ seems to be the most appropriate model.

9.3 Experimental Data

In addition to an evaluation of the parameter estimates with respect to prior knowledge and of the model simulations and predictions, in this section we will explicitly use the residuals for model validation. Recall that the residuals are defined as

$$\varepsilon(t) := y(t) - \widehat{y}(t|\vartheta) \tag{9.4}$$

The residuals do play a key role in the model validation process, since they reflect the difference between the measured output $y(t)$ and the predicted model output $\widehat{y}(t|\vartheta)$, given an estimate of ϑ. In the following, we therefore introduce some common residual tests and finalize this section with a real-world example.

9.3.1 Graphical Inspection

A first test on the residuals should be based on graphical inspection. Plotting the residuals, at an appropriate scale, will directly show some peculiarities, as outliers, drift and periodicity, see Fig. 5.4 and Fig. 7.5 in previous chapters. Let us also demonstrate this graphical inspection to the moving object of Example 7.3.

Example 9.6 Moving object (*constant velocity*): Recall that the following observations on an object moving in a straight line with constant velocity v, as presented in Table 9.3, were available.

Table 9.3 Moving object data

Time t (s)	1	2	3	4	10	12	18
Distance $y(t)$ (ft)	9	15	19	20	45	55	78

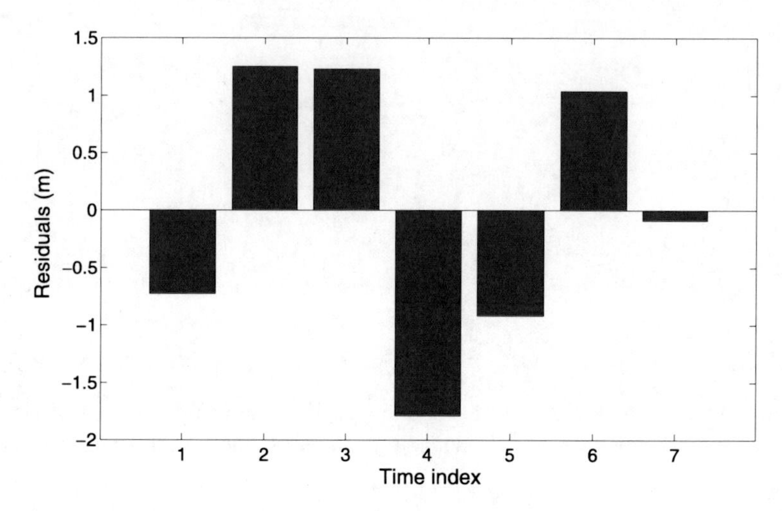

Fig. 9.6 Residuals

The proposed model for the moving object was

$$y(t) = s_0 + vt + e(t)$$

with final estimates $\widehat{s_0} = 5.7027$ ft and $\widehat{v} = 4.0215$ ft/s. In Fig. 9.6 the residuals $\varepsilon(t) := y(t) - \phi(t)^T\widehat{\vartheta}$, with $\vartheta = [s_0\ v]^T$, are plotted.

Notice from Fig. 9.6 that the observation at time index 4 is possibly an outlier. Furthermore, the residuals show some periodicity, but the time series is far too short to come up with firm statements. Hence, this clearly illustrates the problem of model validation for small data sets.

However, apart from the model validation problem related to small data sets, it is never clear beforehand whether drift and periodicity in the residuals originate from an invalid model; the experimental data may contain these characteristics as well. Hence, analysis of the experimental data, using, for instance, linear regression and correlation techniques, and examination of the sensor system may help to solve this dilemma. Furthermore, some basic properties of the prediction error sequence, as $\max_t|\varepsilon(t)| = \|\varepsilon\|_\infty$ and $\frac{1}{N}\sum^N \varepsilon(t)^2 = \|\varepsilon\|_2$, may also help to validate the model. For instance, when $\|\varepsilon\|_\infty$ is large, most likely outliers are present in the data, and thus, for an appropriate validation of the model, these should be removed. The 2-norm of the residuals can be used to compare models, and, under the assumption that the system is time-invariant, it indicates the expected magnitude of prediction errors. Hence, on the basis of these statistics, interpreted as quantities calculated from a set of data, one may or may not accept the model as valid.

9.3.2 Correlation Tests

Ideally, the residuals or prediction errors related to dynamic models should not depend on the inputs or previous residuals. If that is not the case, there is room for model improvement. For instance, in case of a general transfer function model structure, the exogenous part $G(q)u(t)$ can be extended with delayed inputs or the noise model $H(q)e(t)$ modified. To check the dependencies, it is very natural to study the correlations between residuals and past inputs. Let N data points of the input and residuals, respectively, be given. Then, the cross-correlation function, see also Sect. 4.1, between input and residuals is given by

$$r_{u\varepsilon}(l) = \frac{1}{N-l}\sum_{i=1}^{N-l} u(i)\varepsilon(i+l) \tag{9.5}$$

Hence, if the cross-correlations are small, this indicates that the residuals, and thus the model output $y(t)$, do not contain any further information originated from past inputs. In particular, it should be noted that significant correlation for negative l indicates output feedback in the input.

In a similar way, we can use the auto-correlation function for investigating the correlations among the residuals. The autocorrelation function is given by

$$r_{\varepsilon\varepsilon}(l) = \frac{1}{N-l}\sum_{i=1}^{N-l} \varepsilon(i)\varepsilon(i+l) \tag{9.6}$$

As mentioned before, the auto-correlation function can be used to test whether the residuals are white and thus do not contain any further information that can be used to improve the model predictions. A popular test for whiteness of the residuals, implicitly assuming that the residuals are normally distributed and within a range of M data points, is

$$\frac{N}{\widehat{r}_{\varepsilon\varepsilon}(0)^2}\sum_{l=1}^{M} \widehat{r}_{\varepsilon\varepsilon}(l)^2 \leq \chi^2_\alpha(M) \tag{9.7}$$

with $\chi^2_\alpha(M)$ the α-level of the $\chi^2(M)$-distribution (see Appendix B). Hence, if this inequality holds, we may conclude that the residuals are serially uncorrelated over a range of M data points.

For a formal test on the statistically independence between residuals and inputs, we could check if the following holds for the estimated cross-correlations:

$$\left|\widehat{r}_{u\varepsilon}(l)\right| \leq \sqrt{\frac{P_1}{N}}N_\alpha \tag{9.8}$$

where $P_1 = \sum_{i=-\infty}^{\infty} r_{\varepsilon\varepsilon}(l) r_{uu}(l)$, and N_α denotes the α-level of the standard normal distribution, $N(0,1)$. Notice that, since the right-hand side of (9.8) does not depend on l, it is a constant. Apart from this formal test, we could also investigate the scatter plot of the pairs $(\varepsilon(t), u(t-l))$.

Let us demonstrate the correlation tests on different models of a mass-spring-damper system with known input and output.

Example 9.7 Mass-spring-damper: Let in discrete-time, the model of a mass-spring-damper system be given by

$$x(t) = \begin{bmatrix} 1 & 0.5 \\ -0.5 & 0.5 \end{bmatrix} x(t-1) + \begin{bmatrix} 0 \\ 0.5 \end{bmatrix} u(t-1)$$

$$y(t) = [1 \quad 0] x(t)$$

The corresponding discrete-time transfer function is

$$G(q) = \frac{0.25q^{-2}}{1 - 1.5q^{-1} + 0.75q^{-2}}$$

If, however, we select an ARX(1, 1, 0) and ARX(2, 1, 0) model structure, we obtain the discrete-time transfer functions

$$G_1(q) = \frac{0.1337}{1 - 0.9082q^{-1}}$$

and

$$G_2(q) = \frac{0.0762}{1 - 1.6979q^{-1} + 0.8735q^{-2}}$$

respectively. The corresponding residuals and correlation functions are presented in Figs. 9.7 and 9.8.

From Fig. 9.7 significant correlations between the residuals can be seen, indicating an inappropriate model structure. Furthermore, the cross-correlation function between input and residuals shows a clear peak at lag 2, which refers to a deficiency in the time lag of the model. Increasing the model complexity toward an ARX(2, 1, 0) model removes the significant correlations between the residuals, but the peak at lag 2 in the cross-correlation function remains, as expected. Obviously, an ARX(2, 1, 2) model, with a similar structure as the transfer function $G(q)$ derived from the discrete-time state-space model, gives a perfect fit.

However, unlike the previous example, in practice always some noise is present. For an evaluation of a noisy data case, the heating system (Example 2.2) is considered again.

Example 9.8 Heating system: The following responses to a random binary signal with switching probability (p_0) of 0.2 and 0.5, respectively, have been measured from a simple heating system (see Figs. 9.9 and 9.10). Recall that the input of the

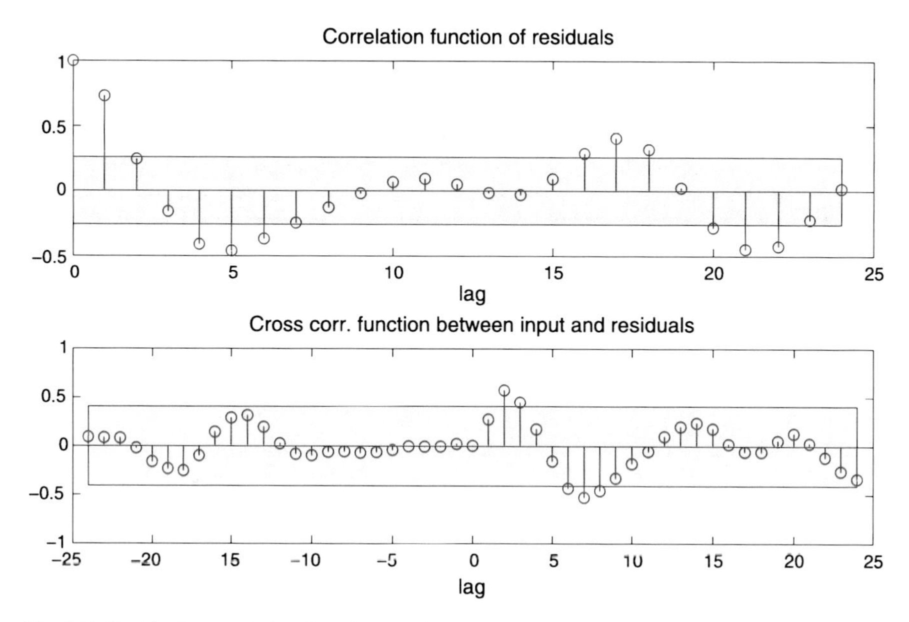

Fig. 9.7 Residuals, correlation functions, and α-levels (see (9.7) and (9.8)) related to ARX(1, 1, 0) model

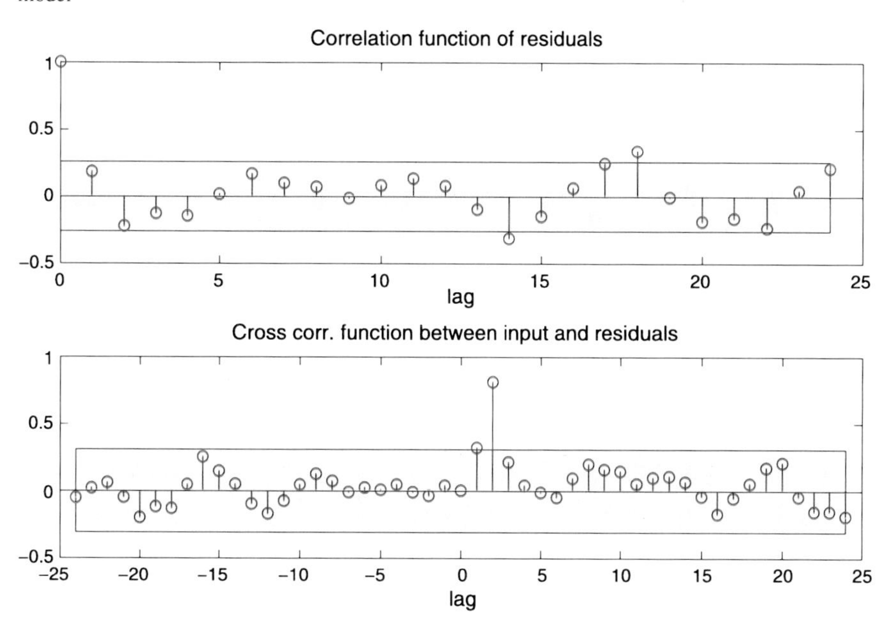

Fig. 9.8 Residuals, correlation functions, and α-levels related to ARX(2, 1, 0) model

system is the voltage applied to the heating element and the output, also in voltage, is measured with a thermistor. The maximum allowable magnitude of the input is 10 V, and the sampling interval is 0.08 s.

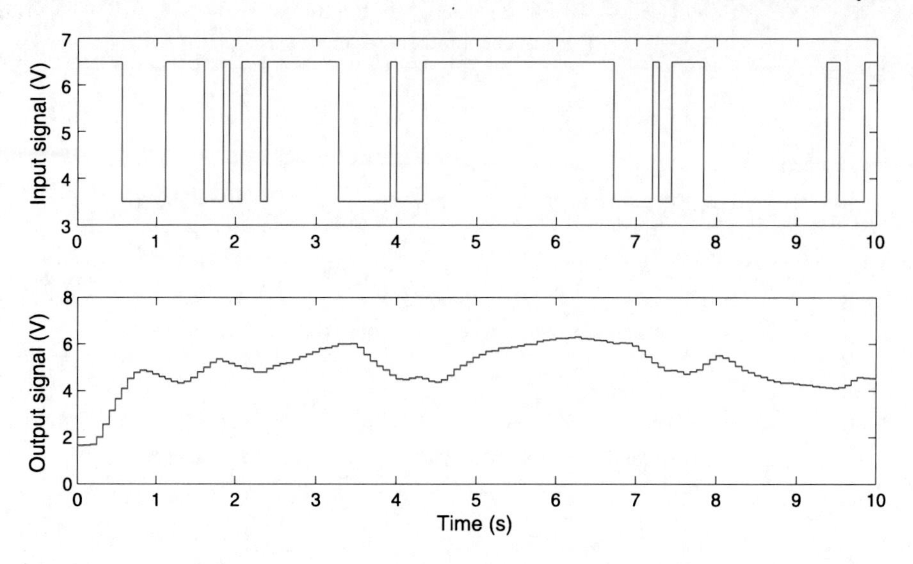

Fig. 9.9 Input and measured output signals for $p_0 = 0.2$

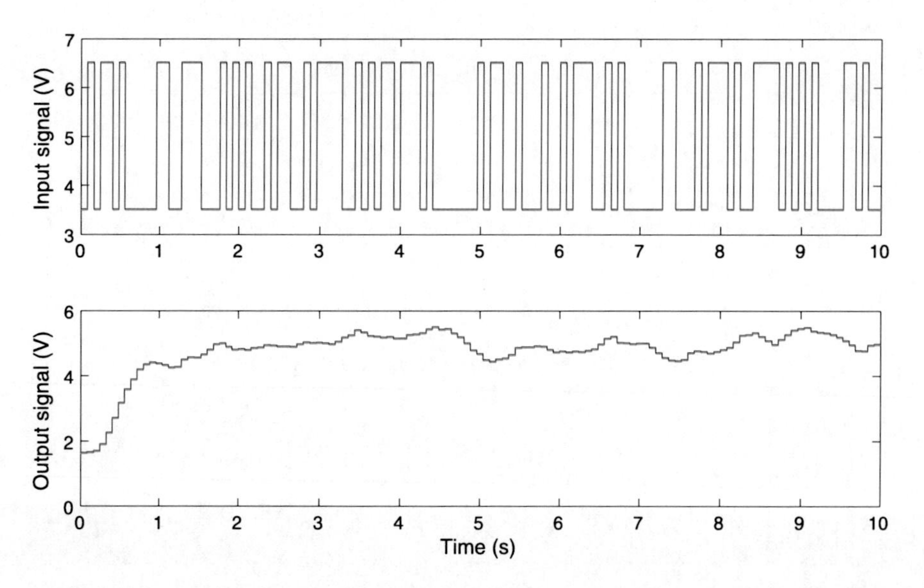

Fig. 9.10 Input and measured output signals for $p_0 = 0.5$

In both figures, the effect of the initial conditions is clearly visible. Furthermore, apart from the difference in switching probability, a similar behavior is seen. In an identification step, after neglecting the first 2 s of the data set and after detrending both the input and output signals, we found from the first data set, with $p_0 = 0.2$,

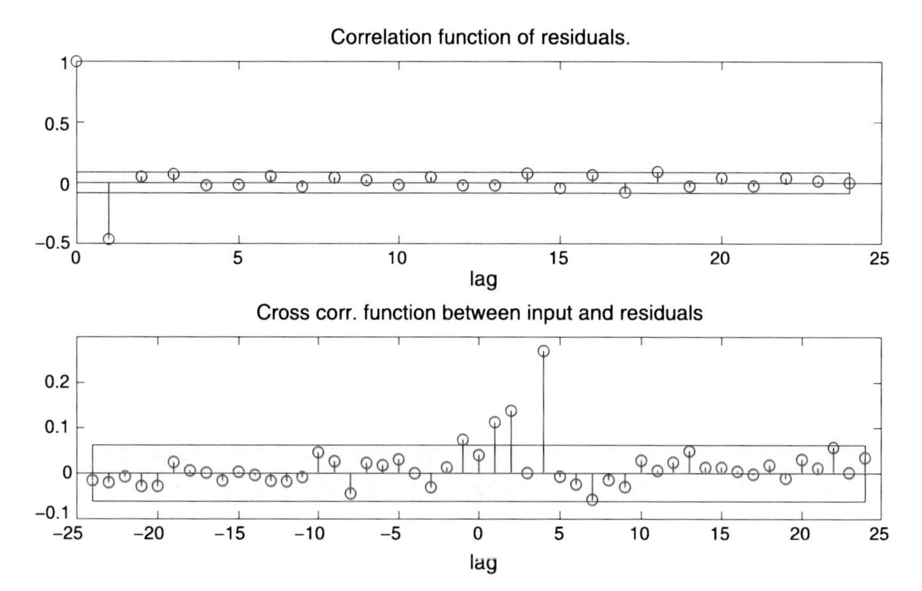

Fig. 9.11 Auto- and cross-correlation functions related to an ARX(2, 1, 3) model for $p_0 = 0.2$

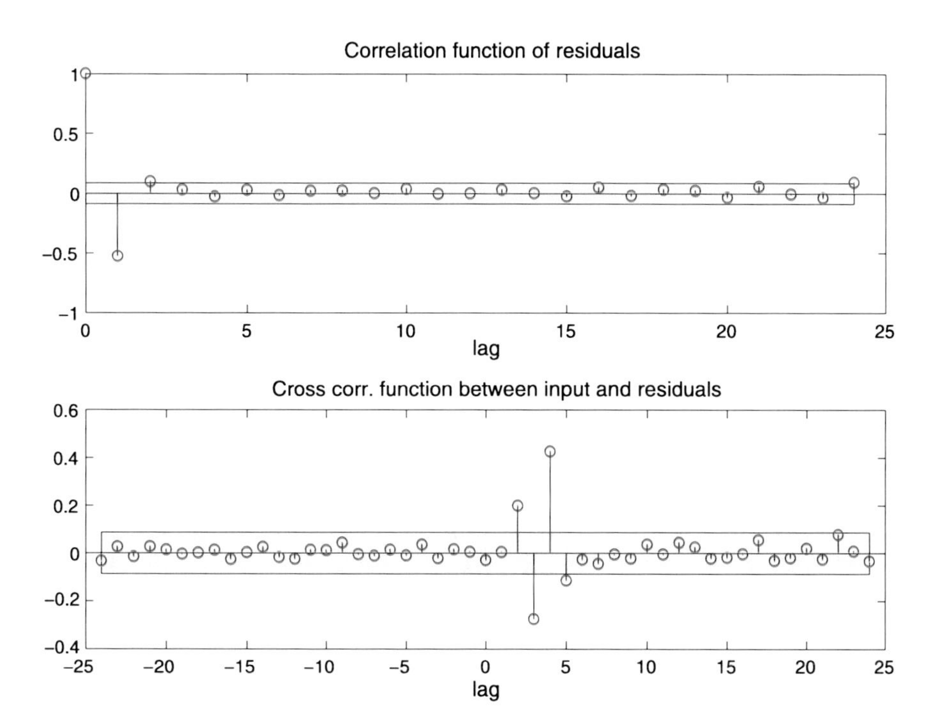

Fig. 9.12 Auto- and cross-correlation functions related to an ARX(2, 1, 3) model for $p_0 = 0.5$

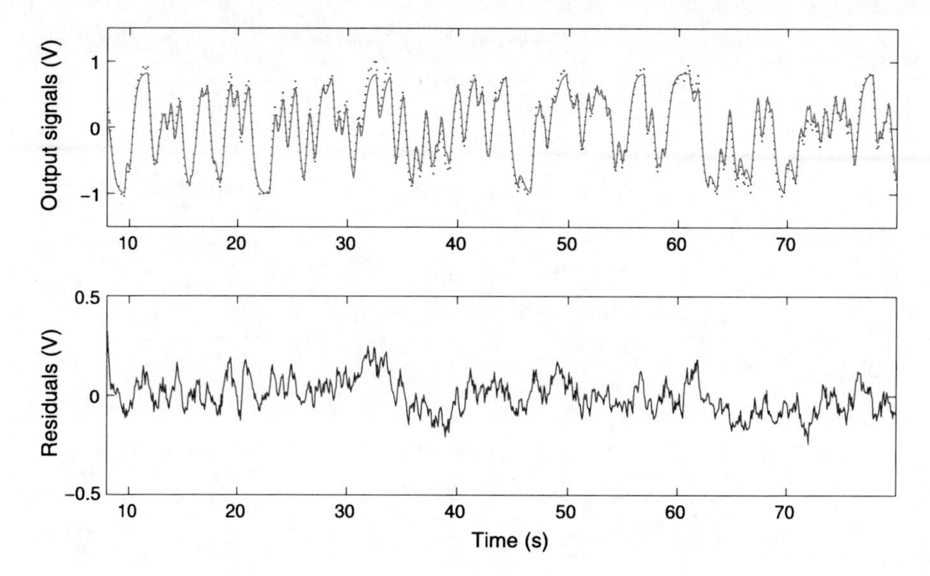

Fig. 9.13 Simulated (*solid line*) and measured (*dotted line*) output signals for $p_0 = 0.2$

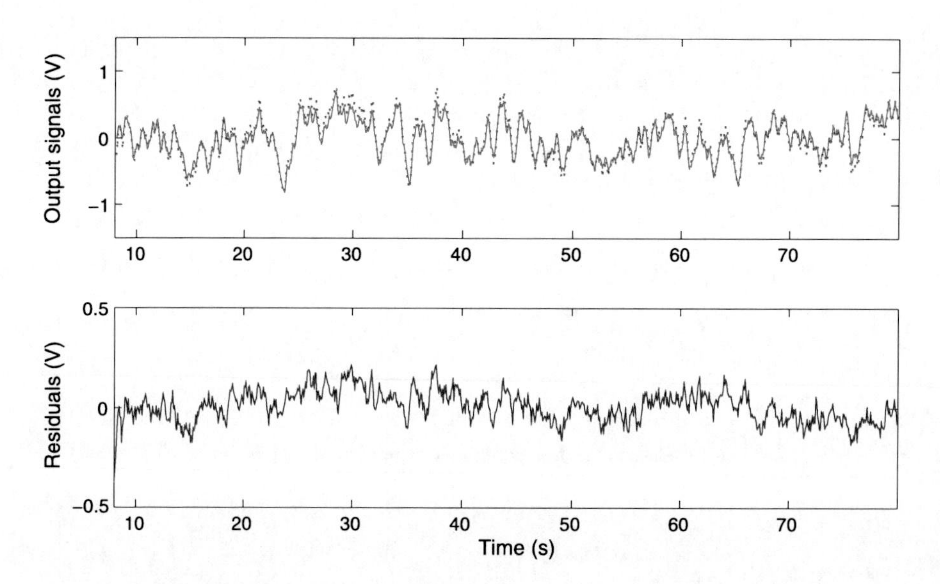

Fig. 9.14 Simulated (*solid line*) and measured (*dotted line*) output signals for $p_0 = 0.5$

the following ARX model:

$$G(q) = \frac{0.0507q^{-3}}{1 - 1.437q^{-1} + 0.520q^{-2}}$$

with a loss function value of 0.00181512 and an FPE function value of 0.00182725. In what follows, we fix this model structure and test it on the two data sets. The correlation functions related to both data sets are presented in Figs. 9.11 and 9.12. In both cases significant auto-correlations between the residuals at lag 1 can be seen. Furthermore, for lag 2–5, also significant cross-correlations between input and residuals are visible, indicating an additional time lag of 2, i.e., 0.16 seconds. Hence, at this point we conclude that there is some model deficiency.

At last, we evaluate the model behavior in time domain by simulating the ARX(2, 1, 3) model for both data sets. The results are presented in Figs. 9.13 and 9.14. Since the model shows a good behavior with respect to the observed data, our overall conclusion is that the model is appropriate, at least for short-term predictions. Thus, as yet, the ARX(2, 1, 3) model passes the model validation test.

Finally, in this subsection, we will demonstrate the use of predictions and experimental data in a *cross-validation* step by a real-world example.

Example 9.9 Storage facility (based on [KD09]): A discrete-time nonlinear model describing the temperature dynamics in a storage room with a respiring product and suitable for incorporation in a model-based control strategy is given by (see [KPL03])

$$\begin{aligned} T_p(t) = {} & \left(p_1 + \frac{p_2}{p_3 + p_4 u(t-1)} + \frac{p_5}{p_6 + p_7 u(t-1)}\right) T_p(t-1) \\ & + \frac{p_8 + p_9 u(t-1)}{p_3 + p_4 u(t-1)} T_e(t-1) + \frac{p_{10} + p_{11} u(t-1)}{p_6 + p_7 u(t-1)} X_e(t-1) \\ & + \left(p_{12} + \frac{p_{13}}{p_6 + p_7 u(t-1)}\right) \end{aligned} \tag{9.9}$$

where $T_p(t)$ is measured. The variable T_p denotes the temperature of the produce (°C), T_e is the external temperature (°C), X_e is the external absolute humidity (kg/kg), and $p = [p_1, \dots, p_{13}]^T$ the parameter vector. Finally, the control input u denotes the product of fresh inlet ratio and ventilation rate and is bounded by $0 \le u \le 1$. In Fig. 9.15 a schematic representation of the storage facility with corresponding variables is presented. The variables T_{in} (air temperature in channel), T_a (air temperature in bulk), X_{in} (absolute humidity in channel), and X_a (absolute humidity in bulk), as shown in the figure, do not appear in (9.9) as a result of a model reduction step based on singular perturbation analysis of the full system, see [KPL03] for details on this. In this model reduction step, quasi-steady states of air temperature and humidity were substituted in the heat balance of the product. This substitution finally leads to the rational terms, as in (9.9), and it enforces that the product temperature in (9.9) depends only on the external temperature T_e and external absolute humidity X_e.

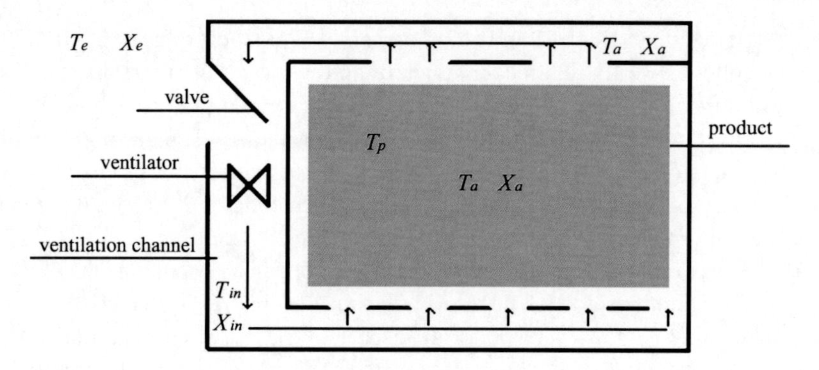

Fig. 9.15 Schematic representation of the storage facility

Rearranging (9.9) into a linear regression (see also Sect. 6.2.5 on model reparameterization) leads to

$$\begin{aligned} T_p(t) = \big[&u(t-1)T_p(t)\ u(t-1)^2T_p(t)\ T_p(t-1)\ u(t-1)T_p(t-1) \\ &u(t-1)^2T_p(t-1)\ T_e(t-1)\ u(t-1)T_e(t-1)\ u(t-1)^2T_e(t-1) \\ &X_e(t-1)\ u(t-1)X_e(t-1)\ u(t-1)^2X_e(t-1)\ u(t-1) \\ &u(t-1)^2\ 1\big][\theta_1 \cdots \theta_{14}]^T \end{aligned} \tag{9.10}$$

with $\theta = \varphi(p)$. After estimating $\theta = [\theta_1, \ldots, \theta_{14}]^T$ the model (9.10) can be rewritten in the nonlinear predictor form

$$\begin{aligned} \widehat{T}_p(t) = &\frac{1}{1-\widehat{\theta}_1 u(t-1) - \widehat{\theta}_2 u(t-1)^2}\big[\widehat{\theta}_3 T_p(t-1) + \widehat{\theta}_4 u(t-1)T_p(t-1) \\ &+ \widehat{\theta}_5 u(t-1)^2 T_p(t-1) + \widehat{\theta}_6 T_e(t-1) + \widehat{\theta}_7 u(t-1)T_e(t-1) \\ &+ \cdots + \widehat{\theta}_{13} u(t-1)^2\big] + \widehat{\theta}_{14} \end{aligned} \tag{9.11}$$

In this example, our focus was not so much on the reconstruction of the physical parameters p. Hence, a nonlinear estimation step from $\theta = \varphi(p)$ can be avoided. Our focus was on the performance of the predictors, (9.9) and (9.11). Both predictors were evaluated for two different data sets in terms of the mean square error (MSE) of the prediction errors. Notice, however, that if (9.10) is rewritten in the predictor form (9.11), a constraint on the estimated parameters $\widehat{\theta}_1$ and $\widehat{\theta}_2$ must be added, because for the whole range of u, the denominator of (9.11) should not be equal to zero. Hence, the constraint is given by

$$1 - \widehat{\theta}_1 u(t-1) - \widehat{\theta}_2 u(t-1)^2 \neq 0, \quad 0 \leq u \leq 1$$

If, however, the constraint is violated, the solution is rejected. In that case, the prediction at time instant t can simply be considered as infeasible.

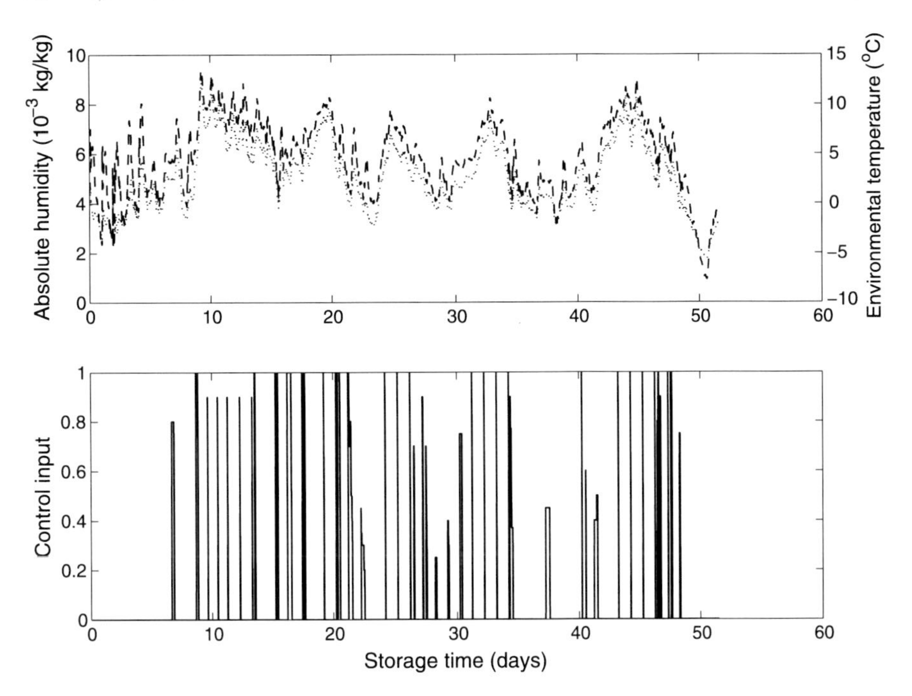

Fig. 9.16 Disturbance (T_e: - -; X_e: . . .) and control (u) inputs of calibration data set

Given input–output data, the parameters in (9.9) were estimated with a nonlinear least-squares (NLS) (see Sect. 5.2.2) method, while the parameters in (9.11) were estimated with direct estimation methods. The direct estimation methods applied here were the truncated least-squares (tLS) (see Sect. 5.1.6) and Generalized Total Least-Squares (GTLS) (see Sect. 5.1.7 for TLS) method. For details on GTLS, we refer to [VHV89]. In this specific application, two data-sets with measured variables, that is, T_p, T_e, X_e, and u, of about 50 days with a sampling interval of 15 minutes were available. The data were obtained from the same location during the same season, but for a different period within the season. All parameters were assumed to be constant during the whole season. The parameters were calibrated from one data set in the so-called calibration period. The prediction performance was subsequently evaluated using an open-loop prediction over the same data set and cross-validated over the second data set in the so-called validation period. See Figs. 9.16–9.17 for inputs to the system in these periods. Notice from Fig. 9.17 that in the first 15 days of this period the room was hardly ventilated, which significantly affected the product temperature, as we will see later on.

The MSE of predictor (9.9) with the parameters ($\widehat{p}$) estimated by an NLS algorithm and the MSE of predictor (9.11) with the parameters ($\widehat{\theta}$) estimated by the truncated LS and GTLS method, using data set 1 for calibration and data set 2 for validation, are presented in Table 9.4. From Table 9.4 it can be seen that the truncated LS method, assuming an equation-error structure in (9.10), gives good results for

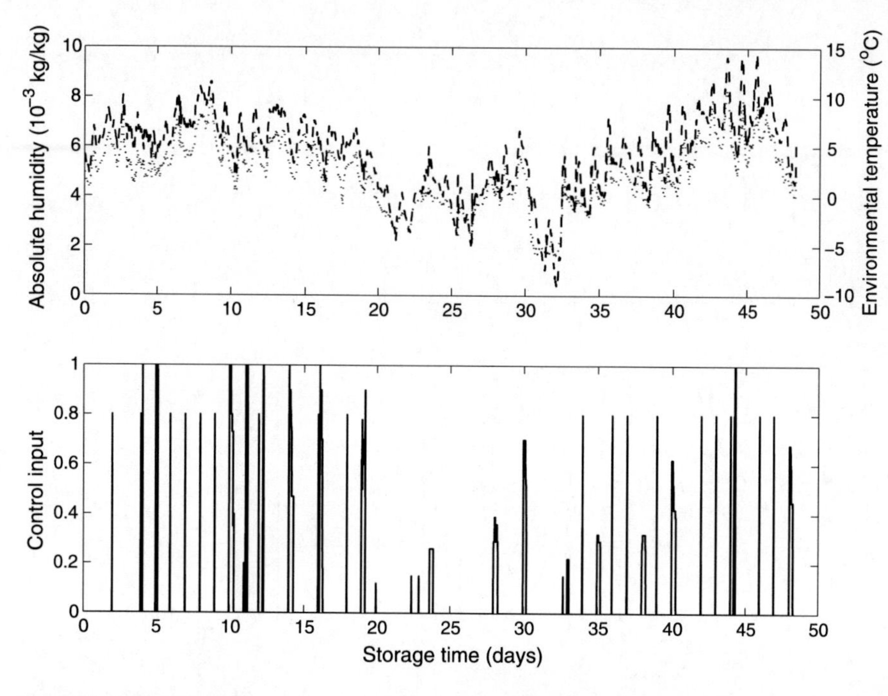

Fig. 9.17 Disturbance (T_e: - -; X_e: ...) and control (u) inputs of validation data set

Table 9.4 MSE of predictors with parameters estimated by NLS, truncated LS (tLS), and GTLS in the calibration (data set 1) and the validation (data set 2) period

	NLS	tLS	GTLS
Calibration	0.027	0.019	0.022
Validation	0.113	0.031	0.053

both the calibration and the validation periods. Although we know that the data matrix contains errors, it is not very likely that the results of GTLS estimation become significantly better as the error is probably close to the accuracy of the measurement device. Notice that the original model (9.9), in combination with an NLS parameter estimation method, shows the least predictive performance of all predictors for both periods. The predicted and measured temperatures using NLS, truncated LS, and GTLS estimation for both data sets are shown in Figs. 9.18 and 9.19.

Let us now switch the data sets. Consequently, data set 2 is used for estimation of $\widehat{p}$ and $\widehat{\theta}$, and data set 1 is used for validation. Furthermore, the same estimation, validation, and cross-validation procedures were performed. The results are given in Table 9.5 and Figs. 9.20 and 9.21.

After switching the data sets, several points are noticeable from Table 9.5. First, it is clear that the predictor with GTLS estimates has the best performance. Furthermore, the original predictor with NLS estimates has a better performance in the

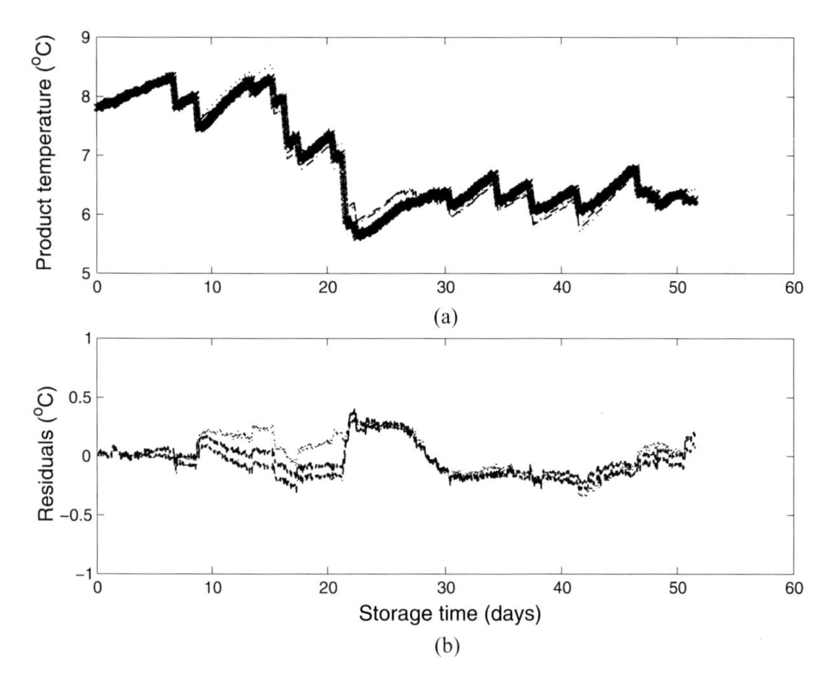

Fig. 9.18 (**a**) Measured (∗) and predicted product temperatures (NLS: …; tLS: -.-.; GTLS: - - -), and (**b**) residuals in the calibration period (data set 1)

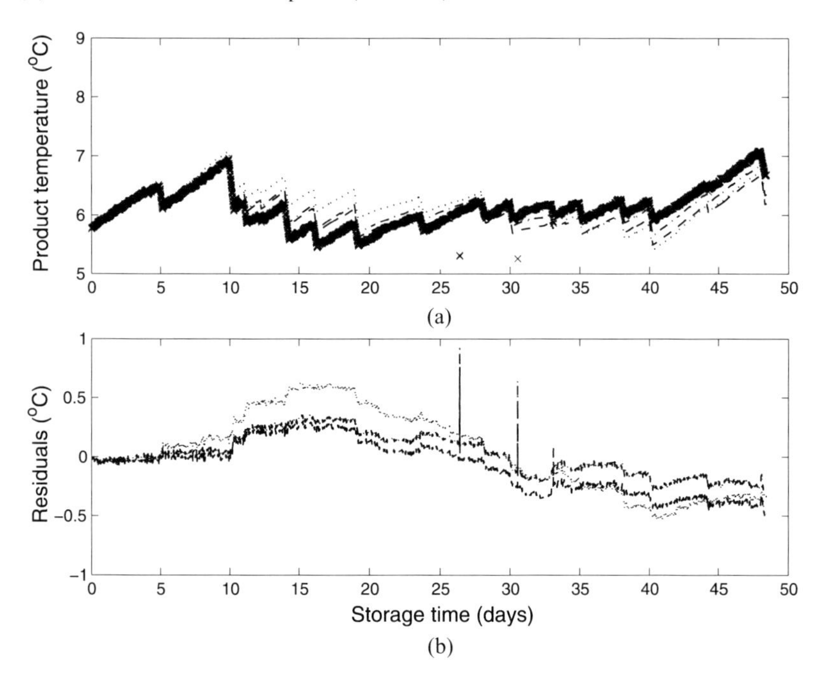

Fig. 9.19 (**a**) Measured (∗) and predicted product temperatures (NLS: …; tLS: -.-.; GTLS: - - -), and (**b**) residuals in the validation period (data set 2)

Table 9.5 MSE of predictors with parameters estimated by NLS, truncated LS (tLS), and GTLS in the calibration (data set 2) and the validation (data set 1) period

	NLS	tLS	GTLS
Calibration	0.097	0.035	0.017
Validation	0.036	0.766	0.016

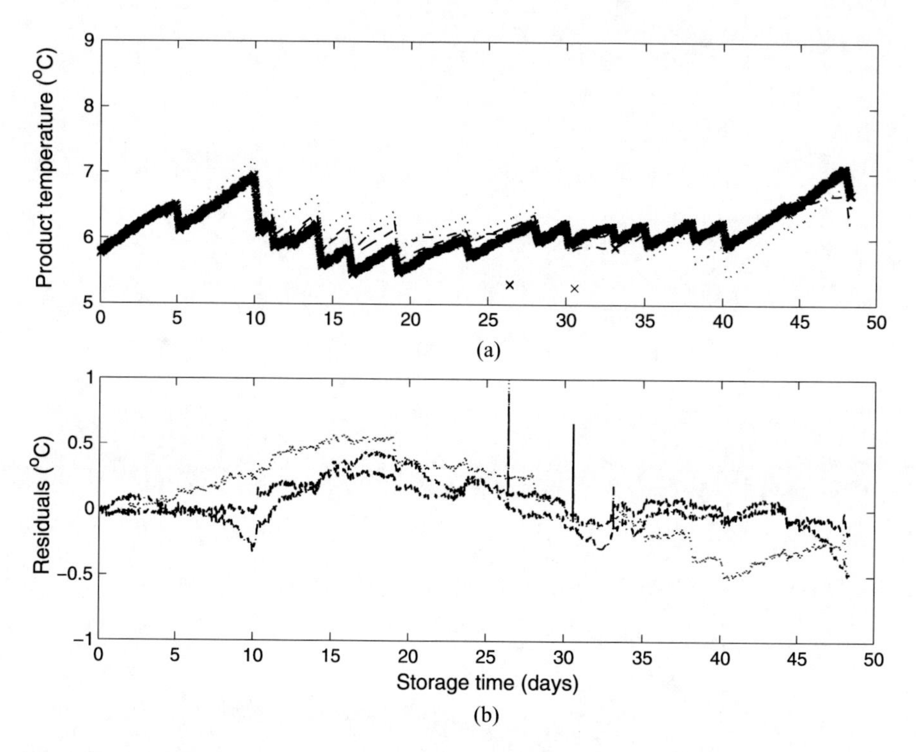

Fig. 9.20 (**a**) Measured ($*$) and predicted product temperatures (NLS: …; tLS: -.-.; GTLS: - - -), and (**b**) residuals in the calibration period (data set 2)

validation period than in the calibration period. Finally, the predictor (9.11) with the truncated LS estimates performs poorly. If the prediction performance of the predictor with the truncated LS estimates is further analyzed, it can be seen in Fig. 9.21 that from day 0 to 20 the predictor has very poor performance, but from day 25 till the end of the period it performs quite well. A possible explanation is that the truncated LS estimates are very sensitive to lack of information in the data set. If the calibration data set is informative enough (Fig. 9.18), that is, the data span the whole range, then this predictor performs properly (see Table 9.4 and Fig. 9.19).

Summarizing, the predictor with GTLS estimates has a good performance in each of the four cases and clearly outperforms the original predictor with NLS estimates. Using truncated LS estimation, as an alternative to the GTLS procedure, has a good performance only if the calibration data set is informative enough.

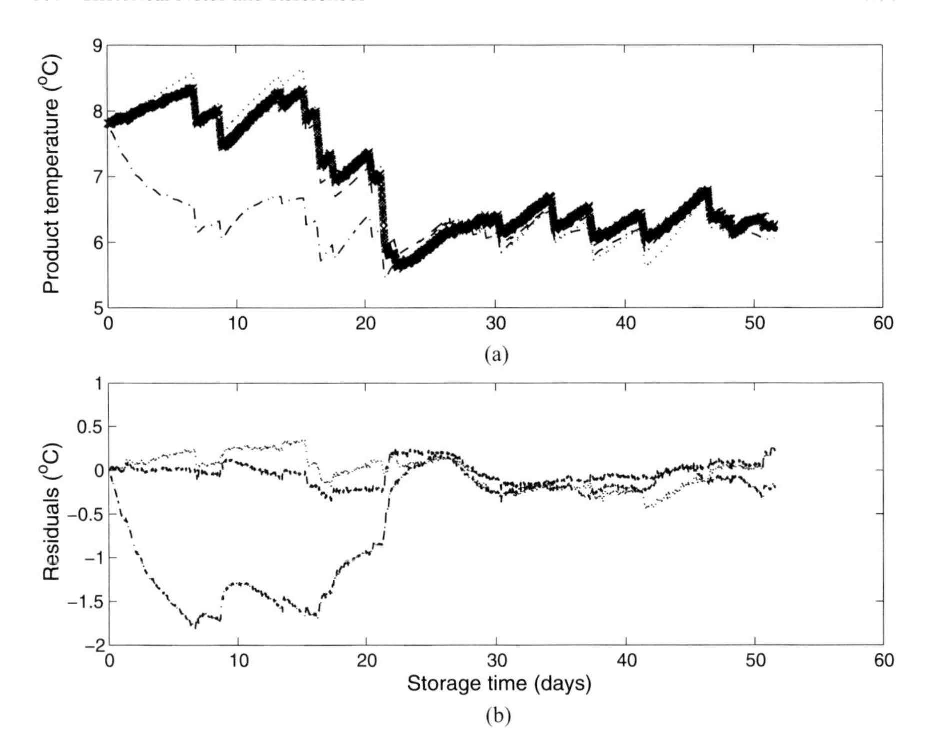

Fig. 9.21 (**a**) Measured (∗) and predicted product temperatures (NLS: ...; tLS: -.-.; GTLS: - - -), and (**b**) residuals in the validation period (data set 1)

9.4 Historical Notes and References

Model validation is a crucial step in the modeling process. Hence, many papers have appeared on this subject. For additional information on procedures and tests in time and frequency domain, we refer to [BBM86, LG97, CDC04, ZT00, KB94, MN95]. In particular, we mention [RH78] for his work on nonstationary signals and [HB94, BZ95, SBD99, MB00] for their work on nonlinear systems. For model validation within a set-membership context, and thus suited for small data sets, we refer to [BBC90, Lju99a, FG06]. A nice overview of model validation techniques for simulation models is given by [Sar84].

If, however, the model does not pass the model validation test, in the worst case new experiments have to be designed (see Fig. 1.7). A new experiment design should also be considered when the practical identifiability of model parameters is low. Consequently, experiment design plays a key role in the identification process. In the past, in addition to some books [GP77, Zar79, WP97], many articles on this subject have appeared, see, for instance, [KS73, NGS77, Bit77, Zar81, QN82, deS87, SMMH94, BP02, RWGF07, Luo07] for linear systems and [DI76, WP87, PW88, WP90, DGW96, BSK^{+}02, BG03, JDVCJB06] for nonlinear systems, to mention a few. Recently, the optimal input design problem has also been tackled by optimizing the parametric sensitivities using optimal singular control [SK01, KS02, KS03,

SK04, SVK06]. For low-dimensional systems, this approach allows analytical solutions and thus insight into the design procedure.

9.5 Outlook

Let us finish with a concise outlook on the developments in system identification for the next decade, 2010–2020.

The *curse of dimensionality* is an issue in many problems, such as in nonlinear parameter estimation, optimal experiment/input design, and in identifiability analysis. Hence, it is expected that, in combination with ever increasing computer power, new methods that circumvent this problem to a large extent will be found.

Identification for control, but then within a dynamic optimization context for nonlinear systems, the development of automatic identification procedures using some *multimodel approach*, the design of *measurement networks for complex systems*, and the application of *advanced particle filtering techniques* are other issues that need (further), and most likely will receive, attention in the near future.

Nowadays, there is a trend to focus on specific model classes, such as, for example, polynomial, Wiener, Hammerstein, and rational models, and to aim at further development of identification procedures using this specific system knowledge. This trend will diversify the system identification methods that can be applied. However, the proven "least-squares" concept, as can be found in many sections of this book, will remain the basis for most of these special cases.

In the last decade, system identification has become a mature discipline. Hence, system identification will be more and more involved in new developments, with increasing complexity, in industry and society. This creates a big challenge to identify complex processes, in engineering and in biological and economical applications, for more insight and better control/management.

Looking back, we see a very active system identification community. Hence, it is expected that in the near future this community may be able to tackle the abovementioned problems effectively, with a focus on errors and—in line with the first words of this book—learning from mistakes.

9.6 Problems

Problem 9.1 A rather popular cross-validation technique is the so-called leave-one-out cross-validation (LOOCV). Instead of splitting a data set in one calibration and one validation set, as in Example 9.6, LOOCV uses a single observation from the original data set as the validation data set and the remaining observations as the training data set for calibration. This procedure is repeated so that each observation in the sample is used once as the validation data. Consequently, for large data sets, leave-one-out cross-validation is usually very expensive from a computational point of view because of the large number of repetitions. Hence, in what follows, we will test it on a small data set, as presented in Example 9.5.

(a) Apply LOOCV to the process data (Table 9.2) using a polynomial model as in Example 9.5.
(b) Repeat the procedure for the sine function and the constant.
(c) Compare and evaluate the results.
(d) Finally, evaluate the results of an LOOCV with respect to the validation results obtained in Example 9.5.

Problem 9.2 As model validation is best illustrated by examples using real data, repeat the procedures as demonstrated in Examples 9.1 and 9.3, and in Examples 9.4 and 9.7.

Appendix A
Matrix Algebra

A.1 Basic Definitions

A matrix A is a rectangular array whose elements a_{ij} are arranged in rows and columns. If there are m rows and n columns, we say that we have an $m \times n$ matrix

$$A = \begin{bmatrix} a_{11} & a_{12} & \cdots & a_{1n} \\ a_{21} & a_{22} & \cdots & a_{2n} \\ \vdots & \vdots & . & \vdots \\ a_{m1} & a_{m2} & \cdots & a_{mn} \end{bmatrix} = (a_{ij})$$

When $m = n$, the matrix is said to be square. Otherwise, it is rectangular. A triangular matrix is a special kind of square matrix where the entries either below or above the main diagonal are zero. Hence, we distinguish between a lower and an upper triangular matrix. A lower triangular matrix L is of the form

$$L = \begin{bmatrix} l_{11} & 0 & \cdots & 0 \\ l_{21} & l_{22} & \cdots & 0 \\ \vdots & \vdots & \ddots & \vdots \\ l_{n1} & a_{n2} & \cdots & l_{nn} \end{bmatrix}$$

Similarly, an upper triangular matrix can be formed. The elements a_{ii} or l_{ii}, with $i = 1, \dots, n$ are called the diagonal elements. When $n = 1$, the matrix is said to be a column matrix or vector. Furthermore, one distinguishes between row vectors in case $m = 1$, submatrices, diagonal matrices (a_{ii}), zero or null matrices $\mathbf{O} = (0_{ij})$, and the identity matrix I, a diagonal matrix of the form $(a_{ii}) = 1$. To indicate an n-dimensional identity matrix, sometimes the notation I_n is used. A matrix is called symmetric if $(a_{ij}) = (a_{ji})$.

K.J. Keesman, *System Identification*,
Advanced Textbooks in Control and Signal Processing,
DOI 10.1007/978-0-85729-522-4,

A.2 Important Operations

In this book, calculation is only based on real-valued matrices, hence $A \in \mathbb{R}^{m\times n}$. As for the scalar case ($m = 1, n = 1$), addition, subtraction (both element-wise), and multiplication of matrices is defined. If A is an $m \times n$ and B an $n \times p$ matrix, then the product AB is a matrix of dimension $m \times p$ whose elements c_{ij} are given by

$$c_{ij} = \sum_{k=1}^{n} a_{ik} b_{kj}$$

Consequently, the ijth element is obtained by, in turn, multiplying the elements of the ith row of A by the jth column of B and summing over all terms. However, it should be noted that for the matrices A and B of appropriate dimensions, in general

$$AB \neq BA$$

Hence, generally, premultiplication of matrices yields different results than post-multiplication.

The transpose of the matrix $A = (a_{ij})$, denoted by A^T and defined as $A^T = (a_{ij})^T := (a_{ji})$, is another important operation, which in general changes the dimension of the matrix. The following holds:

$$(AB)^T = B^T A^T$$

For vectors $x, y \in \mathbb{R}^n$, however,

$$x^T y = y^T x$$

which is called the inner or scalar product. If the inner product is equal to zero, i.e., $x^T y = 0$, the two vectors are said to be orthogonal. In addition to the inner product of two vectors, also the matrix inner product has been introduced. The inner product of two real matrices A and B is defined as

$$\langle A, B\rangle := \mathrm{Tr}\big(A^T B\big)$$

Other important operations are the outer or dyadic product (AB^T) and matrix inversion (A^{-1}), which is only defined for square matrices. In the last operation one has to determine the determinant of the matrix, denoted by $\det(A)$ or simply $|A|$, which is a scalar.

The determinant of an $n \times n$ matrix A is defined as

$$|A| := a_{i1}c_{i1} + a_{i2}c_{i2} + \cdots + a_{in}c_{in}$$

Herein, the *cofactors* c_{ij} of A are defined as follows:

$$c_{ij} := (-1)^{i+j} |A_{ij}|$$

where $|A_{ij}|$ is the determinant of the submatrix obtained when the ith row and the jth column are deleted from A. Thus, the determinant of a matrix is defined in terms of the determinants of the associated submatrices. Let us demonstrate the calculation of the determinant to a 3×3 matrix.

Example A.1 Determinant: Let

$$A = \begin{bmatrix} a_{11} & a_{12} & a_{13} \\ a_{21} & a_{22} & a_{23} \\ a_{31} & a_{32} & a_{33} \end{bmatrix}$$

Then,

$$|A| = a_{11} \begin{vmatrix} a_{22} & a_{23} \\ a_{32} & a_{33} \end{vmatrix} - a_{12} \begin{vmatrix} a_{21} & a_{23} \\ a_{31} & a_{33} \end{vmatrix} + a_{13} \begin{vmatrix} a_{21} & a_{22} \\ a_{31} & a_{32} \end{vmatrix}$$

After some algebraic manipulation using the same rules for the subdeterminants, we obtain

$$|A| = a_{11}(a_{22}a_{33} - a_{32}a_{23}) - a_{12}(a_{21}a_{33} - a_{31}a_{23}) + a_{13}(a_{21}a_{32} - a_{31}a_{22})$$

When the determinant of a matrix is equal to zero, the matrix is singular, and the inverse A^{-1} does not exist. If $\det(A) \neq 0$, the inverse exists, and the matrix is said to be regular. Whether a matrix is invertible or not can also be checked by calculating the rank of a matrix. The column rank of a matrix A is the maximal number of linearly independent columns of A. Likewise, the row rank is the maximal number of linearly independent rows of A. Since the column rank and the row rank are always equal, they are simply called the rank of A. Thus, an $n \times n$ matrix A is invertible when the rank is equal to n.

The inverse of a square $n \times n$ matrix is calculated from

$$A^{-1} = \frac{1}{|A|} \operatorname{adj}(A)$$

$$\begin{bmatrix} \frac{c_{11}}{|A|} & \frac{c_{21}}{|A|} & \cdots & \frac{c_{n1}}{|A|} \\ \frac{c_{12}}{|A|} & \frac{c_{22}}{|A|} & \cdots & \frac{c_{n2}}{|A|} \\ \vdots & \vdots & . & \vdots \\ \frac{c_{1n}}{|A|} & \frac{c_{2n}}{|A|} & \cdots & \frac{c_{nn}}{|A|} \end{bmatrix}$$

where $\operatorname{adj}(A)$ denotes the adjoint of the matrix A and is obtained after transposing the $n \times n$ matrix C with elements c_{ij}, the cofactors of A.

The following properties are useful:

1. $(AB)^{-1} = B^{-1}A^{-1}$
2. $(AB)(B^{-1}A^{-1}) = A(BB^{-1})A^{-1} = AIA^{-1} = I$
3. $(ABC)^{-1} = C^{-1}B^{-1}A^{-1}$

4. $(A^T)^{-1} = (A^{-1})^T$
5. $|A|^{-1} = 1/|A|$.

A square matrix is said to be an orthogonal matrix if

$$AA^T = I$$

so that $A^{-1} = A^T$.

If, however, the matrix is rectangular, the matrix inverse does not exist. For these cases, the so-called *generalized* or *pseudo*-inverse has been introduced. The pseudo-inverse A^+ of an $m \times n$ matrix A, also known as the Moore–Penrose pseudo-inverse, is given by

$$A^+ = \left(A^T A\right)^{-1} A^T$$

provided that the inverse $(A^T A)^{-1}$ exists. Consequently,

$$\left(A^+ A\right) = \left(A^T A\right)^{-1} A^T A = I$$

and thus A^+ of this form is also called the *left semi-inverse* of A. This Moore–Penrose pseudo-inverse forms the heart of the ordinary least-squares solution to a linear regression problem, where $m = N$ (number of measurements), and $n = p$ (number of parameters). The generalized inverse is not unique. For the case $m < n$, where the inverse $(A^T A)^{-1}$ does not exist, one could use

$$A^+ = A^T \left(AA^T\right)^{-1}$$

so that

$$\left(AA^+\right) = AA^T \left(AA^T\right)^{-1} = I$$

if $(AA^T)^{-1}$ exists. Application of this generalized inverse or *right semi-inverse* plays a key role in so-called *minimum-length solutions*. Finally, for the cases where $(A^T A)^{-1}$ and $(AA^T)^{-1}$ do not exist, the generalized inverse can be computed via a limiting process

$$A^+ = \lim_{\delta \to 0} \left(A^T A + \delta I\right)^{-1} A^T = \lim_{\delta \to 0} A^T \left(AA^T + \delta I\right)^{-1}$$

which is related to Tikhonov regularization.

A.3 Quadratic Matrix Forms

Define the vector $x := [x_1, x_2, \ldots, x_n]^T$; then a *quadratic form* in x is given by

$$x^T Qx$$

where $Q = (q_{ij})$ is a symmetric $n \times n$ matrix. Following the rules for matrix multiplication, the scalar $x^T Qx$ is calculated as

$$\begin{aligned} x^T Qx = q_{11}x_1^2 &+ 2q_{12}x_1x_2 + \cdots + 2q_{1n}x_1x_n \\ &+ q_{22}x_2^2 + \cdots + 2q_{2n}x_2x_n \\ &+ \cdots + q_{nn}x_n^2 \end{aligned}$$

Hence, if Q is diagonal, the quadratic form reduces to a weighted inner product, which is also called the weighted Euclidean squared norm of x. In shorthand notation, $\|x\|_{2,Q}^2$; see Sect. A.4 for a further introduction of vector and matrix norms. Consequently, the weighted squared norm represents a weighted sum of squares.

An $n \times n$ real symmetric matrix Q is called *positive definite* if

$$x^T Qx > 0$$

for all nonzero vectors x. For an $n \times n$ positive definite matrix Q, all diagonal elements are positive, that is, $q_{ii} > 0$ for $i = 1, \ldots, n$. A positive definite matrix is invertible. In case $x^T Qx \geq 0$, we call the matrix *semi-positive definite*.

A.4 Vector and Matrix Norms

Let us introduce the norm of a vector $x \in \mathbb{R}^n$, as introduced in the previous section, in some more detail, where the norm is indicated by the double bar. A vector norm on $\mathbb{R}^n$ for $x, y \in \mathbb{R}^n$ satisfies the following properties:

$$\begin{aligned} \|x\| &\geq 0 \quad \big(\|x\| = 0 \iff x = 0\big) \\ \|x + y\| &\leq \|x\| + \|y\| \\ \|\alpha x\| &= |\alpha|\|x\| \end{aligned}$$

Commonly used vector norms are the 1-, 2-, and ∞-norm, which are defined as

$$\begin{aligned} \|x\|_1 &:= |x_1| + \cdots + |x_n| \\ \|x\|_2 &:= \big(x_1^2 + \cdots + x_n^2\big)^{\frac{1}{2}} \\ \|x\|_\infty &:= \max_{1 \leq i \leq n} |x_i| \end{aligned}$$

where the subscripts on the double bar are used to indicate a specific norm. Hence, the 2-norm, also known as the Euclidean (squared) norm, is frequently used to indicate a length of a vector. The weighted Euclidean norm for diagonal matrix Q, as already introduced in Sect. A.3, is then defined as

$$\|x\|_{2,Q} := \big(q_{11}x_1^2 + \cdots + q_{nn}x_n^2\big)^{\frac{1}{2}}$$

Sometimes this norm is also denoted as $\|x\|_Q$, thus without an explicit reference to the 2-norm. However, in the following, we will use the notation $\|x\|_{2,Q}$ for a weighted 2-norm to avoid confusion. This idea of norms can be further extended to matrices $A, B \in \mathbb{R}^{m\times n}$ with the same kind of properties as presented above. For the text in this book, it suffices to introduce one specific matrix norm, the so-called *Frobenius* norm $\|\cdot\|_F$,

$$\|A\|_F = \sqrt{\sum_{i=1}^{m}\sum_{j=1}^{n}|a_{ij}|^2} = \sqrt{\mathrm{Tr}\big(A^T A\big)}$$

where the trace (denoted by $\mathrm{Tr}(\cdot)$) of a square $n \times n$ matrix is the sum of its diagonal elements. The Frobenius norm is used in the derivation of a total least-squares solution to an estimation problem with noise in both the regressors and regressand, the dependent variable.

A.5 Differentiation of Vectors and Matrices

Differentiation of vector and matrix products is important when deriving solutions to optimization problems. Let us start by considering the inner product of two n-dimensional vectors a and x,

$$x^T a = x_1 a_1 + x_2 a_2 + \cdots + x_n a_n$$

Then, the partial derivatives with respect to a_i are given by

$$\frac{\partial(x^T a)}{\partial a_i} = x_i$$

Consequently, after stacking all the partial derivatives, we obtain x, and thus vector differentiation can be summarized as

$$\frac{\partial(x^T a)}{\partial a} = x, \qquad \frac{\partial(x^T a)}{\partial a^T} = x^T$$

In general terms, a vector differentiation operator is defined as

$$\frac{\mathrm{d}}{\mathrm{d}x} := \left[\frac{\partial}{\partial x_1}, \dots, \frac{\partial}{\partial x_n}\right]^T$$

Applying to any scalar function $f(x)$ to find its derivative with respect to x, we obtain

$$\frac{\mathrm{d}}{\mathrm{d}x} f(x) = \left[\frac{\partial f(x)}{\partial x_1}, \dots, \frac{\partial f(x)}{\partial x_n}\right]^T$$

Vector differentiation has the following properties:

1. $\frac{\mathrm{d}}{\mathrm{d}x}a^T x = \frac{\mathrm{d}}{\mathrm{d}x}x^T a = a$.
2. $\frac{\mathrm{d}}{\mathrm{d}x}x^T x = 2x$.
3. $\frac{\mathrm{d}}{\mathrm{d}x}x^T Ax = x^T A^T + x^T A$, and thus for $A^T = A$, $\frac{\mathrm{d}}{\mathrm{d}x}x^T Ax = 2Ax$.

The matrix differentiation operator is defined as

$$\frac{\mathrm{d}}{\mathrm{d}A} := \begin{bmatrix} \frac{\partial}{\partial a_{11}} & \cdots & \frac{\partial}{\partial a_{1n}} \\ \vdots & . & \vdots \\ \frac{\partial}{\partial a_{m1}} & \cdots & \frac{\partial}{\partial a_{mn}} \end{bmatrix}$$

The derivative of a scalar function $f(A)$ with respect to A is given by

$$\frac{\mathrm{d}}{\mathrm{d}A}f(A) = \begin{bmatrix} \frac{\partial f(A)}{\partial a_{11}} & \cdots & \frac{\partial f(A)}{\partial a_{1n}} \\ \vdots & . & \vdots \\ \frac{\partial f(A)}{\partial a_{m1}} & \cdots & \frac{\partial f(A)}{\partial a_{mn}} \end{bmatrix}$$

For the special case $f(A) = u^T Av$ with u an $m \times 1$ constant vector, v an $n \times 1$ constant vector, and A an $m \times n$ matrix,

$$\frac{\mathrm{d}}{\mathrm{d}A}u^T Av = uv^T$$

Example A.2 Derivative of cost function: Let a cost function $J_W(\vartheta)$, with ϑ a parameter vector, be defined as

$$\begin{aligned} J_W(\vartheta) &:= \|y - \Phi\vartheta\|_{2,Q}^2 \\ &= (y - \Phi\vartheta)^T Q(y - \Phi\vartheta) \\ &= y^T Qy - y^T Q\Phi\vartheta - \vartheta^T \Phi^T Qy + \vartheta^T \Phi^T Q\Phi\vartheta \\ &= y^T Qy - 2\vartheta^T \Phi^T Qy + \vartheta^T \Phi^T Q\Phi\vartheta \end{aligned}$$

with Q a symmetric positive definite weighting matrix. Then, following the rules of vector differentiation,

$$\frac{\mathrm{d}}{\mathrm{d}\vartheta}J_W = -2\Phi^T Qy + 2\Phi^T Q\Phi\vartheta$$

After setting $\frac{\mathrm{d}}{\mathrm{d}\vartheta}J_W = 0$, a necessary condition for finding a minimum, a weighted least-squares estimate of ϑ is found.

A.6 Eigenvalues and Eigenvectors

Eigenvalue decomposition of a square matrix is given by

$$Au = \lambda u$$

where u is an eigenvector, and λ the associated eigenvalue. The combination (λ, u) is called an eigenpair. For an n-dimensional square matrix A, n (not necessarily different) eigenvalues exist. The eigenpairs (λ, u) of a square matrix A can be computed (in principle) by first determining the roots λ of the characteristic polynomial

$$c(\lambda) = \det(\lambda I_n - A)$$

Subsequently, the eigenvectors u can be found by solving the associated linear equations

$$(\lambda I_n - A)u = 0$$

Let us illustrate this by an example.

Example A.3 Eigenvalues and eigenvectors: Let

$$A = \begin{bmatrix} 1 & 0 \\ 0 & 0 \end{bmatrix}$$

Then

$$\det\left(\begin{bmatrix} \lambda & 0 \\ 0 & \lambda \end{bmatrix} - \begin{bmatrix} 1 & 0 \\ 0 & 0 \end{bmatrix}\right) = \det\left(\begin{bmatrix} \lambda - 1 & 0 \\ 0 & \lambda \end{bmatrix}\right) = (\lambda - 1)\lambda = 0$$

$$\Longrightarrow \quad \lambda_1 = 1: \quad \begin{bmatrix} 1 & 0 \\ 0 & 1 \end{bmatrix}\begin{bmatrix} u_{11} \\ u_{21} \end{bmatrix} = \begin{bmatrix} u_{11} \\ u_{21} \end{bmatrix} \quad \Longrightarrow \quad u_{11} \text{ free}, \quad u_{21} = 0$$

$$\Longrightarrow \quad u_1 = \begin{bmatrix} 1 \\ 0 \end{bmatrix}$$

$$\Longrightarrow \quad \lambda_2 = 0: \quad \begin{bmatrix} 1 & 0 \\ 0 & 1 \end{bmatrix}\begin{bmatrix} u_{21} \\ u_{22} \end{bmatrix} = \begin{bmatrix} 0 \\ 0 \end{bmatrix} \quad \Longrightarrow \quad u_{21} = 0, \quad u_{22} \text{ free}$$

$$\Longrightarrow \quad u_2 = \begin{bmatrix} 0 \\ 1 \end{bmatrix}$$

When

$$A = \begin{bmatrix} 1 & 0 \\ 0 & 1 \end{bmatrix} \quad \Longrightarrow \quad \lambda_{1,2} = 1, \quad u_1 = \begin{bmatrix} 1 \\ 0 \end{bmatrix} \quad \text{and} \quad u_2 = \begin{bmatrix} 0 \\ 1 \end{bmatrix}$$

or

$$A = \begin{bmatrix} 2 & 0 \\ 0 & 1 \end{bmatrix} \quad \Longrightarrow \quad \lambda_1 = 2, \quad u_1 = \begin{bmatrix} 1 \\ 0 \end{bmatrix} \quad \text{and} \quad \lambda_2 = 1, \quad u_2 = \begin{bmatrix} 0 \\ 1 \end{bmatrix}$$

Finally, when

$$A = \begin{bmatrix} 2 & 1 \\ 1 & 1 \end{bmatrix} \implies \lambda_1 = 2.618, \quad u_1 = \begin{bmatrix} -0.8507 \\ -0.5257 \end{bmatrix}$$

$$\lambda_2 = 0.382, \quad u_2 = \begin{bmatrix} -0.5257 \\ -0.8507 \end{bmatrix}$$

Two matrices A and B are said to be similar if and only if there exists a non-singular matrix P such that $B = P^{-1}AP$. The matrix function $f(A) = P^{-1}AP$ is called the similarity transformation of A. It appears that an $n \times n$ matrix A is similar to the diagonal matrix with the eigenvalues of A on the diagonal, provided that A has n distinct eigenvalues.

In the case with n linearly independent eigenvectors, we can generalize $Au = \lambda u$ to $AU = UD$, and thus

$$A = UDU^{-1}$$

also called the eigendecomposition or spectral decomposition of matrix A. From this it also follows that

$$U^{-1}AU = D$$

with $U = [u_1 \; u_2 \; \cdots \; u_n]$ that is a matrix formed by the eigenvectors, sometimes called the eigenmatrix, and

$$D = \begin{bmatrix} \lambda_1 & 0 & \cdots & 0 \\ 0 & \lambda_2 & \cdots & 0 \\ \vdots & \vdots & \ddots & \vdots \\ 0 & 0 & \cdots & \lambda_n \end{bmatrix}$$

If this holds, it is said that A is diagonalizable. This property is important in having a geometrical interpretation of the eigenvectors and also in analyzing the behavior and error propagation of linear dynamic models. Note from the above that only diagonalizable matrices can be factorized in terms of eigenvalues and eigenvectors.

The following list (derived from [HK01]) summarizes some useful eigen properties

1. If (λ, u) is an eigenpair of A, then so is (λ, ku) for any $k \neq 0$.
2. If (λ_1, u_1) and (λ_2, u_2) are eigenpairs of A with $\lambda_1 \neq \lambda_2$, then u_1 and u_2 are linearly independent. In other words, eigenvectors corresponding to distinct eigenvalues are linearly independent.
3. A and A^T have the same eigenvalues.
4. If A is diagonal, upper triangular, or lower triangular, then its eigenvalues are its diagonal entries.
5. The eigenvalues of a symmetric matrix are real.
6. Eigenvectors corresponding to distinct eigenvalues of a symmetric matrix are orthogonal.

7. $\det(A)$ is the product of the absolute values of the eigenvalues of A.
8. A is nonsingular if and only if 0 is not an eigenvalue of A. This implies that A is singular if and only if 0 is an eigenvalue of A.
9. Similar matrices have the same eigenvalues.
10. An $n \times n$ matrix A is similar to a diagonal matrix if and only if A has n linearly independent eigenvectors. In that case, it is said that A is diagonalizable.

Especially, properties (5) and (6) are important in the evaluation of the symmetric, positive definite covariance matrices and the corresponding ellipsoidal uncertainty regions (see Appendix B). Finally, it is worth noting that eigenvalues do also play an important role in the calculation of the spectral norm of a square matrix. The spectral or 2-norm of a real matrix A is the square root of the largest eigenvalue of the positive semi-definite matrix $A^T A$,

$$\|A\|_2 = \sqrt{\lambda_{\max}\left(A^T A\right)}$$

with is different from the entry-wise Frobenius norm, introduced before. However, the following inequality holds: $\|A\|_2 \leq \|A\|_F \leq \sqrt{n}\|A\|_2$.

A.7 Range and Kernel of a Matrix

The range of a matrix A, denoted by $\text{ran}(A)$, also called the column space of A, is the set of all possible linear combinations of the column vectors of A. Consequently, for a matrix A of dimension $m \times n$,

$$\text{ran}(A) = \left\{y \in \mathbb{R}^m : y = Ax \text{ for some } x \in \mathbb{R}^n\right\}$$

In addition to the range of a matrix, the kernel or null space (also nullspace) of a matrix A is defined as the set of all vectors x for which $Ax = 0$. In mathematical notation, for an $m \times n$ matrix A,

$$\ker(A) = \left\{x \in \mathbb{R}^n : Ax = 0, x \neq 0\right\}$$

Let us demonstrate these properties of a matrix by some examples.

Example A.4 If

$$A = \begin{bmatrix} 2 & 1 \\ 1 & 1 \end{bmatrix}$$

then given the column vectors $v_1 = \left[\begin{smallmatrix} 2 \\ 1 \end{smallmatrix}\right]$ and $v_2 = \left[\begin{smallmatrix} 1 \\ 1 \end{smallmatrix}\right]$, a linear combination of v_1 and v_2 is any vector of the form

$$\beta_1 \begin{bmatrix} 2 \\ 1 \end{bmatrix} + \beta_2 \begin{bmatrix} 1 \\ 1 \end{bmatrix} = \begin{bmatrix} 2\beta_1 + \beta_2 \\ \beta_1 + \beta_2 \end{bmatrix}$$

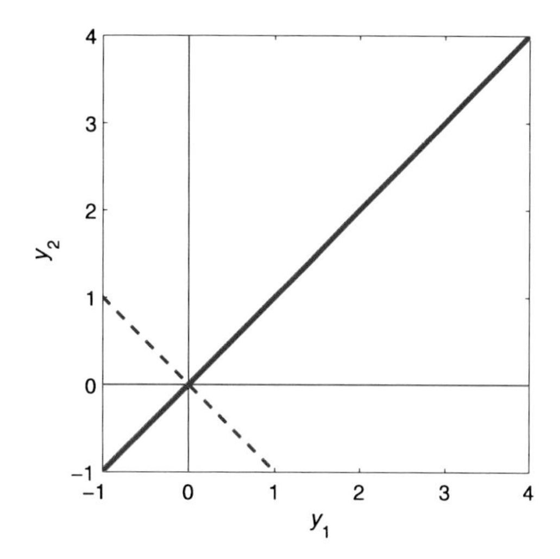

Fig. A.1 Range of $A=[1 \quad 1]^T$ (*bold solid line*) and kernel of $A^T=[1 \quad 1]$ (*dashed line*)

Hence, in this case with real constants $\beta_1, \beta_2 \in \mathbb{R}$, the range of A is the two-dimensional plane $\mathbb{R}^2$. If, however,

$$A = \begin{bmatrix} 1 \\ 1 \end{bmatrix}$$

then the range of A is precisely the set of vectors $[y_1 y_2]^T \in \mathbb{R}^2$ that satisfy the equation $y_1 = y_2$. Consequently, this set is a line through the origin in the two-dimensional space (see Fig. A.1). It can then be easily verified that the kernel of A is empty, but $\ker(A^T)$ with

$$A^T = \begin{bmatrix} 1 & 1 \end{bmatrix}$$

is a line with normalized slope, thus with unit length: $\left[\begin{smallmatrix} -0.7071 \\ 0.7071 \end{smallmatrix}\right]$, as $x_1 + x_2 = 0$ with $\|x\| = 1$. The kernel of A^T is also presented in Fig. A.1 by the dashed line that satisfies the equation $x_1 = -x_2$. Hence, for the linear transformation $A^T x$ with $A^T = [1\ 1]$, every point on this line maps to the origin. For more information on these properties and their application in estimation problems, we refer to, for instance, [GVL89].

A.8 Exponential of a Matrix

The matrix exponential, denoted by e^A or $\exp(A)$, is a matrix function on the square matrix A, analogous to the ordinary exponential function. Let A be an $n \times n$ real or complex matrix. The exponential of A is the $n \times n$ matrix given by the power series

$$e^A = \sum_{k=0}^{\infty} \frac{1}{k!} A^k$$

This series always converges, and thus e^A is well defined. However, using this power series is not a very practical way to calculate the matrix exponential. In the famous paper of Moler and Van Loan [MVL78], 19 ways to compute the matrix exponential are discussed. Here, it suffices to illustrate this matrix function by an example.

Example A.5 Exponential of a matrix:

$$A = \begin{bmatrix} 2 & 1 \\ 1 & 1 \end{bmatrix} \implies e^A = \begin{bmatrix} 10.3247 & 5.4755 \\ 5.4755 & 4.8492 \end{bmatrix}$$

This result is based on MATLAB's *expm*. It is important to realize that this result is completely different from the case where the exponential function is taken element-wise, which in fact is *not* the exponential of a matrix. For a comparison, using MATLAB's *exp* gives

$$\begin{bmatrix} 7.3891 & 2.7183 \\ 2.7183 & 2.7183 \end{bmatrix} \approx \begin{bmatrix} e^2 & e \\ e & e \end{bmatrix}$$

A.9 Square Root of a Matrix

A matrix B is said to be a square root of the square matrix A, if the matrix product BB is equal to A. Furthermore, B is the unique square root for which every eigenvalue has nonnegative real part. The square root of a diagonal matrix D, for example, is found by taking the square root of all the entries on the diagonal. For a general diagonalizable matrix A such that $D = U^{-1}AU$, we have $A = UDU^{-1}$. Consequently,

$$A^{1/2} = UD^{1/2}U^{-1}$$

since $UD^{1/2}U^{-1}UD^{1/2}U^{-1} = UDU^{-1} = A$. This square root is real if A is positive definite. For nondiagonalizable matrices, we can calculate the Jordan normal form followed by a series expansion. However, this is not an issue here, as in most system identification applications, in particular filtering, the square root of a real, symmetric, and positive definite covariance matrix is calculated. Let us demonstrate the calculation of the square root of the matrix used in Example A.3.

Example A.6 Square root of a matrix:

$$A = \begin{bmatrix} 2 & 1 \\ 1 & 1 \end{bmatrix} \implies U = \begin{bmatrix} -0.8507 & 0.5257 \\ -0.5257 & -0.8507 \end{bmatrix}, \quad D = \begin{bmatrix} 2.618 & 0 \\ 0 & 0.382 \end{bmatrix}$$

and thus

$$A^{1/2} = \begin{bmatrix} -0.8507 & 0.5257 \\ -0.5257 & -0.8507 \end{bmatrix} \begin{bmatrix} \sqrt{2.618} & 0 \\ 0 & \sqrt{0.382} \end{bmatrix} \begin{bmatrix} -0.8507 & 0.5257 \\ -0.5257 & -0.8507 \end{bmatrix}$$

$$= \begin{bmatrix} 1.3416 & 0.4472 \\ 0.4472 & 0.8944 \end{bmatrix}$$

as in this special case $U = U^{-1}$.

Let us also evaluate this result analytically. Then, for an unknown matrix $B = \begin{bmatrix} b_1 & b_2 \\ b_3 & b_4 \end{bmatrix}$, we obtain

$$A = \begin{bmatrix} 2 & 1 \\ 1 & 1 \end{bmatrix} = \begin{bmatrix} b_1 & b_2 \\ b_3 & b_4 \end{bmatrix} \begin{bmatrix} b_1 & b_2 \\ b_3 & b_4 \end{bmatrix} = \begin{bmatrix} b_1^2 + b_2 b_3 & b_1 b_2 + b_2 b_4 \\ b_1 b_3 + b_3 b_4 & b_2 b_3 + b_4^2 \end{bmatrix}$$

Consequently, the solution set is given by

$$\begin{bmatrix} b_1 \\ b_2 \\ b_3 \\ b_4 \end{bmatrix} \in \left\{ \begin{bmatrix} 1 \\ 1 \\ 1 \\ 0 \end{bmatrix}, \begin{bmatrix} -1 \\ -1 \\ -1 \\ 0 \end{bmatrix}, \begin{bmatrix} \frac{3}{5}\sqrt{5} \\ \frac{1}{5}\sqrt{5} \\ \frac{3}{5}\sqrt{5} \\ \frac{2}{5}\sqrt{5} \end{bmatrix}, \begin{bmatrix} -\frac{3}{5}\sqrt{5} \\ -\frac{1}{5}\sqrt{5} \\ -\frac{3}{5}\sqrt{5} \\ -\frac{2}{5}\sqrt{5} \end{bmatrix} \right\}$$

Hence, from this matrix equality four solutions appear. However, only the third solution results in a matrix for which every eigenvalue has nonnegative real part, that is, $\lambda_1 = 1.6180$ and $\lambda_2 = 0.6180$. The other solutions result in a matrix B with at least one negative eigenvalue. Thus, the matrix B with entries as defined by the third solution is the principal square root of A.

A.10 Choleski Decomposition

The Choleski decomposition or Choleski factorization is a decomposition of a symmetric, positive definite matrix A into the product of a lower triangular matrix L and its transpose L^T, for A real and thus with all entries real. Hence, A can be decomposed as

$$A = LL^T$$

where L is a lower triangular matrix with strictly positive diagonal, real entries. Notice that the Choleski decomposition is an example of a square root of a matrix. The Choleski decomposition is unique: given a symmetric, positive definite matrix A, there is only one lower triangular matrix L with strictly positive diagonal entries such that $A = LL^T$. In general, Choleski factorizations for positive semidefinite matrices are not unique.

Example A.7 Choleski decomposition:

$$A = \begin{bmatrix} 2 & 1 \\ 1 & 1 \end{bmatrix} \implies L = \begin{bmatrix} 1.4142 & 0 \\ 0.7071 & 0.7071 \end{bmatrix}, \quad L^T = \begin{bmatrix} 1.4142 & 0.7071 \\ 0 & 0.7071 \end{bmatrix}$$

such that $A = LL^T$.

Let us evaluate this result analytically. Then, for $L = \left[\begin{smallmatrix} l_1 & 0 \\ l_3 & l_2 \end{smallmatrix}\right]$,

$$A = \begin{bmatrix} 2 & 1 \\ 1 & 1 \end{bmatrix} = \begin{bmatrix} l_1 & 0 \\ l_3 & l_2 \end{bmatrix} \begin{bmatrix} l_1 & l_3 \\ 0 & l_2 \end{bmatrix} = \begin{bmatrix} l_1^2 & l_1 l_3 \\ l_1 l_3 & l_2^2 + l_3^2 \end{bmatrix}$$

$$\implies l_1 = \sqrt{2}, \quad l_3 = \frac{1}{2}\sqrt{2}, \quad l_2 = \frac{1}{2}\sqrt{2}$$

A.11 Modified Choleski (UD) Decomposition

In addition to the Choleski decomposition, as presented in the previous subsection, there is also a unique, real, lower triangular matrix, $\tilde{L}$, and a real, positive diagonal matrix, $\tilde{D}$, such that

$$A = \tilde{L}\tilde{D}\tilde{L}^T$$

where $\tilde{L}$ has unit diagonal elements. This decomposition is known as the modified Choleski decomposition, [vRLS73]. Given the Choleski and modified Choleski decomposition, we obtain that $L = \tilde{L}\tilde{D}^{1/2}$. However, A can also be decomposed as $A = UDU^T$, known as the modified Choleski (UD) decomposition, with U a unit upper triangular matrix and D a diagonal matrix with nonnegative elements.

Example A.8 Modified Choleski decomposition: For $L = \left[\begin{smallmatrix} 1 & 0 \\ l & 1 \end{smallmatrix}\right]$ and $D = \left[\begin{smallmatrix} d_1 & 0 \\ 0 & d_2 \end{smallmatrix}\right]$, we obtain

$$A = \begin{bmatrix} 2 & 1 \\ 1 & 1 \end{bmatrix} = \begin{bmatrix} 1 & 0 \\ l & 1 \end{bmatrix} \begin{bmatrix} d_1 & 0 \\ 0 & d_2 \end{bmatrix} \begin{bmatrix} 1 & l \\ 0 & 1 \end{bmatrix} = \begin{bmatrix} d_1 & d_1 l \\ d_1 l & d_2 \end{bmatrix}$$

$$\implies d_1 = 2, \quad d_2 = 1, \quad l = 0.5$$

Similarly, we can decompose A as UDU^T with $U = \left[\begin{smallmatrix} 1 & u \\ 0 & 1 \end{smallmatrix}\right]$ and D as before, leading to $d_1 = 1$, $d_2 = 1$, $u = 1$.

A.12 QR Decomposition

Any real square matrix A may be decomposed as $A = QR$, where Q is an orthogonal matrix (i.e., $Q^T Q = I$), and R is an upper triangular matrix, also called right triangular matrix. If A is nonsingular, then the factorization is unique if the diagonal elements of R are positive. However, this so-called QR decomposition or QR factorization also exists for rectangular matrices; see an example below.

Example A.9 QR decomposition: Let us start with a square matrix. For example,

$$A = \begin{bmatrix} 2 & 1 \\ 1 & 1 \end{bmatrix} = \begin{bmatrix} q_1 & q_2 \\ q_3 & q_4 \end{bmatrix} \begin{bmatrix} r_1 & r_2 \\ 0 & r_3 \end{bmatrix} = \begin{bmatrix} q_1 r_1 & q_1 r_2 + q_2 r_3 \\ q_3 r_1 & q_3 r_2 + q_4 r_3 \end{bmatrix}$$

with

$$\begin{bmatrix} q_1 & q_3 \\ q_2 & q_4 \end{bmatrix} \begin{bmatrix} q_1 & q_2 \\ q_3 & q_4 \end{bmatrix} = \begin{bmatrix} q_1^2 + q_3^2 & q_1 q_2 + q_3 q_4 \\ q_1 q_2 + q_3 q_4 & q_2^2 + q_4^2 \end{bmatrix} = \begin{bmatrix} 1 & 0 \\ 0 & 1 \end{bmatrix}$$

Hence, we do have seven equations with seven unknowns. The solution set is given by

$$\begin{bmatrix} q_1 \\ q_2 \\ q_3 \\ q_4 \\ r_1 \\ r_2 \\ r_3 \end{bmatrix} \in \left\{ \begin{bmatrix} \frac{2}{5}\sqrt{5} \\ -\frac{1}{5}\sqrt{5} \\ \frac{1}{5}\sqrt{5} \\ \frac{2}{5}\sqrt{5} \\ \sqrt{5} \\ \frac{3}{5}\sqrt{5} \\ \frac{1}{5}\sqrt{5} \end{bmatrix}, \begin{bmatrix} \frac{2}{5}\sqrt{5} \\ \frac{1}{5}\sqrt{5} \\ \frac{1}{5}\sqrt{5} \\ -\frac{2}{5}\sqrt{5} \\ \sqrt{5} \\ \frac{3}{5}\sqrt{5} \\ -\frac{1}{5}\sqrt{5} \end{bmatrix}, \begin{bmatrix} -\frac{2}{5}\sqrt{5} \\ -\frac{1}{5}\sqrt{5} \\ -\frac{1}{5}\sqrt{5} \\ \frac{2}{5}\sqrt{5} \\ -\sqrt{5} \\ -\frac{3}{5}\sqrt{5} \\ \frac{1}{5}\sqrt{5} \end{bmatrix}, \begin{bmatrix} -\frac{2}{5}\sqrt{5} \\ \frac{1}{5}\sqrt{5} \\ -\frac{1}{5}\sqrt{5} \\ -\frac{2}{5}\sqrt{5} \\ -\sqrt{5} \\ -\frac{3}{5}\sqrt{5} \\ -\frac{1}{5}\sqrt{5} \end{bmatrix} \right\}$$

Thus, the QR decomposition is not unique. However, there is one unique combination that gives positive values for the diagonal elements r_1 and r_3, and this one is given by the entries as defined by the first solution.

As a second case, let A be an $n \times m$ matrix with $n = 3$ and $m = 2$. For instance,

$$A = \begin{bmatrix} 2 & 1 \\ 1 & 1 \\ 1 & 2 \end{bmatrix} \implies Q = \begin{bmatrix} -0.8165 & 0.4924 & 0.3015 \\ -0.4082 & -0.1231 & -0.9045 \\ -0.4082 & -0.8616 & 0.3015 \end{bmatrix},$$

$$R = \begin{bmatrix} -2.4495 & -2.0412 \\ 0 & -1.3540 \\ 0 & 0 \end{bmatrix}$$

such that $A = QR = Q\left[\begin{smallmatrix} R_1 \\ 0 \end{smallmatrix}\right] = [Q_1\ Q_2]\left[\begin{smallmatrix} R_1 \\ 0 \end{smallmatrix}\right] = Q_1 R_1$, where R_1 is an $n \times n$ upper triangular matrix, Q_1 is an $m \times n$ matrix, Q_2 is an $m \times (m - n)$ matrix, and Q_1 and Q_2 both have orthogonal columns.

A.13 Singular Value Decomposition

Recall that eigenvalue decomposition is limited to square matrices only. The *singular value decomposition* (SVD) is an important factorization of a rectangular matrix. It has several applications in signal processing and statistics. In particular, in this book, SVD is used in relation with least-squares fitting, identifiability, and total least-squares solutions. In what follows, we focus on the $N \times p$ regressor matrix Φ. The SVD technique decomposes Φ into

$$\Phi = USV^T$$

where U and V are orthogonal matrices of dimensions $N \times N$ and $p \times p$, respectively, such that $U^T U = I_N$ and $V^T V = I_p$. The $N \times p$ singular value matrix S has the following structure:

$$S = \begin{bmatrix} \sigma_1 & 0 & \cdots & 0 \\ 0 & \sigma_2 & \cdots & 0 \\ \vdots & \vdots & \vdots & \vdots \\ 0 & 0 & \cdots & \sigma_p \\ \hdashline & 0_{(N-p)\times p} & & \end{bmatrix}$$

where $0_{(N-p)\times p}$ denotes an $(N - p) \times p$ zero or null matrix. An intuitive explanation of this result is that the columns of V form a set of orthonormal "input" or "analyzing" basis vector directions for Φ, the columns of U form a set of orthonormal "output" basis vector directions for Φ, and the matrix S contains the singular values that can be thought of as a scalar "gain" by which each corresponding input is multiplied to give a corresponding output. If the SVD of Φ is calculated and $\sigma_1 \geq \cdots \geq \sigma_r > \sigma_{r+1} = \cdots = \sigma_p = 0$, then the rank of Φ is equal to r. Hence, there exists a clear link between the rank of a matrix and its singular values.

For a further interpretation of the singular vectors and values, notice that the SVD of $\Phi^T \Phi$ is given by

$$\begin{aligned} \Phi^T \Phi &= VS^T U^T USV^T \\ &= VS^T SV^T \end{aligned}$$

Since $V^T V = I_p$, we have $V^T = V^{-1}$, and thus

$$(\Phi^T \Phi)V = VS^T S$$

with $S^T S$ a $p \times p$ diagonal matrix. Consequently, with $\Lambda = S^T S$, the right singular vector matrix V in the SVD of Φ can be calculated as the eigenmatrix of $\Phi^T \Phi$, and σ_i in S as the square root of the corresponding eigenvalues. Similarly, the left singular vector matrix U can be calculated from an eigenvalue decomposition of $\Phi \Phi^T$.

A.14 Projection Matrices

A square matrix A is said to be idempotent if $A^2 = A$. Idempotent matrices have the following properties:

1. $A^r = A$ for r being a positive integer.
2. $I - A$ is idempotent.
3. If A_1 and A_2 are idempotent matrices and $A_1 A_2 = A_2 A_1$, then $A_1 A_2$ is idempotent.
4. A is a projection matrix.

If, however, an idempotent matrix is also symmetric, then we call it an orthogonal projection matrix. Hence, for orthogonal projection in real spaces, it holds that the projection matrix is idempotent and symmetric, i.e., $A^2 = A$ and $A = A^T$. In linear regression with least-squares estimate $\widehat{\vartheta} = (\Phi^T \Phi)^{-1} \Phi^T y$, where Φ the regressor matrix, we use the matrix $P = \Phi(\Phi^T \Phi)^{-1} \Phi^T$ to calculate the predicted model output, $\widehat{y} = \Phi \widehat{\vartheta} = \Phi(\Phi^T \Phi)^{-1} \Phi^T y$, from the output vector y. Since

$$\begin{aligned}\left(\Phi\left(\Phi^T\Phi\right)^{-1}\Phi^T\right)^2 &= \Phi\left(\Phi^T\Phi\right)^{-1}\Phi^T\Phi\left(\Phi^T\Phi\right)^{-1}\Phi^T \\ &= \Phi\left(\Phi^T\Phi\right)^{-1}\Phi^T\end{aligned}$$

and

$$\begin{aligned}\left(\Phi\left(\Phi^T\Phi\right)^{-1}\Phi^T\right)^T &= \Phi\left(\Phi^T\Phi\right)^{-T}\Phi^T \\ &=_{[(\Phi^T\Phi)=(\Phi^T\Phi)^T]} \Phi\left(\Phi^T\Phi\right)^{-1}\Phi^T\end{aligned}$$

P defines an orthogonal projection in a real space, as does $I - P$.

Appendix B
Statistics

B.1 Random Entities

B.1.1 Discrete/Continuous Random Variables

A *discrete random variable* ξ is defined as a discrete-valued function $\xi(j)$ with probability of occurrence of the jth value given by $p(j)$, where

$p(j)$: probability density function of the random variable $\xi(j)$

The mean value or first moment of $p(j)$, which is defined as the expected value of ξ ($E[\xi]$) and also denoted by $\bar{\xi}$, is given by

$$E[\xi] = \bar{\xi} = \sum_j \xi(j) p(j)$$

This concept can be extended to *continuous random variables*, where, for simplicity, we write ξ and $p(\xi)$. In the following, some useful operations on ξ for given $p(\xi)$ are defined. First, we define the *expectation*,

$$E[\xi] = \mu := \int_{\infty}^{\infty} \xi p(\xi)\, \mathrm{d}\xi$$

which can be interpreted as the center of the probability density function (pdf). Hence, for $a, b \in \mathbb{R}$,

(i) $E[a] = a$,
(ii) $E[a\xi + b] = aE[\xi] + b$.

Second, we define the *variance*,

$$\operatorname{Var} \xi := E\big[\big(\xi - E[\xi]\big)^2\big]$$

which can be interpreted as the dispersion of the probability density function. The following properties hold:

K.J. Keesman, *System Identification*,
Advanced Textbooks in Control and Signal Processing,
DOI 10.1007/978-0-85729-522-4,

(i) $\operatorname{Var} a\xi + b = a^2 \operatorname{Var} \xi$.
(ii) $\operatorname{Var} \xi = E[\xi^2] - (E[\xi])^2$.

For two, either discrete or continuous random variables, ξ and ψ, the *covariance* is defined as

$$\operatorname{Cov}(\xi, \psi) = E\big[(\xi - E[\xi])(\psi - E[\psi])\big]$$

which represents the dependence between two random variables. The following properties hold:

(i) $\operatorname{Cov}(\xi, \psi) = E[\xi\psi] - E[\xi]E[\psi]$.
(ii) ξ and ψ independent: $\operatorname{Cov}(\xi, \psi) = 0$ and $E[\xi\psi] = E[\xi]E[\psi]$.
(iii) $\operatorname{Var} a\xi + b\psi = a^2 \operatorname{Var} \xi + b^2 \operatorname{Var} \psi + 2ab \operatorname{Cov}(\xi, \psi)$.

Then, the *normalized* covariance or *correlation* is given by

$$\rho = \frac{\operatorname{Cov}(\xi, \psi)}{\sqrt{\operatorname{Var} \xi}\sqrt{\operatorname{Var} \psi}}$$

We state the following properties:

(i) $-1 < \rho < 1$.
(ii) ξ and ψ are linearly related if $\rho = \pm 1$.
(iii) ξ and ψ independent if $\rho = 0$.

B.1.2 Random Vectors

A random vector is a column vector whose elements are random variables. The expectation of $\xi \in \mathbb{R}^r$ is given by

$$E[\xi] = \mu_\xi = \begin{bmatrix} E[\xi_1] \\ E[\xi_2] \\ \vdots \\ E[\xi_r] \end{bmatrix} = \begin{bmatrix} \mu_1 \\ \mu_2 \\ \vdots \\ \mu_r \end{bmatrix}$$

The covariance of ξ is given by

$$\begin{aligned} P = P_\xi &= E\big[(\xi - \mu_\xi)(\xi - \mu_\xi)^T\big] \\ &= E \begin{bmatrix} (\xi_1 - \mu_1)^2 & (\xi_1 - \mu_1)(\xi_2 - \mu_2) & \cdots & (\xi_1 - \mu_1)(\xi_r - \mu_r) \\ \vdots & (\xi_2 - \mu_2)^2 & & \vdots \\ \vdots & & \ddots & \vdots \\ \vdots & & \cdots & (\xi_r - \mu_r)^2 \end{bmatrix} \end{aligned}$$

In the following examples, the covariance matrix is calculated and visualized for some cases.

Example B.1 Covariance matrix: Let for a given vector ξ, ξ_i and ξ_j for $i \neq j$ be uncorrelated with constant variance σ^2 and zero mean. Then

$$P = \begin{bmatrix} \sigma^2 & 0 & \cdots & 0 \\ 0 & \sigma^2 & \cdots & 0 \\ \vdots & & \ddots & \vdots \\ 0 & 0 & \cdots & \sigma^2 \end{bmatrix}$$

because $\mu_\xi = 0$, $E[\xi_i \ \xi_j] = 0$ for $i \neq j$, and $E[\xi_i^2] = \sigma^2$ for $i = j$.

Example B.2 Covariance matrix: Let two random vectors $\xi_1 = [1\ 2\ 2\ 3]^T$ and $\xi_2 = [1\ 1\ 2\ 2]^T$ be given. Consequently, $E[\xi_1] = \bar{\xi}_1 = 2$ and $E[\xi_2] = \bar{\xi}_2 = 1.5$, so that the normalized vectors are given by $\tilde{\xi}_1 = \xi_1 - E[\xi_1] = [-1\ 0\ 0\ 1]^T$ and $\tilde{\xi}_2 = \xi_2 - E[\xi_2] = [-0.5\ -0.5\ 0.5\ 0.5]^T$. Define

$$\tilde{\Xi} := [\tilde{\xi}_1\ \tilde{\xi}_2] = \begin{bmatrix} -1 & -0.5 \\ 0 & -0.5 \\ 0 & 0.5 \\ 1 & 0.5 \end{bmatrix}$$

Thus,

$$P = \frac{1}{3}\tilde{\Xi}^T\tilde{\Xi} = \begin{bmatrix} \frac{2}{3} & \frac{1}{3} \\ \frac{1}{3} & \frac{1}{3} \end{bmatrix} = \begin{bmatrix} \operatorname{Var}\tilde{\xi}_1 & \operatorname{Cov}(\tilde{\xi}_1, \tilde{\xi}_2) \\ \operatorname{Cov}(\tilde{\xi}_2, \tilde{\xi}_1) & \operatorname{Var}\tilde{\xi}_2 \end{bmatrix}$$

The eigenvalues and eigenvectors of P are given by

$$\lambda_{1,2} = \frac{1}{2} \pm \frac{1}{6}\sqrt{5}$$

and

$$u_1 = \begin{bmatrix} -0.8507 \\ -0.5257 \end{bmatrix} \quad \text{and} \quad u_2 = \begin{bmatrix} 0.5257 \\ -0.8507 \end{bmatrix}$$

Given any positive definite and symmetric covariance matrix, the so-called *uncertainty ellipsoid* can be constructed (see, for instance, [Bar74]). This ellipsoid represents an isoline connecting points of equal probability, which could be specified if we would accept, for example, a specific symmetric, unimodal multivariate distribution of the data. For our case, the uncertainty ellipse, with center $\bar{\xi} = [2\ 1.5]^T$ and form matrix P, is given by

$$\left\{\xi \in \mathbb{R}^2 : \left(\xi - \bar{\xi}\right)^T P^{-1}\left(\xi - \bar{\xi}\right) = 1\right\}$$

and presented in Fig. B.1. Another interpretation for the uncertainty ellipse is that, assuming that the given data have a bivariate Gaussian distribution (see explanation below), the ellipse in Fig. B.1 is the smallest area that contains a fixed probability mass. Notice from Fig. B.1 that the (orthogonal) main axes are defined by the

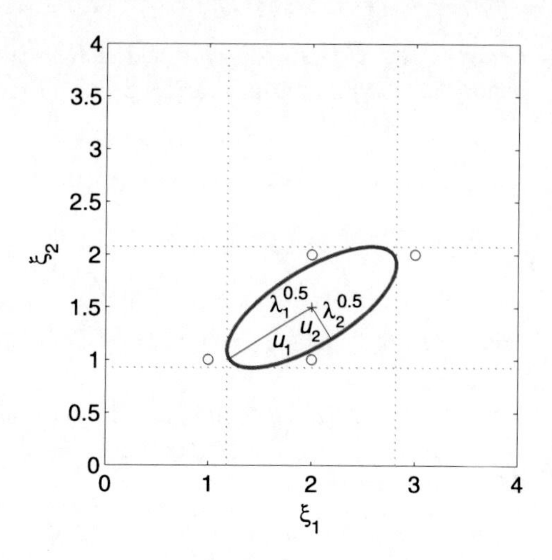

Fig. B.1 Uncertainty ellipse with data points (o)

eigenvectors u_1 and u_2. In addition to this, the lengths of the semi-axes are given by the square root of the corresponding eigenvalues λ_1 and λ_2 multiplied by a scaling factor. Consequently, the eigenvectors and eigenvalues of a covariance matrix and a distribution-related scaling factor define the uncertainty ellipse.

Example B.3 Forecast errors, see [DK05]: Forecast and observation files for location "De Bilt" from 1 March 2001 until 1 March 2002 were provided by the weather agency "HWS". These forecast files contain data from 0 until 31 hours ahead with an hourly interval. Every six hours, a new forecast file was delivered. Observation files were received daily containing hourly data from the previous 24 hours. From these data the average forecast error (i.e., observation minus forecast) of the temperature and the covariance matrix Q of the forecast error are obtained. The covariance matrix is graphically represented in Fig. B.2.

The normally distributed probability density function (pdf) of the random vector $\xi \in \mathbb{R}^r$ is given by

$$p(\xi) = \frac{1}{(2\pi)^{r/2}|P|^{1/2}} \exp\left\{-\frac{1}{2}(\xi - \mu_\xi)^T P^{-1}(\xi - \mu_\xi)\right\}$$

where $(\xi - \mu_\xi)^T P^{-1}(\xi - \mu_\xi)$ defines an *ellipsoid* with center μ_ξ and principal axes that are the eigenvectors of the $r \times r$ matrix P. Clearly,

$$E[\xi] = \mu_\xi$$
$$E\left[(\xi - \mu_\xi)(\xi - \mu_\xi)^T\right] = P$$

In short-hand notation, $\xi \sim N(\mu_\xi, P)$. It is also common practice to say that ξ has a Gaussian distribution.

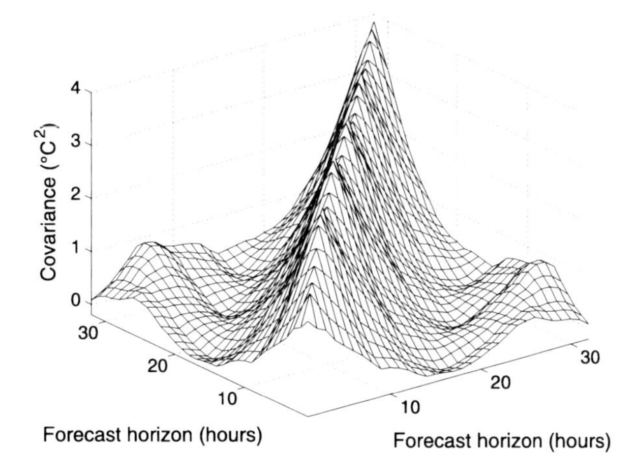

Fig. B.2 Covariance matrix of the short-term forecast error

The Gaussian distribution is of paramount importance in statistics, as stated by the Central Limit Theorem. Let $\xi_1, \xi_2, \dots, \xi_n$ be a sequence of n independent and identically distributed (iid) random variables, each having finite values of expectation μ and variance $\sigma^2 > 0$. The central limit theorem states that, as the sample size n increases, the distribution of the sample average of these random variables approaches the normal distribution with mean μ and variance σ^2/n, irrespective of the shape of the common distribution of the individual terms ξ_i. The central limit theorem is formally stated as follows.

Theorem B.1 *Let $\xi_1, \xi_2, \dots$ be independent, identically distributed random variables having mean μ and finite nonzero variance σ^2.*

Let $S_n = \xi_1 + \xi_2 + \cdots + \xi_n$. Then

$$\lim_{n\to\infty} P\left(\frac{S_n - n\mu}{\sigma\sqrt{n}} \le x\right) = \Phi(x)$$

where $\Phi(x)$ is the probability that a standard normal random variable is less than x.

Another distribution, which is especially of paramount importance in hypothesis testing, is the so-called chi-square or χ^2-distribution. The chi-square distribution (also called chi-squared distribution) with M degrees of freedom is the distribution of a sum of the squares of M independent standard normal random variables. In Fig. B.3 the cumulative distribution function, which is the integral of the chi-square probability density function, is presented.

The chi-square distribution is commonly used in the so-called chi-square tests for goodness of fit of an observed distribution to a theoretical one. A chi-square test is any statistical hypothesis test in which the sampling distribution of the test statistic (as in (9.7)) has a chi-square distribution when the null hypothesis is true, or any in which this is asymptotically true.

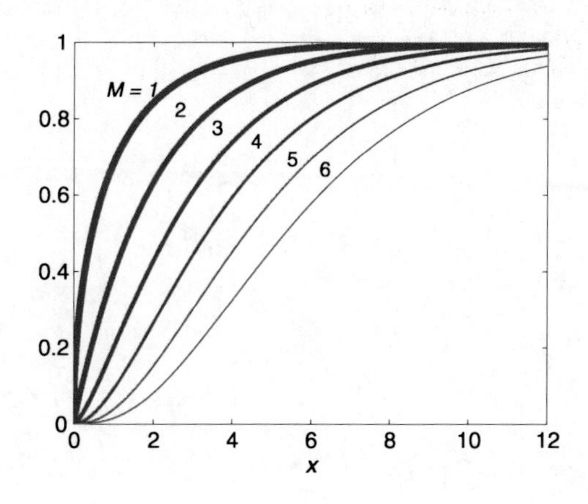

Fig. B.3 Cumulative distribution function related to $\chi^2(M)$-distributions with $M = 1, \ldots, 6$

B.1.3 Stochastic Processes

A statistical phenomenon that evolves in time according to probabilistic laws is called a *stochastic process*, see [BJ70]. Let ξ and ψ be elements of the same stochastic process, denoted by $\xi(t)$ and $\xi(t+\tau)$. Then, the *autocorrelation function* is defined by

$$r_{\xi\xi}(\tau, t) := E\big[\xi(t)\xi(t+\tau)\big]$$

Some properties of the autocorrelation function are:

(i) For $\tau = 0$, $r_{\xi\xi}(\tau, t)$ has its maximum.
(ii) $r_{\xi\xi}(\tau, t) = r_{\xi\xi}(-\tau, t)$.

In the analysis of sequences or time series, a series is said to be *stationary in the wide sense* if the first and second moments, i.e., mean and (co)variances, are not functions of time. In what follows, we will call these series simply stationary. A very useful property of stationary series is that they are *ergodic*, in other words, we can estimate the mean and (co)variances by averages over time. This implies that the autocorrelation function can be estimated from

$$r_{\xi\xi}(\tau) = \lim_{T\to\infty} \frac{1}{2T} \int_{-T}^{T} u(t)u(t+\tau)\,\mathrm{d}t$$

Notice that this function is now, under the assumption of ergodicity, only a function of lag τ and not of t. Let us illustrate the autocorrelation function by an example using real-world data.

Example B.4 Autocorrelation function of wastewater data: Consider the following measurements of the ammonium concentration entering a wastewater treatment plant (upper graph, Fig. B.4) and associated normalized autocorrelation function (bottom graph, Fig. B.4).

It suffices here to say that this autocorrelation function clearly demonstrates the previously mentioned properties of the autocorrelation function and that strong correlations between subsequent ammonium concentrations are observed.

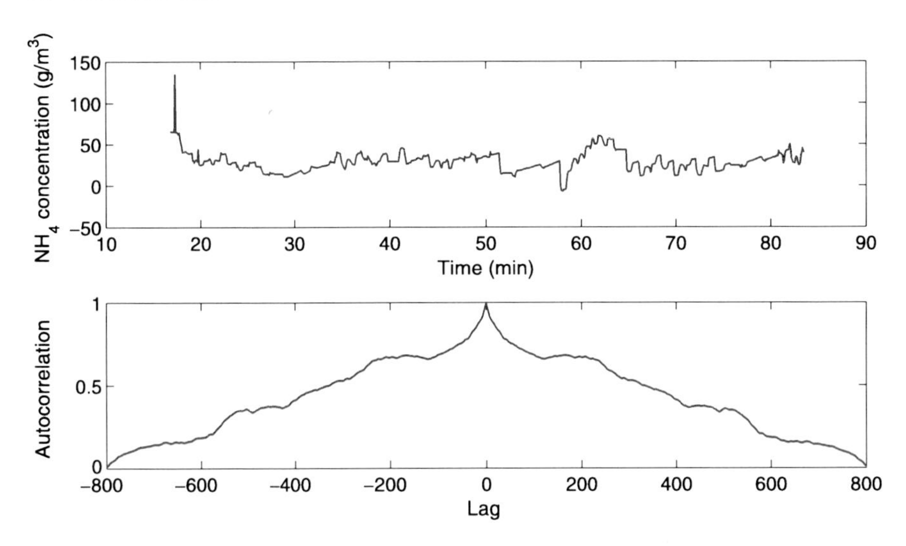

Fig. B.4 Measured ammonium concentrations with corresponding autocorrelation function

In a similar way, the cross-correlation function between $\xi(t)$ and $\psi(t)$ is defined as

$$r_{\xi\psi}(\tau,t) := E\big[\xi(t)\psi(t+\tau)\big]$$

which under ergodic conditions becomes

$$r_{\xi\psi}(\tau) = \lim_{T\to\infty} \frac{1}{2T} \int_{-T}^{T} \xi(t)\psi(t+\tau)\,\mathrm{d}t$$

Appendix C
Laplace, Fourier, and z-Transforms

C.1 Laplace Transform

The Laplace transform is one of the best-known and most widely used integral transforms. It is commonly used to produce an easily solvable algebraic equation from an ordinary differential equation. Furthermore, the Laplace transform is often interpreted as a transformation from the time domain, in which inputs and outputs are functions of time, to the frequency domain, where the same inputs and outputs are functions of complex angular frequency, or radians per unit time. For LTI systems, the Laplace transform provides an alternative functional description that often simplifies the analysis of the behavior of the system. The most commonly applied Laplace transform is defined as

$$\mathscr{L}\big[f(t)\big] \equiv F(s) := \int_0^\infty f(t)\mathrm{e}^{-st}\,\mathrm{d}t$$

It is a linear operator on a function $f(t)$ (original) with real argument t that transforms it to a function $F(s)$ (image) with complex argument s. Let us illustrate this unilateral or one-sided Laplace transform by two simple examples.

Example C.1 Laplace transform: Given $f(t) = \mathrm{e}^{-at}$ with $a, t \in \mathbb{R}_+$. Then,

$$\begin{aligned}
\mathscr{L}\big[\mathrm{e}^{-at}\big] &= \int_0^\infty \mathrm{e}^{-at}\mathrm{e}^{-st}\,\mathrm{d}t \\
&= \int_0^\infty \mathrm{e}^{-(a+s)t}\,\mathrm{d}t \\
&= \left[\frac{1}{-(a+s)}\mathrm{e}^{-(a+s)t}\right]_0^\infty \\
&= \frac{1}{-(a+s)}[0-1] \\
&= \frac{1}{(a+s)}
\end{aligned}$$

K.J. Keesman, *System Identification*,
Advanced Textbooks in Control and Signal Processing,
DOI 10.1007/978-0-85729-522-4, © Springer-Verlag London Limited 2011

Example C.2 Laplace transform: Consider the function $f(t-\tau)$ with time delay $\tau \in \mathbb{R}$ and $f(t-\tau)=0$ for $t<\tau$. Then,

$$\begin{aligned}
\mathscr{L}\left[f(t-\tau)\right] &= \int_0^\infty f(t-\tau)\mathrm{e}^{-st}\,\mathrm{d}t \\
&=_{[t'=t-\tau]} \int_{-\tau}^\infty f(t')\mathrm{e}^{-s(t'+\tau)}\,\mathrm{d}t' \\
&= \mathrm{e}^{-s\tau}\int_{-\tau}^\infty f(t')\mathrm{e}^{-st'}\,\mathrm{d}t' \\
&=_{[f(t')=0 \text{ for } t'<0]} \mathrm{e}^{-s\tau}\int_0^\infty f(t')\mathrm{e}^{-st'}\,\mathrm{d}t' \\
&= \mathrm{e}^{-s\tau}\mathscr{L}\left[f(t)\right]
\end{aligned}$$

A very powerful property of the Laplace transform is given in the following example, without showing all the intermediate steps.

Example C.3 Laplace transform: Laplace transformation of the convolution integral $y(t)=\int_{-\infty}^{t} g(t-\tau)u(\tau)\,\mathrm{d}\tau$, with $y(t)$, $u(t)$, and $g(t)$ appropriate (integrable) time functions, leads to

$$Y(s)=G(s)U(s)$$

which defines an algebraic relationship between transformed output signal $Y(s)$ and transformed input signal $U(s)$.

Finally, for the transformation from one model representation in the frequency domain to the time domain and vice versa, as depicted in Fig. 2.1, the following property is essential.

Example C.4 Laplace transform: The Laplace transform of a derivative can be found after integrating the expression for the definition of the Laplace transform, as given above, by parts. Hence,

$$\int_0^\infty f(t)\mathrm{e}^{-st}\,\mathrm{d}t = \left[\frac{-f(t)\mathrm{e}^{-st}}{s}\right]_0^\infty + \frac{1}{s}\int_0^\infty f'(t)\mathrm{e}^{-st}\,\mathrm{d}t$$

After evaluating the limits and multiplying by s we obtain

$$\begin{aligned}
&s\mathscr{L}\left[f(t)\right] = f(0) + \mathscr{L}\left[f'(t)\right] \\
&\quad\Longrightarrow \quad \mathscr{L}\left[f'(t)\right] = s\mathscr{L}\left[f(t)\right] - f(0)
\end{aligned}$$

The Laplace transform has the useful property that not only the ones shown above but many relationships and operations over the originals $f(t)$ correspond to simpler relationships and operations over the images $F(s)$.

C.2 Fourier Transform

The Fourier transform shows a close similarity to the Laplace transform. The continuous Fourier transform is equivalent to evaluating the bilateral Laplace transform with complex argument $s = j\omega$, with ω in rad/s. The result of a Fourier transformation of a real-valued function $(f(t))$ is often called the frequency domain representation of the original function. In particular, it describes which frequencies are present in the original function. There are several common conventions for defining the Fourier transform of an integrable function $f(t)$. In this book, with angular frequency $\omega = 2\pi\xi$ in rad/s and frequency ξ in Hertz, we use

$$\mathscr{F}\big[f(t)\big] \equiv F(\omega) := \int_{-\infty}^{\infty} f(t)\mathrm{e}^{-j\omega t}\,\mathrm{d}t$$

for every real number t. The most important property for further use in this book is illustrated by the following example.

Example C.5 Fourier transform: Fourier transformation of the convolution integral $y(t) = \int_{-\infty}^{t} g(t-\tau)u(\tau)\,\mathrm{d}\tau$, with $y(t)$, $u(t)$, and $g(t)$ integrable functions, leads to

$$Y(\omega) = G(\omega)U(\omega)$$

which, as in the case of the Laplace transform, defines an algebraic relationship between transformed output signal $Y(\omega)$ and transformed input signal $U(\omega)$.

In this book, in addition to the continuous Fourier transform, the Discrete Fourier Transform (DFT) of the sampled, continuous-time signal $f(t)$ for $t = 1, 2, \ldots, N$ is used as well and is given by

$$F_N(\omega) = \frac{1}{\sqrt{N}} \sum_{t=1}^{N} f(t)\mathrm{e}^{-j\omega t}$$

where $\omega = 2\pi k/N$, $k = 1, 2, \ldots, N$. In this definition, N/k is the period associated with the specific frequency ω_k. The absolute square value of $F(\omega_k)$, $|F(2\pi k/N)|^2$, is a measure of the energy contribution of this frequency to the energy of the signal. The plot of values of $|F(\omega)|^2$ as a function of ω is called the periodogram of the signal $f(t)$.

C.3 z-Transform

The z-transform converts a discrete time-domain signal, which in general is a sequence of real numbers, into a complex frequency domain representation. The z-transform is like a discrete equivalent of the Laplace transform. The unilateral

Table C.1 Transforms of commonly used functions

Function	Time domain $x(t)$	Laplace s-domain $X(s) = \mathscr{L}[x(t)]$	z-domain $X(z) = \mathscr{Z}[x(t)]$
Unit impulse	$\delta(t)$	1	–
Ideal delay	$x(t-\tau)$	$e^{-s\tau} X(s)$	$z^{-\tau/T_s} X(z)$
Unit step	$H_s(t)$	$\frac{1}{s}$	$\frac{T_s}{z-1}$
Unit pulse	$\frac{1}{T_s}[H_s(t) - H_s(t-T_s)]$	$\frac{1}{T_s} \frac{1-e^{-sT_s}}{s}$	1
Ramp	$t H_s(t)$	$\frac{1}{s^2}$	$\frac{T_s^2(z+1)}{2(z-1)^2}$
Exp. decay	$e^{-\alpha t} H_s(t)$	$\frac{1}{(s+\alpha)}$	$\frac{1}{\alpha} \frac{1-e^{-\alpha T_s}}{2-e^{-\alpha T_s}}$

or one-sided z-transform is simply the Laplace transform of an ideally sampled signal after the substitution $z = e^{sT_s}$, with T_s the sampling interval. The z-transform can also be seen as a generalization of the Discrete Fourier transform (DFT), where the DFT can be found by evaluating the z-transform $F(z)$ at $z = e^{j\omega}$. The two-sided z-transform of a discrete-time signal $f(t)$ is the function $F(z)$ defined as

$$\mathscr{Z}\big[f(t)\big] \equiv F(z) := \sum_{t=-\infty}^{\infty} f(t) z^{-t}$$

where $t \in \mathbb{Z}$, and z is, in general, a complex number. In this book, and basically for causal signals, the unilateral z-transform is used as well and is given by

$$\mathscr{Z}\big[f(t)\big] \equiv F(z) := \sum_{t=0}^{\infty} f(t) z^{-t}$$

Again, a very relevant property of the z-transform is illustrated in the following.

Example C.6 z-transform: z-transformation of the convolution sum

$$y(t) = \sum_{k=0}^{t} g(t-k) u(k)$$

with $y(t)$, $u(t)$, and $g(t)$ discrete-time functions, gives

$$Y(z) = G(z) U(z)$$

which defines a similar relationship between transformed output signal $Y(z)$ and transformed input signal $U(z)$, as in the case of Laplace or Fourier transformation.

For the approximate conversion from Laplace to z-domain and vice versa, the following relationships can be used:

$$s = \frac{2}{T_s} \frac{z-1}{z+1} \quad \text{(Tustin transformation)}$$

$$z = \frac{2 + sT_s}{2 - sT_s}$$

with T_s the sampling interval.

Finally, we will show some relationships between the transforms. Let $H_s(t)$ be the Heaviside step function, and $\delta(t)$ the Dirac delta function with t a real number (usually, but not necessarily, time) and T_s the sampling interval. Then, some basic time functions with their transforms are presented in Table C.1.

Appendix D
Bode Diagrams

D.1 The Bode Plot

In literature, the Bode plot is also referred to as the logarithmic or corner plot. The Bode plot allows graphical instead of analytical interpretations of signals and LTI systems in the frequency domain, for example, of $F(\omega)$. Because of the logarithmic transformation of functions, multiplication and division are reduced to addition and subtraction. Recall that in the frequency domain, complex numbers appear, even if the original function is real-valued. Hence, in what follows, we use $F(j\omega)$ instead of $F(\omega)$. Because of the complex numbers, the Bode plot consists of two plots, i.e., 20 times the logarithm of the magnitude (Lm) in decibels (dB) and the phase shift in degrees, as functions of the angular frequency ω. Notice then that if $|F(j\omega)|$ increases by tenfold, or one decade, then the log magnitude increases by 20. To simplify the interpretation of Bode plots of transfer functions, four basic types of terms are specified and analyzed beforehand. In general, the numerator and denominator of transfer functions of LTI systems consists of these four basic types, which are

$$K$$

$$(j\omega)^{\pm n}$$

$$(1 + j\omega T)^{\pm m}$$

$$e^{\pm j\omega\tau}$$

Because of the logarithmic transformation, a large class of Bode plots of the entire transfer function can be analyzed by simply adding the contribution of each of these simple terms.

K.J. Keesman, *System Identification*,
Advanced Textbooks in Control and Signal Processing,
DOI 10.1007/978-0-85729-522-4,

D.2 Four Basic Types

D.2.1 Constant or K Factor

For a real positive constant K, it holds that $\mathrm{Lm}(K) = 20\log|K|$ appears as a horizontal line that raises or lowers the log magnitude curve of the entire transfer function by a fixed amount. Clearly, because of the constant value, there is no contribution to the phase shift.

D.2.2 $(j\omega)^{\pm n}$ Factor

The log magnitude and phase shift of $(j\omega)^{\pm n}$ are given by

$$\mathrm{Lm}(j\omega)^{\pm n} = \pm 20n\log\omega$$
$$\angle(j\omega)^{\pm n} = \pm\frac{n\pi}{2}$$

Hence, the magnitude plot consists of a straight line whose slope is $\pm 20n$ dB/decade and goes through 0 dB at $\omega = 1$. The phase shift is a constant with a value of $\pm\frac{n\pi}{2}$. These results have been obtained using the following rules for complex numbers, i.e., given $z = a + bi \in \mathbb{C}$ with $a, b \in \mathbb{R}$, $|z| = \sqrt{\mathrm{Re}^2 z + \mathrm{Im}^2 z} = \sqrt{a^2 + b^2}$, and $\angle z = \arg z = \arctan\frac{\mathrm{Im}\,z}{\mathrm{Re}\,z} = \arctan\frac{b}{a}$.

D.2.3 $(1 + j\omega T)^{\pm m}$ Factor

Let us first consider the case with $m = 1$ and negative exponent. Then,

$$\begin{aligned}
\mathrm{Lm}(1 + j\omega T)^{-1} &= 20\log\left|\frac{1}{1 + j\omega T}\right| \\
&= 20\log\left|\frac{1 - j\omega T}{1 + \omega^2 T^2}\right| \\
&= 20\log\sqrt{\frac{1}{(1 + \omega^2 T^2)^2} + \frac{\omega^2 T^2}{(1 + \omega^2 T^2)^2}} \\
&= 20\log\sqrt{\frac{1}{(1 + \omega^2 T^2)}} \\
&= -20\log\sqrt{1 + \omega^2 T^2} \\
\angle(1 + j\omega T)^{-1} &= \arctan\frac{-\omega T}{1} = -\arctan\omega T
\end{aligned}$$

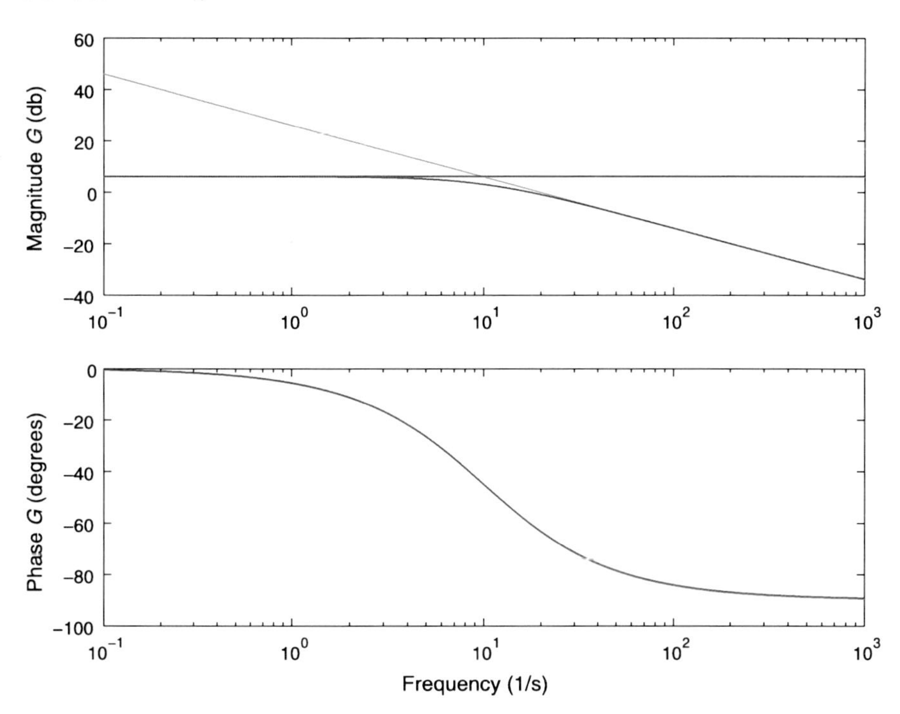

Fig. D.1 Bode plot for $G(j\omega) = \frac{20}{j\omega+10}$ with asymptotes in the magnitude plot

For very small values of ωT, the log magnitude becomes

$$\mathrm{Lm}(1 + j\omega T)^{-1} = -20 \log 1 = 0$$

Consequently, for low frequencies, the log magnitude becomes a line at 0 dB. On the contrary, if $\omega T \gg 1$,

$$\mathrm{Lm}(1 + j\omega T)^{-1} \approx \mathrm{Lm}(j\omega T)^{-1} = -20 \log \omega T$$

This defines a line through zero dB at $\omega = 1/T$ and with a -20 dB/decade slope for $\omega T > 1$. The intersection of both lines for $\omega T < 1$ and $\omega T > 1$ is at $\omega = 1/T$. The point is called the *corner frequency*. Let us first demonstrate this by an example.

Example D.1 Bode plot: Let a transfer function of an LTI system be given by

$$G(j\omega) = \frac{20}{j\omega + 10} = \frac{2}{0.1 j\omega + 1}$$

The corresponding Bode plot is presented in Fig. D.1

Taking $\omega = 0$ gives the static gain, which in this case is equal to 2. The corner frequency is at $\omega = 1/T = 10$. Furthermore, we obtain the asymptotes for $\omega T \ll 1$: $|G| = 20 \log 2 = 6.0206$ (horizontal line) and for $\omega T \gg 1$: $|G| = 20/\omega$ (line with

-20 dB/decade slope). These asymptotes are also presented in Fig. D.1. Notice that the error between the exact curve of the magnitudes and the asymptotes is greatest at the corner frequency.

If we now consider the case with $m = 1$ and positive exponent, we find that the corner frequency is still at $\omega = 1/T$ except that the asymptote for $\omega T \gg 1$ has a slope of $+20$ dB/decade. This is the only, but significant, difference, since this implies that for increasing frequency, the magnitude increases as well. When we consider a transfer function with two or more factors of this form, we can simply add the contribution of each. Consider, for example, the transfer function

$$G(j\omega) = \frac{1}{j\omega T_1 + 1} \frac{1}{j\omega T_2 + 1}$$

Then, the magnitude and phase are given by

$$\begin{aligned} \mathrm{Lm}\, G(j\omega) &= -20 \log \sqrt{1 + \omega^2 T_1^2} - 20 \log \sqrt{1 + \omega^2 T_2^2} \\ \angle G(j\omega) &= -\arctan \omega T_1 - \arctan \omega T_2 \end{aligned}$$

If we assume that $T_2 < T_1$, then the contribution of both terms is 0 dB for $\omega < 1/T_1$. For $1/T_1 < \omega < 1/T_2$, the first factor contributes -20 dB/decade. Up to frequency $\omega = 1/T_2$, the second factor has no contribution to the asymptotic behavior of the plot. However, at $\omega = 1/T_2$, which is another corner frequency, and increasing frequencies, the second factor is approximated by $-20 \log |\omega T_2|$. This again defines a straight line with a slope of -20 dB/decade. However, since this reinforces the first factor, the magnitude follows a straight line with slope of -40 dB/decade. If the exponent of the second factor of $G(j\omega)$ were positive, a horizontal line for $\omega > 1/T_2$ would appear.

D.2.4 $e^{\pm j\omega\tau}$ Factor

The log magnitude and phase shift of $e^{\pm j\omega\tau}$ is given by

$$\begin{aligned} \mathrm{Lm}\, e^{\pm j\omega\tau} &= 20 \log \left| e^{\pm j\omega\tau} \right| \\ &= 20 \log \left| \cos(\pm\omega\tau) + j \sin(\pm\omega\tau) \right| \\ &= 20 \log 1 = 0 \\ \angle e^{\pm j\omega\tau} &= \arctan \frac{\sin(\pm\omega\tau)}{\cos(\pm\omega\tau)} \\ &= \pm\omega\tau \end{aligned}$$

Hence, the magnitude plot consists of a horizontal line at 0 dB, because the magnitude is equal to one. The phase shift is proportional to the frequency ω.

In conclusion, we can state that given a (sampled) Bode plot related to, for example, an empirical transfer function estimate (ETFE), we are able to approximately recover the underlying factors of the complete transfer function. Thus, we can identify the entire continuous-time transfer function, provided that the transfer function consists of the four basic types mentioned above.

Appendix E
Shift Operator Calculus

E.1 Forward- and Backward-shift Operator

Shift operator calculus is a convenient tool for manipulating linear difference equations with constant coefficients, and for details, we refer to [Åst70] and the references therein. In the development of shift operator calculus, systems are viewed as operators that map input signals to output signals. To specify an operator, it is necessary to define its range. In particular, the class of input signals must be specified and how the operator acts on the signals. In shift operator calculus, all (sampled) signals are considered as double infinite sequences $f(t): t = \dots, -1, 0, 1, \dots$ with t the time index. In what follows, the sampling interval is chosen as the time unit.

The *forward-shift* operator, denoted by q, is defined by

$$qf(t) = f(t+1)$$

If the norm of a signal is defined as $\|f\|^2 := \sum_{t=-\infty}^{\infty} f^2(t)$, 2-norm, then it directly follows that the shift operator is bounded with unit norm. The inverse of the forward-shift operator is called the *backward-shift* operator or delay operator. It is denoted by q^{-1}. Consequently,

$$q^{-1}f(t) = f(t-1)$$

This inverse of q exists simply because all the signals are considered as double infinite sequences. Shift operator calculus allows compact descriptions of discrete-time systems. Furthermore, relationship between system variables can be easily derived, because the manipulation of difference equations is reduced to purely algebraic manipulation. A similar result holds for differential equations after applying integral transforms, as the Laplace or Fourier transforms (see Appendix C).

Let us illustrate this to a general difference equation of the form

$$y(t+n_a) + a_1 y(t+n_a-1) + \cdots + a_{n_a} y(t) = b_0 u(t+n_b) + \cdots + b_{n_b} u(t) \quad \text{(E.1)}$$

K.J. Keesman, *System Identification*,
Advanced Textbooks in Control and Signal Processing,
DOI 10.1007/978-0-85729-522-4, © Springer-Verlag London Limited 2011

where $n_a \geq n_b$ to guarantee causality. We call $d = n_a - n_b$ the pole excess of the system. Application of the shift operator gives

$$\left(q^{n_a} + a_1 q^{n_a - 1} + \cdots + a_{n_a}\right) y(t) = \left(b_0 q^{n_b} + \cdots + b_{n_b}\right) u(t)$$

Define

$$\begin{aligned} A(q) &:= q^{n_a} + a_1 q^{n_a - 1} + \cdots + a_{n_a} \\ B(q) &:= b_0 q^{n_b} + b_1 q^{n_b - 1} + \cdots + b_{n_b} \end{aligned}$$

Then, in compact notation, the difference equation can be written as

$$A(q) y(t) = B(q) u(t) \tag{E.2}$$

Shifting (E.1) to the left, i.e., after substituting $k + n_a$ by κ and using $d = n_a - n_b$, we obtain

$$y(\kappa) + a_1 y(\kappa - 1) + \cdots + a_{n_a} y(\kappa - n_a) = b_0 u(\kappa - d) + \cdots + b_{n_b} u(\kappa - d - n_b)$$

With

$$\begin{aligned} A^*\left(q^{-1}\right) &= 1 + a_1 q^{-1} + \cdots + a_{n_a} q^{-n_a} \\ B^*(q) &= b_0 + b_1 q^{-1} + \cdots + b_{n_b} q^{-n_b} \end{aligned}$$

we obtain

$$A^*\left(q^{-1}\right) y(t) = B^*\left(q^{-1}\right) u(t - d)$$

and

$$A^*\left(q^{-1}\right) = q^{-n_a} A(q)$$

Thus, it follows from the definition of the shift operator that the difference equation (E.1) can be multiplied by powers of q, which implies a forward shift of time. The equations for shifted times can also be multiplied by real numbers and added. This operation implies the multiplication of a polynomial in q by another polynomial in q. Hence, if (E.2) holds, then also

$$C(q) A(q) y(t) = C(q) B(q) u(t)$$

with $C(q)$ a polynomial in q. Multiplication of polynomials is illustrated by the next example.

Example E.1 Multiplication of polynomials: Let

$$A(q) = q^2 + a_1 q + a_2, \quad C(q) = q - 1$$

Then

$$\begin{aligned} C(q) A(q) y(t) &= (q - 1)\left(q^2 + a_1 q + a_2\right) y(t) \\ &= \left(q^3 + a_1 q^2 + a_2 q - q^2 - a_1 q - a_2\right) y(t) \end{aligned}$$

$$
\begin{aligned}
&= \big(q^3 + (a_1 - 1)q^2 + (a_2 - a_1)q - a_2\big)y(t) \\
&= y(t+3) + (a_1 - 1)y(t+2) + (a_2 - a_1)y(t+1) - a_2 y(t) \\
&= (q-1)\big(y(t+2) + a_1 y(t+1) + a_2 y(t)\big) \\
&= C(q)\big[A(q)y(t)\big]
\end{aligned}
$$

To obtain a convenient algebra, division by a polynomial in q has also to be defined. It is possible to develop an operator algebra that allows division by an arbitrary polynomial in q if it is assumed that there is some k_0 such that all sequences are zero for $k \leq k_0$. Consequently, this shift operator calculus allows the normal manipulations of multiplication, division, addition, and subtraction. Let us illustrate this algebra by an example.

Example E.2 Shift operator calculus: Consider the difference equation

$$x(t+2) = ax(t+1) + bu(t)$$

with output equation $y(t) = x(t)$. Hence,

$$y(t+2) - ay(t+1) = bu(t)$$

and thus

$$\big(q^2 - aq\big)y(t) = bu(t)$$

Under the assumption that $y(1) = 0$ and after premultiplication both sides by $1/(q^2 - aq)$,

$$y(t) = \frac{b}{(q^2 - aq)}u(t) = \frac{bq^{-2}}{1 - aq^{-1}}u(t)$$

After long division we obtain

$$y(t) = bq^{-2}\big(1 + aq^{-1} + a^2 q^{-2} + \cdots\big)\, u(t)$$

Hence, multiplication and division can be applied on polynomials in q in a normal way. Similarly, addition and subtraction rules hold. However, it should be realized that q is not a variable but an operator. As in Example E.2, the basic assumption, which holds throughout this book, is that all initial conditions for the difference equation are zero.

E.2 Pulse Transfer Operator

The introduction of shift operator calculus also allows input–output relationships to be conveniently expressed as rational functions, as illustrated by Example E.2. For instance, since division by polynomials in q is defined, from (E.2) we can derive that

$$y(t) = \frac{1}{A(q)} B(q)u(t)$$

The function $B(q)/A(q)$ is called the *pulse-transfer* operator. This operator can be easily obtained from a linear time-invariant (LTI) system description as follows. Let an LTI system be described by

$$\begin{aligned} x(t+1) = qx(t) &= Ax(t) + Bu(t) \\ y(t) &= Cx(t) + Du(t) \end{aligned} \tag{E.3}$$

with A, B, C, D matrices of appropriate dimensions. From this discrete-time state-space description we obtain

$$(qI - A)x(t) = Bu(t)$$

and thus

$$y(t) = \left[C(qI - A)^{-1}B + D\right]u(t)$$

For $u(t)$ a unit pulse, the pulse-transfer operator of the LTI system (E.3) is given by

$$G(q) = \left[C(qI - A)^{-1}B + D\right] \tag{E.4}$$

In the backward-shift operator,

$$G^*(q^{-1}) = \left[C(I - q^{-1}A)^{-1}q^{-1}B + D\right] = G(q)$$

The pulse-transfer operator of the system of (E.3) is thus a matrix whose elements are rational functions in q. For single-input single-output (SISO) systems with B and C vectors and D a scalar,

$$\begin{aligned} G(q) &= \left[C(qI - A)^{-1}B + D\right] \\ &= \frac{C\,\mathrm{adj}(qI - A)B}{\det(qI - A)} \\ &= \frac{B(q)}{A(q)} \end{aligned} \tag{E.5}$$

If the state vector is of dimension n and if the polynomials $A(q)$ and $B(q)$ do not have common factors, then $A(q)$ is of degree n. It directly follows from (E.5) that the polynomial $A(q)$ is the characteristic polynomial of the system matrix A. We call $A(q)$ *monic* if its zeroth coefficient is 1 or the identity matrix in the multiinput multioutput case. The *poles* of a system are the zeros of the denominator of $G(q)$, the characteristic polynomial $A(q)$. The system *zeros* are obtained from $B(q) = 0$. Time delay in a system gives rise to poles at the origin. From Example E.2 we notice that this system has two poles, one in a and one at the origin, as a result of a single time delay. The pulse-transfer operator of a discrete-time LTI system $G(q)$ is also called the discrete-time transfer function. The z-transform of Sect. 3.2.1 maps a semi-infinite time sequence into a function of a complex variable ($z = e^{j\omega}$ with ω the frequency, see also Appendix C). Notice then the difference in range for the z-transform and the shift-operator calculus. In the operator calculus double-infinite sequences are considered. In practice, this means that the main difference between z and q, apart from being a complex variable or an operator, respectively, is that the z-transform takes the initial values explicitly into account.

Example E.3 Shift operator calculus: Consider the LTI system with matrices

$$A = \begin{bmatrix} 1 & 0 \\ 0 & 0 \end{bmatrix}, \qquad B = \begin{bmatrix} 1 \\ 0 \end{bmatrix}, \qquad C = [\,1\;0\,], \qquad D = 0$$

Then

$$\begin{aligned} G(q) &= [\,1\;0\,] \begin{bmatrix} q-1 & 0 \\ 0 & q \end{bmatrix}^{-1} \begin{bmatrix} 1 \\ 0 \end{bmatrix} \\ &= \frac{q}{q(q-1)} \end{aligned}$$

Hence, after pole-zero cancelation, the pulse-transfer operator or discrete-time transfer function becomes

$$G(q) = \frac{1}{q-1}$$

with no zero and a single pole in 1. Consequently, this system is a simple single integrator.

Appendix F
Recursive Least-squares Derivation

F.3 Least-squares Method

Recall that in this book the linear regression at each time instant t is written as

$$y(t) = \phi(t)^T \vartheta + e(t)$$

where $y(t)$ is the measurement, $\phi(t)$ the regressor vector, and $e(t)$ the prediction error (in this case also called the equation error) for $t = 1, 2, \dots, N$, and ϑ is the unknown $p \times 1$ parameter vector. The ordinary least-squares method minimizes the criterion function

$$J(\vartheta) = \sum_{t=1}^{N} \big(y(t) - \phi(t)^T \vartheta\big)^2$$

with respect to ϑ. Notice that the criterion function is quadratic in ϑ, and thus it can be minimized analytically, as shown before. The ordinary least-squares estimate is then given by

$$\widehat{\vartheta}(N) = \left[\sum_{t=1}^{N} \phi(t)\phi^T(t)\right]^{-1} \sum_{t=1}^{N} \phi(t)y(t) \tag{F.1}$$

provided that the inverse exists.

The estimate can also be written in a recursive form. Here we follow the derivation as presented in Ljung and Söderström [LS83]. Define the $p \times p$ matrix

$$\bar{K}(t) := \sum_{k=1}^{t} \phi(k)\phi^T(k)$$

Then from (F.1) we obtain

$$\sum_{k=1}^{t-1} \phi(k)y(k) = \bar{K}(t-1)\widehat{\vartheta}(t-1)$$

K.J. Keesman, *System Identification*,
Advanced Textbooks in Control and Signal Processing,
DOI 10.1007/978-0-85729-522-4,

From the definition of $\bar{K}(t)$ it follows that

$$\bar{K}(t-1) = \bar{K}(t) - \phi(t)\phi^T(t)$$

Thus,

$$\begin{aligned}
\widehat{\vartheta}(t) &= \bar{K}^{-1}(t)\left[\sum_{k=1}^{t-1} \phi^k y(k) + \phi(t)y(t)\right] \\
&= \bar{K}^{-1}(t)\left[\bar{K}(t-1)\widehat{\vartheta}(t-1) + \phi(t)y(t)\right] \\
&= \bar{K}^{-1}(t)\left[\bar{K}(t)\widehat{\vartheta}(t-1) + \phi(t)\left(-\phi^T(t)\widehat{\vartheta}(t-1) + y(t)\right)\right] \\
&= \widehat{\vartheta}(t-1) + \bar{K}^{-1}(t)\phi(t)\left[y(t) - \phi^T(t)\widehat{\vartheta}(t-1)\right]
\end{aligned} \tag{F.2}$$

and

$$\bar{K}(t) = \bar{K}(t-1) + \phi(t)\phi^T(t) \tag{F.3}$$

Consequently, (F.2) and (F.3) together form a recursive algorithm for the estimation of $\vartheta(t)$, given the previous estimate $\widehat{\vartheta}(t-1)$, $\bar{K}(t-1)$ and $\phi(t)$, $y(t)$. Notice that we do not need previous values of ϕ and $y(t)$; the actual values suffice to estimate the current value of $\widehat{\vartheta}(t)$. However, there is still a need to invert a $p \times p$ matrix at each time step.

F.4 Equivalent Recursive Form

Let us first define

$$K(t) := \frac{1}{t}\bar{K}(t)$$

so that (F.3) becomes

$$\begin{aligned}
K(t) &= \frac{1}{t}\left[\bar{K}(t-1) + \phi(t)\phi^T(t)\right] = \frac{t-1}{t}K(t-1) + \frac{1}{t}\phi(t)\phi^T(t) \\
&= K(t-1) + \frac{1}{t}\left[\phi(t)\phi^T(t) - K(t-1)\right]
\end{aligned} \tag{F.4}$$

Subsequently, we introduce

$$P(t) = \bar{K}^{-1}(t) = \frac{1}{t}R^{-1}(t)$$

so that $P(t)$ can be updated directly, instead of using (F.3). However, to accomplish this, we need the so-called *matrix inversion lemma*, which is given by

$$[A + BCD]^{-1} = A^{-1} - A^{-1}B\left[DA^{-1}B + C^{-1}\right]^{-1}DA^{-1}$$

with matrices A, B, C, and D of appropriate dimensions, such that the product BCD and the sum $A + BCD$ exist. Furthermore, the inverses A^{-1} and C^{-1} must exist. Given the definition of $P(t)$, (F.4) can be written as

$$
\begin{aligned}
tK(t) &= \left[(t-1)\ K(t-1) + \phi(t)\phi^T(t)\right] \\
\Longrightarrow\quad \left[tK(t)\right]^{-1} &= \left[(t-1)\ K(t-1) + \phi(t)\phi^T(t)\right]^{-1} \\
\Longrightarrow\quad P(t) &= \left[P^{-1}(t-1) + \phi(t)\phi^T(t)\right]^{-1}
\end{aligned}
$$

Consequently, after setting $P^{-1}(t-1) = A$, $\phi(t) = B$, $C = 1$, and $\phi^T(t) = D$, we obtain

$$
\begin{aligned}
P(t) &= \left[P^{-1}(t-1) + \phi(t)\cdot 1 \cdot \phi^T(t)\right]^{-1} \\
&= P(t-1) - P(t-1)\phi(t)\left[\phi^T(t)P(t-1)\phi(t) + 1\right]^{-1}\phi^T(t)P(t-1) \\
&= P(t-1) - \frac{P(t-1)\phi(t)\phi^T(t)P(t-1)}{1 + \phi^T(t)P(t-1)\phi(t)}
\end{aligned}
$$

Instead of inverting a $p \times p$ matrix, we only need the division by a scalar. From this equation we also find that

$$
\begin{aligned}
P(t)\phi(t) &= P(t-1)\phi(t) - \frac{P(t-1)\phi(t)\phi^T(t)P(t-1)\phi(t)}{1 + \phi^T(t)P(t-1)\phi(t)} \\
&= \frac{P(t-1)\phi(t)}{1 + \phi^T(t)P(t-1)\phi(t)}
\end{aligned}
$$

Define $L(t) := P(t)\phi(t)$; then the so-called *recursive least-squares* (RLS) algorithm is given by

$$
\begin{aligned}
\widehat{\vartheta}(t) &= \widehat{\vartheta}(t-1) + L(t)\left[y(t) - \phi^T(t)\widehat{\vartheta}(t-1)\right] \\
L(t) &= \frac{P(t-1)\phi(t)}{1 + \phi^T(t)P(t-1)\phi(t)} \\
P(t) &= P(t-1) - \frac{P(t-1)\phi(t)\phi^T(t)P(t-1)}{1 + \phi^T(t)P(t-1)\phi(t)}
\end{aligned}
$$

Recall that this algorithm has been derived from an un-weighted least-squares criterion function. If, however, a time-varying weight $R(t)$ is included, such that

$$
J_W(\vartheta) = \sum_{t=1}^{N} \frac{1}{R(t)}\left(y(t) - \phi(t)^T\vartheta\right)^2
$$

we will obtain

$$
\widehat{\vartheta}(t) = \widehat{\vartheta}(t-1) + L(t)\left[y(t) - \phi^T(t)\widehat{\vartheta}(t-1)\right]
$$

$$L(t) = \frac{P(t-1)\phi(t)}{R(t) + \phi^T(t)P(t-1)\phi(t)}$$

$$P(t) = P(t-1) - \frac{P(t-1)\phi(t)\phi^T(t)P(t-1)}{R(t) + \phi^T(t)P(t-1)\phi(t)}$$

Notice that the initial values of $\widehat{\vartheta}(0)$ and $P(0)$ must be known in order to run the RLS algorithm. Typical choices are $\widehat{\vartheta}(0) = 0$ and $P(0) = c\,I$ with c a large constant.

Appendix G
Dissolved Oxygen Data

Table G.1 DO data lake "De Poel en 't Zwet"[a]

t (d)	DO (g/m^3)	C_s (g/m^3)	I (W/m^2)
111.8750	9.1860	10.9873	0.0000
111.9170	9.1980	11.0374	0.0000
111.9580	8.9960	11.0374	0.0000
112.0000	8.7930	11.0374	0.0000
112.0420	8.8040	11.0878	0.0000
112.0830	8.8150	11.0878	0.0000
112.1250	8.8270	11.1412	0.0000
112.1670	8.4070	11.1412	0.0000
112.2080	8.4170	11.1412	0.0000
112.2500	8.4280	11.1924	0.0000
112.2920	8.4390	11.1924	15.8900
112.3330	8.4490	11.1924	57.1900
112.3750	8.4600	11.1412	108.0000
112.4170	8.6880	11.0878	165.2000
112.4580	8.9170	10.9873	209.7000
112.5000	9.3650	10.9351	158.9000
112.5420	9.8130	10.8343	130.3000
112.5830	10.4800	10.7350	266.9000
112.6250	10.7100	10.6373	127.1000
112.6670	10.7300	10.6373	85.7800
112.7080	10.7400	10.5902	54.0100

[a]This data set, for the period 21–30 April 1983, was collected by students of the University of Twente.

K.J. Keesman, *System Identification*,
Advanced Textbooks in Control and Signal Processing,
DOI 10.1007/978-0-85729-522-4,

Table G.1 (Continued)

t (d)	DO (g/m^3)	C_s (g/m^3)	I (W/m^2)
112.7500	10.7600	10.6373	31.7700
112.7920	10.9900	10.6373	25.4200
112.8330	10.7800	10.6373	6.3540
112.8750	10.8000	10.6872	0.0000
112.9170	10.8100	10.6872	0.0000
112.9580	10.3800	10.7350	0.0000
113.0000	9.9510	10.7857	0.0000
113.0420	10.1900	10.8343	0.0000
113.0830	9.9760	10.8858	0.0000
113.1250	9.9880	10.8858	0.0000
113.1670	9.5550	10.9351	0.0000
113.2080	9.5670	10.9873	0.0000
113.2500	9.5780	10.9873	0.0000
113.2920	9.1430	11.0374	12.7100
113.3330	9.1540	11.0374	60.3700
113.3750	9.6140	10.9873	98.4900
113.4170	9.8510	10.9351	133.4000
113.4580	10.5400	10.8343	168.4000
113.5000	10.3300	10.7857	225.6000
113.5420	11.0200	10.6373	184.3000
113.5830	11.4800	10.6373	25.4200
113.6250	11.2700	10.6373	15.8900
113.6670	11.5100	10.5902	47.6600
113.7080	11.0700	10.5902	28.5900
113.7500	11.3100	10.5902	9.5320
113.7920	11.7800	10.5435	3.1770
113.8330	11.3400	10.5435	0.0000
113.8750	11.5800	10.5902	0.0000
113.9170	11.1400	10.6373	0.0000
113.9580	11.1500	10.6373	0.0000
114.0000	10.9400	10.6373	0.0000
114.0420	10.9500	10.6872	0.0000
114.0830	10.7300	10.6872	0.0000
114.1250	10.5200	10.7350	0.0000
114.1670	10.0700	10.7857	0.0000
114.2080	10.0800	10.8343	0.0000
114.2500	9.8640	10.8343	0.0000
114.2920	9.4130	10.8343	12.7100
114.3330	9.6560	10.8858	41.3000

Table G.1 (Continued)

t(d)	DO (g/m^3)	C_s (g/m^3)	I (W/m^2)
114.3750	9.8990	10.8858	73.0800
114.4170	9.9110	10.8343	117.6000
114.4580	10.8500	10.7857	139.8000
114.5000	11.5600	10.6872	238.3000
114.5420	12.2800	10.5902	171.6000
114.5830	12.7600	10.4947	251.0000
114.6250	12.7700	10.4008	187.5000
114.6670	11.3900	10.3555	197.0000
114.7080	13.0400	10.3083	168.4000
114.7500	13.5200	10.2639	117.6000
114.7920	13.3000	10.2639	60.3700
114.8330	13.3200	10.2639	19.0600
114.8750	12.8700	10.3555	9.5320
114.9170	12.8800	10.4008	0.0000
114.9580	12.4200	10.4487	0.0000
115.0000	12.2000	10.4947	0.0000
115.0420	11.9800	10.4947	0.0000
115.0830	11.7600	10.4947	0.0000
115.1250	11.5400	10.5435	0.0000
115.1670	11.5500	10.5902	0.0000
115.2080	10.8500	10.6373	0.0000
115.2500	10.6200	10.6872	0.0000
115.2920	10.8800	10.7350	19.0600
115.3330	10.1700	10.7857	50.8400
115.3750	10.6600	10.7857	54.0100
115.4170	10.9100	10.7857	50.8400
115.4580	11.1700	10.7350	114.4000
115.5000	11.1800	10.6373	114.4000
115.5420	11.6700	10.6373	41.3000
115.5830	11.6900	10.6373	19.0600
115.6250	11.4600	10.6373	66.7200
115.6670	9.8720	10.6373	22.2400
115.7080	11.1200	10.5902	82.6100
115.7500	11.5600	10.5435	57.1900
115.7920	11.9900	10.5435	31.7700
115.8330	12.0200	10.5435	9.5320
115.8750	11.8500	10.5435	0.0000
115.9170	11.6700	10.5902	0.0000
115.9580	11.2900	10.5902	0.0000

Table G.1 (Continued)

t (d)	DO (g/m^3)	C_s (g/m^3)	I (W/m^2)
116.0000	11.3200	10.6373	0.0000
116.0420	10.9300	10.6373	0.0000
116.0830	11.3700	10.6373	0.0000
116.1250	11.1900	10.6373	0.0000
116.1670	10.5900	10.6373	0.0000
116.2080	11.0400	10.6872	0.0000
116.2500	10.6400	10.7350	0.0000
116.2920	10.4500	10.7350	15.8900
116.3330	10.6900	10.7350	60.3700
116.3750	10.5100	10.6872	114.4000
116.4170	10.7400	10.6373	149.3000
116.4580	10.7700	10.5902	181.1000
116.5000	11.2200	10.5435	120.7000
116.5420	11.4700	10.4487	155.7000
116.5830	12.3500	10.3555	177.9000
116.6250	12.6000	10.2639	222.4000
116.6670	12.6300	10.2174	212.9000
116.7080	13.0900	10.1737	92.1400
116.7500	12.6900	10.1737	34.9500
116.7920	12.7200	10.2174	19.0600
116.8330	12.5300	10.2174	9.5320
116.8750	12.1200	10.2639	0.0000
116.9170	11.9300	10.3083	0.0000
116.9580	11.9600	10.3555	0.0000
117.0000	11.5500	10.3555	0.0000
117.0420	11.3500	10.4008	0.0000
117.0830	10.9400	10.4008	0.0000
117.1250	10.7400	10.4487	0.0000
117.1670	10.7600	10.4487	0.0000
117.2080	10.3400	10.4487	0.0000
117.2500	9.9170	10.4947	0.0000
117.2920	10.1600	10.4947	0.0000
117.3330	9.9630	10.5435	3.1770
117.3750	9.3080	10.5435	12.7100
117.4170	9.5550	10.5435	9.5320
117.4580	9.5770	10.5902	25.4200
117.5000	10.0500	10.5902	41.3000
117.5420	10.0800	10.5902	22.2400
117.5830	10.1000	10.6373	15.8900

Table G.1 (Continued)

t (d)	DO (g/m^3)	C_s (g/m^3)	I (W/m^2)
117.6250	10.1200	10.6373	12.7100
117.6670	10.1400	10.6373	15.8900
117.7080	10.1700	10.6872	19.0600
117.7500	10.1900	10.7350	12.7100
117.7920	10.2100	10.7350	9.5320
117.8330	10.0000	10.7857	3.1770
117.8750	9.7920	10.8343	0.0000
117.9170	9.8130	10.8343	0.0000
117.9580	9.6010	10.8858	0.0000
118.0000	9.6220	10.9351	0.0000
118.0420	10.3500	10.9351	0.0000
118.0830	10.3700	10.9873	0.0000
118.1250	9.6850	10.9873	0.0000
118.1670	9.4700	11.0374	0.0000
118.2080	9.4900	11.0374	0.0000
118.2500	9.5110	11.0878	0.0000
118.2920	9.5310	11.1412	15.8900
118.3330	9.5510	11.1412	63.5400
118.3750	10.0500	11.0878	88.9600
118.4170	10.0700	11.0374	114.4000
118.4580	10.5700	11.0374	85.7800
118.5000	11.0800	10.9351	181.1000
118.5420	11.1000	10.8343	162.0000
118.5830	11.8500	10.7857	212.9000
118.6250	12.1200	10.7350	247.8000
118.6670	12.3800	10.6373	219.2000
118.7080	12.6500	10.5902	174.7000
118.7500	12.6800	10.5902	123.9000
118.7920	12.7100	10.5435	66.7200
118.8330	12.2400	10.5902	22.2400
118.8750	12.0800	10.6373	0.0000
118.9170	11.9300	10.6872	0.0000
118.9580	11.6500	10.6872	0.0000
119.0000	11.5700	10.7350	0.0000
119.0420	11.4800	10.7350	0.0000
119.0830	11.1800	10.7350	0.0000
119.1250	10.8700	10.7857	0.0000
119.1670	10.5600	10.7857	0.0000
119.2080	11.0900	10.8343	0.0000

Table G.1 (Continued)

t (d)	DO (g/m^3)	C_s (g/m^3)	I (W/m^2)
119.2500	10.3300	10.8343	0.0000
119.2920	10.4300	10.8343	6.3540
119.3330	10.0900	10.8343	12.7100
119.3750	10.1900	10.8343	15.8900
119.4170	10.2800	10.8343	34.9500
119.4580	10.3700	10.8343	57.1900
119.5000	9.8000	10.8343	139.8000
119.5420	10.8200	10.7350	222.4000
119.5830	10.8300	10.6872	152.5000
119.6250	10.8500	10.5902	254.2000
119.6670	11.2700	10.4947	228.8000
119.7080	11.0800	10.4487	184.3000
119.7500	11.5000	10.4008	130.3000
119.7920	11.5200	10.3555	73.0800
119.8330	11.5400	10.4008	25.4200
119.8750	11.3500	10.4008	3.1770
119.9170	10.9600	10.4487	0.0000
119.9580	10.7800	10.4947	0.0000
120.0000	10.5900	10.5435	0.0000

References

[ADSC98] S. Audoly, L. D'Angiò, M.P. Saccomani, C. Cobelli, Global identifiability of linear compartmental models—a computer algebra algorithm. IEEE Trans. Biomed. Eng. **45**(1), 36–47 (1998)

[ÅH84] K.J. Åström, T. Hagglund, Automatic tuning of simple regulators with specifications on phase and amplitude margins. Automatica **20**, 645–651 (1984)

[ÅH88] K.J. Åström, T. Hagglund, *Automatic tuning of PID controllers* (Instrument Society of America, 1988)

[AH01] H. Akcay, P.S.C. Heuberger, Frequency-domain iterative identification algorithm using general orthonormal basis functions. Automatica **37**(5), 663–674 (2001)

[Aka74] H. Akaike, A new look at statistical model identification. IEEE Trans. Autom. Control **AC-19**, 716–723 (1974)

[Akc00] H. Akcay, Continuous-time stable and unstable system modelling with orthonormal basis functions. Int. J. Robust Nonlinear Control **10**(6), 513–531 (2000)

[AMLL02] A. Al Mamun, T.H. Lee, T.S. Low, Frequency domain identification of transfer function model of a disk drive actuator. Mechatronics **12**(4), 563–574 (2002)

[AN99] H. Akcay, B. Ninness, Orthonormal basis functions for modelling continuous-time systems. Signal Process. **77**(3), 216–274 (1999)

[And85] B.D.O. Anderson, Identification of scalar errors-in-variables models with dynamics. Automatica **21**(6), 709–716 (1985)

[Åst70] K.J. Åström, *Introduction to Stochastic Control Theory*. Mathematics in Science and Engineering, vol. 70 (Academic Press, San Diego, 1970)

[BA02] K.P. Burnham, D.R. Anderson, *Model Selection and Multimodel Inference: A Practical Information-theoretic Approach*, 2nd edn. (Springer, Berlin, 2002)

[Bag75] A. Bagchi, Continuous time systems identification with unknown noise covariance. Automatica **11**(5), 533–536 (1975)

[Bai02] E.-W. Bai, A blind approach to the Hammerstein–Wiener model identification. Automatica **38**(6), 967–979 (2002)

[Bai03a] E.-W. Bai, Frequency domain identification of Hammerstein models. IEEE Trans. Autom. Control **48**(4), 530–542 (2003)

[Bai03b] E.-W. Bai, Frequency domain identification of Wiener models. Automatica **39**(9), 1521–1530 (2003)

[Bar74] Y. Bard, *Nonlinear Parameter Estimation* (Academic Press, San Diego, 1974)

[BBB85] D. Bertin, S. Bittanti, P. Bolzern, Prediction-error directional forgetting technique for recursive estimation. Syst. Sci. **11**(2), 33–39 (1985)

[BBC90] G. Belforte, B. Bona, V. Cerone, Identification, structure selection and validation of uncertain models with set-membership error description. Math. Comput. Simul. **32**(5–6), 561–569 (1990)

K.J. Keesman, *System Identification*,
Advanced Textbooks in Control and Signal Processing,
DOI 10.1007/978-0-85729-522-4,

[BBF87] G. Belforte, B. Bona, S. Fredani, Optimal sampling schedule for parameter estimation of linear models with unknown but bounded measurement errors. IEEE Trans. Autom. Control **AC–32**(2), 179–182 (1987)

[BBM86] A. Benveniste, M. Basseville, G. Moustakides, Modelling and monitoring of changes in dynamical systems, in *Proceedings of the IEEE Conference on Decision and Control* (1986), pp. 776–782

[BC94] S. Bittanti, M. Campi, Bounded error identification of time-varying parameters by RLS techniques. IEEE Trans. Autom. Control **39**(5), 1106–1110 (1994)

[BE83] D.R. Brillinger, P.R. Krishnaiah (eds.), *Handbook of Statistics 3: Time Series in the Frequency Domain* (North-Holland, Amsterdam, 1983)

[BEW02] L. Bertino, G. Evensen, H. Wackernagel, Combining geostatistics and Kalman filtering for data assimilation in an estuarine system. Inverse Probl. **18**(1), 1–23 (2002)

[BG02] B. Bamieh, L. Giarre, Identification of linear parameter varying models. Int. J. Robust Nonlinear Control **12**(9), 841–853 (2002)

[BG03] G. Belforte, P. Gay, Optimal input design for set-membership identification of Hammerstein models. Int. J. Control **76**(3), 217–225 (2003)

[Bie77] G.J. Bierman, *Factorization Methods for Discrete Sequential Estimation*. Mathematics in Science and Engineering (Academic Press, San Diego, 1977)

[Bit77] S. Bittanti, On optimal experiment design for parameters estimation of dynamic systems under periodic operation, in *Proceedings of the IEEE Conference on Decision and Control*, 1977, pp. 1126–1131

[BJ70] G.E.P. Box, G.M. Jenkins, *Time Series Analysis: Forecasting and Control* (Holden-Day, Oakland, 1970)

[Bjo96] A. Bjork, *Numerical Methods for Least Squares Problems* (SIAM, Philadelphia, 1996)

[BK70] R. Bellman, K.J. Åström, On structural identifiability. Math. Biosci. **7**, 329–339 (1970)

[Blu72] M. Blum, Optimal smoothing of piecewise continuous functions. IEEE Trans. Inf. Theory **18**(2), 298–300 (1972)

[BM74] G.E.P. Box, J.F. MacGregor, Analysis of closed-loop dynamic-stochastic systems. Technometrics **16**(3), 391–398 (1974)

[BMS+04] T.Z. Bilau, T. Megyeri, A. Sárhegyi, J. Márkus, I. Kollár, Four-parameter fitting of sine wave testing result: Iteration and convergence. Comput. Stand. Interfaces **26**(1), 51–56 (2004)

[Box71] M.J. Box, Bias in nonlinear estimation. J. R. Stat. Soc., Ser. B, Stat. Methodol. **33**(2), 171–201 (1971)

[BP02] G. Belforte, G.A.Y. Paolo, Optimal experiment design for regression polynomial models identification. Int. J. Control **75**(15), 1178–1189 (2002)

[BR97] P. Barone, R. Ragona, Bayesian estimation of parameters of a damped sinusoidal model by a Markov chain Monte Carlo method. IEEE Trans. Signal Process. **45**(7), 1806–1814 (1997)

[BRD97] M. Boutayeb, H. Rafaralahy, M. Darouach, Convergence analysis of the extended Kalman filter used as an observer for nonlinear deterministic discrete-time systems. IEEE Trans. Autom. Control **42**(4), 581–586 (1997)

[Bri81] D.R. Brillinger, *Time Series: Data Analysis and Theory* (Holden-Day, Oakland, 1981)

[BRJ09] E.-W. Bai, J. Reyland Jr., Towards identification of Wiener systems with the least amount of a priori information: IIR cases. Automatica **45**(4), 956–964 (2009)

[BS72] B.D.O. Anderson, S. Chirarattananon, New linear smoothing formulas. IEEE Trans. Autom. Control **17**(1), 160–161 (1972)

[BSK+02] K. Bernaerts, R.D. Servaes, S. Kooyman, K.J. Versyck, J.F. Van Impe, Optimal temperature input design for estimation of the square root model parameters: parameter accuracy and model validity restrictions. Int. J. Food Microbiol. **73**(2–3), 145–157 (2002).

[BY76] B. Beck, P. Young, Systematic identification of do-bod model structure. J. Environ. Eng. Div. ASCE **102**(5 EE5), 909–927 (1976)

[BZ95] S.A. Billings, Q.M. Zhu, Model validation tests for multivariable nonlinear models including neural networks. Int. J. Control **62**(4), 749–766 (1995)

[Car73] N.A. Carlson, Fast triangular formulation of the square root filter. AIAA J. **11**(9), 1259–1265 (1973)

[Car90] N.A. Carlson, Federated square root filter for decentralized parallel processes. IEEE Trans. Aerosp. Electron. Syst. **26**(3), 517–525 (1990)

[CB89] S. Chen, S.A. Billings, Recursive prediction error parameter estimator for nonlinear models. Int. J. Control **49**(2), 569–594 (1989)

[CC94] J.-M. Chen, B.-S. Chen, A higher-order correlation method for model-order and parameter estimation. Automatica **30**(8), 1339–1344 (1994)

[CDC04] M. Crowder, R. De Callafon, Time Domain Control Oriented Model Validation Using Coprime Factor Perturbations, in *Proceedings of the IEEE Conference on Decision and Control*, vol. 2 (2004), pp. 2182–2187

[CG00] J. Chen, G. Gu, *Control-oriented System Identification: an h_∞ Approach* (Wiley, New York, 2000)

[CGCE03] M.J. Chapman, K.R. Godfrey, M.J. Chappell, N.D. Evans, Structural identifiability for a class of non-linear compartmental systems using linear/non-linear splitting and symbolic computation. Math. Biosci. **183**(1), 1–14 (2003)

[Che70] R.T.N. Chen, Recurrence relationship for parameter estimation via method of quasi-linearization and its connection with Kalman filtering. AIAA J. **8(9)**, 1696–1698 (1970)

[CHY02] Y.-Y. Chen, P.-Y. Huang, J.-Y. Yen, Frequency-domain identification algorithms for servo systems with friction. IEEE Trans. Control Syst. Technol. **10**(5), 654–665 (2002)

[CKBR08] J. Chandrasekar, I.S. Kim, D.S. Bernstein, A.J. Ridley, Cholesky-based reduced-rank square-root Kalman filtering, in *Proceedings of the American Control Conference* (2008), pp. 3987–3992

[CM78] F.L. Chernousko, A.A. Melikyan, *Game Problems of Control and Search* (Nauka, Moscow, 1978) (in Russian)

[CSI00] M.-H. Chen, Q.-M. Shao, J.G. Ibrahim, *Monte Carlo Methods in Bayesian Computation* (Springer, New York, 2000)

[CSS08] M.C. Campi, T. Sugie, F. Sakai, An iterative identification method for linear continuous-time systems. IEEE Trans. Autom. Control **53**(7), 1661–1669 (2008)

[CZ95] R.F. Curtain, H.J. Zwart, *An Introduction to Infinite-dimensional Linear Systems Theory* (Springer, Berlin, 1995), p. 698

[DA96] S. Dasgupta, B.D.O. Anderson, A parametrization for the closed-loop identification of nonlinear time-varying systems. Automatica **32**(10), 1349–1360 (1996)

[DdFG01] A. Doucet, N. de Freitas, N. Gordon, *Sequential Monte Carlo Methods in Practice* (Springer, New York, 2001)

[DDk05] H. Deng, M. Doroslovački, Improving convergence of the *pnlms* algorithm for sparse impulse response identification. IEEE Signal Process. Lett. **12**(3), 181–184 (2005)

[deS87] C.W. deSilva, Optimal input design for the dynamic testing of mechanical systems. J. Dyn. Syst. Meas. Control, Trans. ASME **109**(2), 111–119 (1987)

[DGW96] S.K. Doherty, J.B. Gomm, D. Williams, Experiment design considerations for non-linear system identification using neural networks. Comput. Chem. Eng. **21**(3), 327–346 (1996)

[DI76] J.J. DiStefano III, Tracer experiment design for unique identification of nonlinear physiological systems. Am. J. Physiol. **230**(2), 476–485 (1876)

[DI82] J.J. DiStefano III, Algorithms, software and sequential optimal sampling schedule designs for pharmacokinetic and physiologic experiments. Math. Comput. Simul. **24**(6), 531–534 (1982)

[DK05] T.G. Doeswijk, K.J. Keesman, Adaptive weather forecasting using local meteorological information. Biosyst. Eng. **91**(4), 421–431 (2005)

[DK09] T.G. Doeswijk, K.J. Keesman, Linear parameter estimation of rational biokinetic functions. Water Res. **43(1)**, 107–116 (2009)

[DS98] N.R. Draper, H. Smith, *Introduction to Linear Regression Analysis*, 4th edn. Wiley Series in Probability and Statistics (Wiley, New York, 1998)

[DvdH96] H.G.M. Dötsch, P.M.J. van den Hof, Test for local structural identifiability of high-order non-linearly parametrized state space models. Automatica **32**(6), 875–883 (1996)

[dVvdH98] D.K. de Vries, P.M.J. van den Hof, Frequency domain identification with generalized orthonormal basis functions. IEEE Trans. Autom. Control **43**(5), 656–669 (1998)

[DW95] L. Desbat, A. Wernsdorfer, Direct algebraic reconstruction and optimal sampling in vector field tomography. IEEE Trans. Signal Process. **43**(8), 1798–1808 (1995)

[ECCG02] N.D. Evans, M.J. Chapman, M.J. Chappell, K.R. Godfrey, Identifiability of uncontrolled nonlinear rational systems. Automatica **38**(10), 1799–1805 (2002)

[EMT00] A. Esmaili, J.F. MacGregor, P.A. Taylor, Direct and two-step methods for closed-loop identification: A comparison of asymptotic and finite data set performance. J. Process Control **10**(6), 525–537 (2000)

[EO68] L.D. Enochson, R.K. Otnes, Programming and analysis for digital time series data. Technical report, US Dept. of Defense, Shock and Vibration Info. Center, 1968

[ES81] H. El-Sherief, Multivariable system structure and parameter identification using the correlation method. Automatica **17**(3), 541–544 (1981)

[Eve94] G. Evensen, Sequential data assimilation with a nonlinear quasi-geostrophic model using Monte Carlo methods to forecast error statistics. J. Geophys. Res. **99**(C5), 10143–10162 (1994)

[Eyk74] P. Eykhoff, *System Identification: Parameter and State Estimation* (Wiley-Interscience, New York, 1974)

[FBT96] B. Farhang-Boroujeny, T.-T. Tay, Transfer function identification with filtering techniques. IEEE Trans. Signal Process. **44**(6), 1334–1345 (1996)

[FG06] P. Falugi, L. Giarre, Set membership (in)validation of nonlinear positive models for biological systems, in *Proceedings of the IEEE Conference on Decision and Control* (2006)

[FH82] E. Fogel, Y.F. Huang, On the value of information in system identification-bounded noise case. Automatica **18**, 229–238 (1982)

[FL99] U. Forssell, L. Ljung, Closed-loop identification revisited. Automatica **35**(7), 1215–1241 (1999)

[Fre80] P. Freymuth, Sine-wave testing of non-cylindrical hot-film anemometers according to the Bellhouse-Schultz model. J. Phys. E, Sci. Instrum. **13**(1), 98–102 (1980)

[GCH98] S. Grob, P.D.J. Clark, K. Hughes, Enhanced channel impulse response identification for the itu hf measurement campaign. Electron. Lett. **34**(10), 1022–1023 (1998)

[Gel74] A. Gelb, *Applied Optimal Estimation* (MIT Press, Cambridge, 1974)

[GGS01] G.C. Goodwin, S.F. Graebe, M.E. Salgado, *Control System Design* (Prentice Hall, New York, 2001)

[GL09b] J. Gillberg, L. Ljung, Frequency-domain identification of continuous-time ARMA models from sampled data. Automatica **45**(6), 1371–1378 (2009)

[God80] K.R. Godfrey, Correlation methods. Automatica **16**(5), 527–534 (1980)

[GP77] G.C. Goodwin, R.L. Payne, *Dynamic System Identification: Experiment Design and Data Analysis* (Prentice-Hall, New York, 1977)

[GP08] W. Greblicki, M. Pawlak, *Non-Parametric System Identification* (Cambridge University Press, Cambridge, 2008)

[GRC09] F. Giri, Y. Rochdi, F.-Z. Chaoui, An analytic geometry approach to Wiener system frequency identification. IEEE Trans. Autom. Control **54**(4), 683–696 (2009)

[Gre94] W. Greblicki, Nonparametric identification of Wiener systems by orthogonal series. IEEE Trans. Autom. Control **39**(10), 2077–2086 (1994)

[Gre98] W. Greblicki, Continuous-time Wiener system identification. IEEE Trans. Autom. Control **43**(10), 1488–1493 (1998)

[Gre00] W. Greblicki, Continuous-time Hammerstein system identification. IEEE Trans. Autom. Control **45**(6), 1232–1236 (2000)

[GRS96] W.R. Gilks, S. Richardson, D.J. Spiegelhalter, *Markov Chain Monte Carlo in Practice* (Chapman and Hall, London, 1996)

[GS84] G.C. Goodwin, K.S. Sin, *Adaptive Filtering Prediction and Control* (Prentice-Hall, New York, 1984)

[Gui03] R. Guidorzi, *Multivariable System Identification: From Observations to Models* (Bononia University Press, Bologna, 2003)

[GvdH01] M. Gilson, P. van den Hof, On the relation between a bias-eliminated least-squares (BELS) and an IV estimator in closed-loop identification. Automatica **37**(10), 1593–1600 (2001)

[GVL80] G.H. Golub, C.F. Van Loan, An analysis of the total least squares problem. SIAM J. Numer. Anal. **17**(6), 883–893 (1980)

[GVL89] G.H. Golub, C.F. Van Loan, *Matrix Computations*, 2nd edn. (Johns Hopkins University Press, Baltimore, 1989)

[GW74] K. Glover, J.C. Willems, Parametrizations of linear dynamical systems: canonical forms and identifiability. IEEE Trans. Autom. Control **AC-19**(6), 640–646 (1974)

[Har91] N. Haritos, Swept sine wave testing of compliant bottom-pivoted cylinders, in *Proceedings of the First International Offshore and Polar Engineering Conference* (1991), pp. 378–383

[HB94] B.R. Haynes, S.A. Billings, Global analysis and model validation in nonlinear system identification. Nonlinear Dyn. **5**(1), 93–130 (1994)

[HdHvdHW04] P.S.C. Heuberger, T.J. de Hoog, P.M.J. van den Hof, B. Wahlberg, Orthonormal basis functions in time and frequency domain: Hambo transform theory. SIAM J. Control Optim. **42**(4), 1347–1373 (2004)

[HGDB96] H. Hjalmarsson, M. Gevers, F. De Bruyne, For model-based control design, closed-loop identification gives better performance. Automatica **32**(12), 1659–1673 (1996)

[HK01] D.R. Hill, B. Kolman, *Modern Matrix Algebra* (Prentice Hall, New York, 2001)

[HP05] D. Hinrichsen, A.J. Pritchard, *Mathematical Systems Theory I: Modelling, State Space Analysis, Stability and Robustness*. Texts in Applied Mathematics, vol. 48 (Springer, Berlin, 2005)

[HRvS07] C. Heij, A. Ran, F. van Schagen, *Introduction to Mathematical Systems Theory: Linear Systems, Identification and Control* (Birkhäuser, Basel, 2007)

[HS09] M. Hong, T. Söderström, Relations between bias-eliminating least squares, the Frisch scheme and extended compensated least squares methods for identifying errors-in-variables systems. Automatica **45**(1), 277–282 (2009)

[HSZ07] M. Hong, T. Söderström, W.X. Zheng, A simplified form of the bias-eliminating least squares method for errors-in-variables identification. IEEE Trans. Autom. Control **52**(9), 1754–1756 (2007)

[HvdMS02] R.H.A. Hensen, M.J.G. van de Molengraft, M. Steinbuch, Frequency domain identification of dynamic friction model parameters. IEEE Trans. Control Syst. Technol. **10**(2), 191–196 (2002)

[Ips09] I. Ipsen, *Numerical Matrix Analysis: Linear Systems and Least Squares* (SIAM, Philadelphia, 2009)

[Jaz70] A.H. Jazwinski, *Stochastic Processes and Filtering Theory*. Mathematics in Science and Engineering, vol. 64 (Academic Press, San Diego, 1970)

[JDVCJB06] C. Jauberthie, L. Denis-Vidal, P. Coton, G. Joly-Blanchard, An optimal input design procedure. Automatica **42**(5), 881–884 (2006)

[JM05] M.A. Johnson, M.H. Moradi, *PID Control: New Identification and Design Methods* (Springer, London, 2005)

[Joh93] R. Johansson, *System Modeling and Identification* (Prentice Hall, New York, 1993)

[JR04] R. Johansson, A. Robertsson, On behavioral model identification. Signal Process. **84**(7), 1089–1100 (2004)

[JU97] S.J. Julier, J.K. Uhlmann, New Extension of the Kalman Filter to Nonlinear Systems, in *Proceedings of SPIE—The International Society for Optical Engineering*, vol. 3068 (1997), pp. 182–193

[JVCR98] R. Johansson, M. Verhaegen, C.T. Chou, A. Robertsson, Behavioral Model Identification, in *Proceedings of the IEEE Conference on Decision and Control*, 1998, pp. 126–131

[JW68] G.M. Jenkins, D.G. Watts, *Spectral Analysis and Its Applications* (Holden-Day, Oakland, 1968)

[JY79] A. Jakeman, P.C. Young, Joint parameter/state estimation. Electron. Lett. **15**(19), 582–583 (1979)

[Kal60] R.E. Kalman, A new approach to linear filtering and prediction problems. Am. Soc. Mech. Eng. Trans. Ser. D, J. Basic Eng. **82**, 35–45 (1960)

[Kat05] T. Katayama, *Subspace Methods for System Identification*. Communications and Control Engineering (Springer, Berlin, 2005)

[Kau69] H. Kaufman, Aircraft parameter identification using Kalman filtering, in *Proceedings of the National Electronics Conference*, vol. XXV (1969)

[Kay88] S.M. Kay, *Modern Spectral Estimation: Theory and Application*. Prentice-Hall Signal Processing Series (Prentice-Hall, New York, 1988)

[KB61] R.E. Kalman, R.S. Bucy, New results in linear filtering and prediction problems. Am. Soc. Mech. Eng. Trans. Ser. D, J. Basic Eng. **83**, 95–108 (1961)

[KB94] A. Kumar, G.J. Balas, Approach to model validation in the μ framework, in *Proceedings of the American Control Conference*, vol. 3, 1994, pp. 3021–3026

[KD09] K.J. Keesman, T.G. Doeswijk, Direct least-squares estimation and prediction of rational systems: application to food storage. J. Process Control **19**, 340–348 (2009)

[Kee89] K.J. Keesman, On the dominance of parameters in structural models of ill-defined systems. Appl. Math. Comput. **30**, 133–147 (1989)

[Kee90] K.J. Keesman, Membership-set estimation using random scanning and principal component analysis. Math. Comput. Simul. **32**(5–6), 535–544 (1990)

[Kee97] K.J. Keesman, Weighted least-squares set estimation from l_∞ norm bounded-noise data. IEEE Trans. Autom. Control **AC 42**(10), 1456–1459 (1997)

[Kee02] K.J. Keesman, State and parameter estimation in biotechnical batch reactors. Control Eng. Pract. **10**(2), 219–225 (2002)

[Kee03] K.J. Keesman, Bound-based identification: nonlinear-model case, in *Encyclopedia of Life Science Systems Article 6.43.11.2*, ed. by H. Unbehauen. UNESCO EOLSS (2003)

[KH95] Y. Kyongsu, K. Hedrick, Observer-based identification of nonlinear system parameters. J. Dyn. Syst. Meas. Control, Trans. ASME **117**(2), 175–182 (1995)

[KJ97] K.J. Keesman, A.J. Jakeman, Identification for long-term prediction of rainfall-streamflow systems, in *Proceedings of the 11th IFAC Symp. on System Identification*, Fukuoka, Japan, 8–11 July, vol. 3 (1997), pp. 2519–1523

[KK09] K.J. Keesman, N. Khairudin, Linear regressive realizations of LTI state space models, in *Proceedings of the 15th IFAC Symposium on System Identification*, St. Malo, France (2009), pp. 1868–1873

[KKZ77] V.I. Kostyuk, V.E. Kraskevitch, K.K. Zelensky, Frequency domain identification of complex systems. Syst. Sci. **3**(1), 5–12 (1977)

[KM08a] K.J. Keesman, V.I. Maksimov, On reconstruction of unknown characteristics in one system of third order, in *Prikl. Mat. i Informatika: Trudy fakulteta VMiK*

MGU (Applied Mathematics and Informatics: Proc., Computer Science Dept. of Moscow State University), vol. 30 (MAKS Press, Moscow, 2008), pp. 95–116 (in Russian)

[KM08b] K.J. Keesman, V.I. Maksimov, On feedback identification of unknown characteristics: a bioreactor case study. Int. J. Control **81**(1), 134–145 (2008)

[KMVH03] A. Kukush, I. Markovsky, S. Van Huffel, Consistent estimation in the bilinear multivariate errors-in-variables model. Metrika **57**(3), 253–285 (2003)

[Kol93] I. Kollar, On frequency-domain identification of linear systems. IEEE Trans. Instrum. Meas. **42**(1), 2–6 (1993)

[Koo37] T.J. Koopmans, *Linear regression analysis of economic time series*. The Netherlands (1937)

[KPL03] K.J. Keesman, D. Peters, L.J.S. Lukasse, Optimal climate control of a storage facility using local weather forecasts. Control Eng. Pract. **11**(5), 505–516 (2003)

[KR76] R.L. Kashyap, A.R. Rao, *Dynamic Stochastic Models from Empirical Data* (Academic Press, San Diego, 1976)

[KS72] H. Kwakernaak, R. Sivan, *Linear Optimal Control Systems* (Wiley-Interscience, New York, 1972)

[KS73] R.E. Kalaba, K. Spingarn, Optimal inputs and sensitivities for parameter estimation. J. Optim. Theory Appl. **11**(1), 56–67 (1973)

[KS02] K.J. Keesman, J.D. Stigter, Optimal parametric sensitivity control for the estimation of kinetic parameters in bioreactors. Math. Biosci. **179**, 95–111 (2002)

[KS03] K.J. Keesman, J.D. Stigter, Optimal input design for low-dimensional systems: an haldane kinetics example, in *Proceedings of the European Control Conference*, Cambridge, UK (2003), p. 268

[KS04] K.J. Keesman, R. Stappers, Nonlinear set-membership estimation: a support vector machine approach. J. Inverse Ill-Posed Probl. **12**(1), 27–41 (2004)

[Kur77] A.B. Kurzhanski, *Control and Observation Under Uncertainty* (Nauka, Moscow, 1977) (in Russian)

[KvS89] K.J. Keesman, G. van Straten, Identification and prediction propagation of uncertainty in models with bounded noise. Int. J. Control **49**(6), 2259–2269 (1989)

[LB93] D. Ljungquist, J.G. Balchen, Recursive Prediction Error Methods for Online Estimation in Nonlinear State-space Models, in *Proceedings of the IEEE Conference on Decision and Control*, vol. 1 (1993), pp. 714–719

[LB07] Z. Lin, M.B. Beck, On the identification of model structure in hydrological and environmental systems. Water Resources Research **43**(2) (2007)

[LCB+07] L. Lang, W.-S. Chen, B.R. Bakshi, P.K. Goel, S. Ungarala, Bayesian estimation via sequential Monte Carlo sampling-constrained dynamic systems. Automatica **43**(9), 1615–1622 (2007)

[Lee64] R.C.K. Lee, *Optimal Identification, Estimation and Control* (MIT Press, Cambridge, 1964)

[Lev64] M.J. Levin, Estimation of a system pulse transfer function in the presence of noise. IEEE Trans. Autom. Control **9**, 229–335 (1964)

[LG94] L. Ljung, T. Glad, *Modeling of Dynamic Systems* (Prentice Hall, New York, 1994)

[LG97] L. Ljung, L. Guo, The role of model validation for assessing the size of the unmodeled dynamics. IEEE Trans. Autom. Control **42**(9), 1230–1239 (1997)

[LG09] T. Liu, F. Gao, A generalized relay identification method for time delay and non-minimum phase processes. Automatica **45**(4), 1072–1079 (2009)

[Liu94] J.S. Liu, *Monte Carlo Strategies in Scientific Computing* (Springer, New York, 1994)

[Lju81] L. Ljung, Analysis of a general recursive prediction error identification algorithm. Automatica **17**(1), 89–99 (1981)

[Lju87] L. Ljung, *System Identification—Theory for the User* (Prentice Hall, New York, 1987)

[Lju99a] L. Ljung, Comments on model validation as set membership identification, in *Robustness in Identification and Control*. Lecture Notes in Control and Information Sciences, vol. 245 (Springer, Berlin, 1999)

[Lju99b] L. Ljung, *System Identification—Theory for the User*, 2nd edn. (Prentice Hall, New York, 1999)

[LKvS96] L.J.S. Lukasse, K.J. Keesman, G. van Straten, Grey-box identification of dissolved oxygen dynamics in an activated sludge process, in *Proceedings of the 13th IFAC World Congress*, San Francisco, USA, vol. N (1996), pp. 485–490

[LKvS99] L.J.S. Lukasse, K.J. Keesman, G. van Straten, A recursively identified model for short-term predictions of NH_4/NO_3-concentrations in alternating activated sludge processes (1999)

[LL96] W. Li, J.H. Lee, Frequency-domain closed-loop identification of multivariable systems for feedback control. AIChE J. **42**(10), 2813–2827 (1996)

[LP96] L.H. Lee, K. Poolla, Identification of linear parameter-varying systems via LFTs, in *Proceedings of the IEEE Conference on Decision and Control*, vol. 2 (1996), pp. 1545–1550

[LP99] L.H. Lee, K. Poolla, Identification of linear parameter-varying systems using nonlinear programming. J. Dyn. Syst. Meas. Control, Trans. ASME **121**(1), 71–78 (1999)

[LS83] L. Ljung, T. Söderström, *Theory and Practice of Recursive Identification* (MIT Press, Cambridge, 1983)

[Luo07] B. Luo, A dynamic method of experiment design of computer aided sensory evaluation. Adv. Soft Comput. **41**, 504–510 (2007)

[Maj73] J.C. Majithia, Recursive estimation of the mean value of a random variable using quantized data. IEEE Trans. Instrum. Meas. **22**(2), 176–177 (1973)

[Mar87] S.L. Marple, *Digital Spectral Analysis with Applications* (Prentice-Hall, New York, 1987)

[May63] D.Q. Mayne, Optimal non-stationary estimation of the parameters of a linear system with Gaussian inputs. J. Electron. Control **14**, 101–112 (1963)

[May79] P.S. Maybeck, *Stochastic Models, Estimation and Control*, vol. 1 (Academic Press, San Diego, 1979)

[MB82] M. Milanese, G. Belforte, Estimation theory and uncertainty intervals evaluation in presence of unknown but bounded errors. IEEE Trans. Autom. Control **AC 27**(2), 408–414 (1982)

[MB86] J.B. Moore, R.K. Boel, Asymptotically optimum recursive prediction error methods in adaptive estimation and control. Automatica **22**(2), 237–240 (1986)

[MB00] K.Z. Mao, S.A. Billings, Multi-directional model validity tests for non-linear system identification. Int. J. Control **73**(2), 132–143 (2000)

[MCS08] J. Mertl, M. Cech, M. Schlegel, One point relay identification experiment based on constant-phase filter, in *8th International Scientific Technical Conference PROCESS CONTROL 2008*, Kouty nad Desnou, Czech Republic, vol. C037 (2008), pp. 1–9

[MF95] J.F. MacGregor, D.T. Fogal, Closed-loop identification: the role of the noise model and prefilters. J. Process Control **5**(3), 163–171 (1995)

[MG86] R.H. Middleton, G.C. Goodwin, Improved finite word length characteristics in digital control using delta operators. IEEE Trans. Autom. Control **AC–31**(11), 1015–1021 (1986)

[MG90] R.H. Middleton, G.C. Goodwin, *Digital Control and Estimation: A Unified Approach.* (Prentice Hall, New York, 1990)

[Mil95] M. Milanese, Properties of least-squares estimates in set membership identification. Automatica **31**, 327–332 (1995)

[MN95] J.C. Morris, M.P. Newlin, Model validation in the frequency domain, in *Proceedings of the IEEE Conference on Decision and Control*, vol. 4 (1995), pp. 3582–3587

[MNPLE96] M. Milanese, J.P. Norton, H. Piet-Lahanier, E. Walter (eds.), *Bounding Approaches to System Identification* (Plenum, New York, 1996)

[MPV06] D.C. Montgomery, E.A. Peck, G.G. Vining, *Introduction to Linear Regression Analysis*, 4th edn. Wiley Series in Probability and Statistics (Wiley, New York, 2006)

[MR97] C. Maffezzoni, P. Rocco, Robust tuning of PID regulators based on step-response identification. Eur. J. Control **3**(2), 125–136 (1997)

[MRCW01] G. Margaria, E. Riccomagno, M.J. Chappell, H.P. Wynn, Differential algebra methods for the study of the structural identifiability of rational function state-space models in the biosciences. Math. Biosci. **174**(1), 1–26 (2001)

[MV91a] M. Milanese, A. Vicino, Optimal estimation theory for dynamic systems with set membership uncertainty: an overview. Automatica **27**(6), 997–1009 (1991)

[MV91b] M. Moonen, J. Vandewalle, A square root covariance algorithm for constrained recursive least squares estimation. J. VLSI Signal Process. **3**(3), 163–172 (1991)

[MVL78] C. Moler, C. Van Loan, Nineteen dubious ways to compute the exponential of a matrix. SIAM Rev. **20**(4), 801–836 (1978)

[MW79] J.B. Moore, H. Weiss, Recursive prediction error methods for adaptive estimation. IEEE Trans. Syst. Man Cybern. **9**(4), 197–205 (1979)

[MWDM02] I. Markovsky, J.C. Willems, B. De Moor, Continuous-time errors-in-variables filtering, in *Proceedings of the IEEE Conference on Decision and Control*, vol. 3 (2002), pp. 2576–2581

[NGS77] T.S. Ng, G.C. Goodwin, T. Söderström, Optimal experiment design for linear systems with input-output constraints. Automatica **13**(6), 571–577 (1977)

[Nin09] B. Ninness, Some system identification challenges and approaches, in *15th IFAC Symposium on System Identification*, St. Malo, France (2009)

[Nor75] J.P. Norton, Optimal smoothing in the identification of linear time-varying systems. Proc. Inst. Electr. Eng. **122**(6), 663–669 (1975)

[Nor76] J.P. Norton, Identification by optimal smoothing using integrated random walks. Proc. Inst. Electr. Eng. **123**(5), 451–452 (1976)

[Nor86] J.P. Norton, *An Introduction to Identification* (Academic Press, San Diego, 1986)

[Nor87] J.P. Norton, Identification and application of bounded-parameter models. Automatica **23**(4), 497–507 (1987)

[Nor03] J.P. Norton, Bound-based Identification: linear-model case, in *Encyclopedia of Life Science Systems Article 6.43.11.2*, ed. by H. Unbehauen. UNESCO EOLSS (2003)

[NW82] V.V. Nguyen, E.F. Wood, Review and unification of linear identifiability concepts. SIAM Rev. **24**(1), 34–51 (1982)

[OFOFDA96] R. Oliveira, E.C. Ferreira, F. Oliveira, S. Feyo De Azevedo, A study on the convergence of observer-based kinetics estimators in stirred tank bioreactors. J. Process Control **6**(6), 367–371 (1996)

[OWG04] S. Ognier, C. Wisniewski, A. Grasmick, Membrane bioreactor fouling in sub-critical filtration conditions: a local critical flux concept. J. Membr. Sci. **229**, 171–177 (2004)

[Paw91] M. Pawlak, On the series expansion approach to the identification of Hammerstein systems. IEEE Trans. Autom. Control **36**(6), 763–767 (1991)

[PC07] G. Pillonetto, C. Cobelli, Identifiability of the stochastic semi-blind deconvolution problem for a class of time-invariant linear systems. Automatica **43**(4), 647–654 (2007)

[PDAFD00] M. Perrier, S.F. De Azevedo, E.C. Ferreira, D. Dochain, Tuning of observer-based estimators: Theory and application to the on-line estimation of kinetic parameters. Control Eng. Pract. **8**(4), 377–388 (2000)

[Pet75] V. Peterka, Square root filter for real time multivariate regression. Kybernetika **11**(1), 53–67 (1975)

[PH05] R.L.M. Peeters, B. Hanzon, Identifiability of homogeneous systems using the state isomorphism approach. Automatica **41**(3), 513–529 (2005)

[Phi73] P.C.B. Phillips, The problem of identification in finite parameter continuous time models. J. Econom. **1**(4), 351–362 (1973)

[PS97] R. Pintelon, J. Schoukens, Frequency-domain identification of linear timeinvariant systems under nonstandard conditions. IEEE Trans. Instrum. Meas. **46**(1), 65–71 (1997)

[PS01] R. Pintelon, J. Schoukens, *System Identification: A Frequency Domain Approach* (Wiley–IEEE Press, New York, 2001)

[PW88] L. Pronzato, E. Walter, Robust experiment design via maximin optimization. Math. Biosci. **89**(2), 161–176 (1988)

[PW98] J.W. Polderman, J.C. Willems, *Introduction to Mathematical Systems Theory: A Behavioral Approach* (Springer, Berlin, 1998)

[QN82] Z.H. Qureshi, T.S. Ng, Optimal input design for dynamic system parameter estimation: the d//s-optimality case. SIAM J. Control Optim. **20**(5), 713–721 (1982)

[Rak80] H. Rake, Step response and frequency-response methods. Automatica **16**(5), 519–526 (1980)

[RH78] J. Rowland, W. Holmes, Nonstationary signal processing and model validation. IEEE Int. Conf. Acoust. Speech Signal Proc. **3**, 520–523 (1978)

[RSP97] Y. Rolain, J. Schoukens, R. Pintelon, Order estimation for linear time-invariant systems using frequency domain identification methods. IEEE Trans. Autom. Control **42**(10), 1408–1417 (1997)

[RU99] K. Reif, R. Unbehauen, The extended Kalman filter as an exponential observer for nonlinear systems. IEEE Trans. Signal Process. **47**(8), 2324–2328 (1999)

[RWGF07] C.R. Rojas, J.S. Welsh, G.C. Goodwin, A. Feuer, Robust optimal experiment design for system identification. Automatica **43**(6), 993–1008 (2007)

[SAD03] M.P. Saccomani, S. Audoly, L. D'Angiò, Parameter identifiability of nonlinear systems: The role of initial conditions. Automatica **39**(4), 619–632 (2003)

[Sak61] M. Sakaguchi, Dynamic programming of some sequential sampling design. J. Math. Anal. Appl. **2**(3), 446–466 (1961)

[Sak65] D.J. Sakrison, Efficient recursive estimation; application to estimating the parameters of a covariance function. Int. J. Eng. Sci. **3**(4), 461–483 (1965)

[SAML80] G. Salut, J. Aguilar-Martin, S. Lefebvre, New results on optimal joint parameter and state estimation of linear stochastic systems. J. Dyn. Syst. Meas. Control, Trans. ASME **102**(1), 28–34 (1980)

[Sar84] R.G. Sargent, Tutorial on verification and validation of simulation models, in *Winter Simulation Conference Proceedings* (1984), pp. 115–121

[SB94] J.D. Stigter, M.B. Beck, A new approach to the identification of model structure. Environmetrics **5**(3), 315–333 (1994)

[SB04] J.D. Stigter, M.B. Beck, On the development and application of a continuous-discrete recursive prediction error algorithm. Math. Biosci. **191**(2), 143–158 (2004)

[SBD99] R. Smith, A. Banaszuk, G. Dullerud, Model validation approaches for nonlinear feedback systems using frequency response measurements, in *Proceedings of the IEEE Conference on Decision and Control*, vol. 2 (1999), pp. 1500–1504

[SC97] G. Sparacino, C. Cobelli, Impulse response model in reconstruction of insulin secretion by deconvolution: role of input design in the identification experiment. Ann. Biomed. Eng. **25**(2), 398–416 (1997)

[Sch73] F.C. Schweppe, *Uncertain Dynamic Systems* (Prentice-Hall, New York, 1973)

[SD98] W. Scherrer, M. Deistler, A structure theory for linear dynamic errors-in-variables models. SIAM J. Control Optim. **36**(6), 2148–2175 (1998)

[SGM88] M.E. Salgado, G.C. Goodwin, R.H. Middleton, Modified least squares algorithm incorporating exponential resetting and forgetting. Int. J. Control **47**(2), 477–491 (1988)

[SGR+00] A. Stenman, F. Gustafsson, D.E. Rivera, L. Ljung, T. McKelvey, On adaptive smoothing of empirical transfer function estimates. Control Eng. Pract. **9**, 1309–1315 (2000)

[She95] S. Sheikholeslam, Observer-based parameter identifiers for nonlinear systems with parameter dependencies. IEEE Trans. Autom. Control **40**(2), 382–387 (1995)

[SK01] J.D. Stigter, K.J. Keesman, Optimal parametric sensitivity control for a fed batch reactor, in *Proceedings of the European Control Conference 2001*, Porto, Portugal, 4–7 Sep. 2001, pp. 2841–2844

[SK04] J.D. Stigter, K.J. Keesman, Optimal parametric sensitivity control of a fed-batch reactor. Automatica **40**(8), 1459–1464 (2004)

[SL03] S.W. Sung, J.H. Lee, Pseudo-random binary sequence design for finite impulse response identification. Control Eng. Pract. **11**(8), 935–947 (2003)

[SM97] P. Stoica, R.L. Moses, *Introduction to Spectral Analysis* (Prentice-Hall, New York, 1997)

[SMMH94] P. Sadegh, H. Melgaard, H. Madsen, J. Holst, Optimal experiment design for identification of grey-box models, in *Proceedings of the American Control Conference*, vol. 1 (1994), pp. 132–137

[Söd07] T. Söderström, Errors-in-variables methods in system identification. Automatica **43**(6), 939–958 (2007)

[Söd08] T. Söderström, Extending the Frisch scheme for errors-in-variables identification to correlated output noise. Int. J. Adapt. Control Signal Process. **22**(1), 55–73 (2008)

[Sor80] H.W. Sorenson, *Parameter Estimation* (Dekker, New York, 1980)

[Sor85] H.W. Sorenson, *Kalman Filtering: Theory and Application* (IEEE Press, New York, 1985)

[SOS00] L. Sun, H. Ohmori, A. Sano, Frequency domain approach to closed-loop identification based on output inter-sampling scheme, in *Proceedings of the American Control Conference*, vol. 3 (2000), pp. 1802–1806

[SR77] M.W.A. Smith, A.P. Roberts, A study in continuous time of the identification of initial conditions and/or parameters of deterministic system by means of a Kalman-type filter. Math. Comput. Simul. **19**(3), 217–226 (1977)

[SR79] M.W.A. Smith, A.P. Roberts, The relationship between a continuous-time identification algorithm based on the deterministic filter and least-squares methods. Inf. Sci. **19**(2), 135–154 (1979)

[SS83] T. Söderström, P.G. Stoica, *Instrumental Variable Methods for System Identification* (Springer, Berlin, 1983)

[SS87] T. Söderström, P.G. Stoica, *System Identification* (Prentice Hall, New York, 1987)

[SSM02] T. Söderström, U. Soverini, K. Mahata, Perspectives on errors-in-variables estimation for dynamic systems. Signal Process. **82**(8), 1139–1154 (2002)

[SVK06] J.D. Stigter, D. Vries, K.J. Keesman, On adaptive optimal input design: a bioreactor case study. AIChE J. **52**(9), 3290–3296 (2006)

[SVPG99] J. Schoukens, G. Vandersteen, R. Pintelon, P. Guillaume, Frequency-domain identification of linear systems using arbitrary excitations and a nonparametric noise model. IEEE Trans. Autom. Control **44**(2), 343–347 (1999)

[THvdH09] R. Tóth, P.S.C. Heuberger, P.M.J. van den Hof, Asymptotically optimal orthonormal basis functions for LPV system identification. Automatica **45**(6), 1359–1370 (2009)

[TK75] H. Thoem, V. Krebs, Closed loop identification—correlation analysis or parameter estimation [Identifizierung im geschlossenen Regelkreis – Korrelationsanalyse oder Parameterschaetzung?]. Regelungstechnik **23**(1), 17–19 (1975)

[TLH+06] K.K. Tan, T.H. Lee, S. Huang, K.Y. Chua, R. Ferdous, Improved critical point estimation using a preload relay. J. Process Control **16**(5), 445–455 (2006)

[TM03] D. Treebushny, H. Madsen, A new reduced rank square root Kalman filter for data assimilation in mathematical models. Lect. Notes Comput. Sci. **2657**, 482–491 (2003) (including subseries Lecture Notes in Artificial Intelligence and Lecture Notes in Bioinformatics)

[TOS98] M. Takahashi, H. Ohmori, A. Sano, Impulse response identification by use of wavelet packets decomposition, in *Proceedings of the IEEE Conference on Decision and Control*, vol. 1 (1998), pp. 211–214

[Tur85] J.M. Turner, *Recursive Least-squares Estimation and Lattice Filters* (Prentice-Hall, New York, 1985)

[TV72] R. Tomovic, M. Vukobratovic, *General Sensitivity Theory* (American Elsevier, New York, 1972)

[TY90] A.P. Tzes, S. Yurkovich, Frequency domain identification scheme for flexible structure control. J. Dyn. Syst. Meas. Control, Trans. ASME **112**(3), 427–434 (1990)

[vBGKS98] J. van Bergeijk, D. Goense, K.J. Keesman, B. Speelman, Digital filters to integrate global positioning system and dead reckoning. J. Agric. Eng. Res. **70**, 135–143 (1998)

[VD92] M. Verhaegen, P. Dewilde, Subspace model identification. Part 1: The output-error state-space model identification class of algorithms. Int. J. Control **56**, 1187–1210 (1992)

[vdH98] J.M. van den Hof, Structural identifiability of linear compartmental systems. IEEE Trans. Autom. Control **43**(6), 800–818 (1998)

[vdHPB95] P.M.J. van den Hof, P.S.C. Heuberger, J. Bokor, System identification with generalized orthonormal basis functions. Automatica **31**(12), 1821–1834 (1995)

[Ver89] M.H. Verhaegen, Round-off error propagation in four generally-applicable, recursive, least-squares estimation schemes. Automatica **25**(3), 437–444 (1989)

[VGR89] S. Vajda, K.R. Godfrey, H. Rabitz, Similarity transformation approach to identifiability analysis of nonlinear compartmental models. Math. Biosci. **93**(2), 217–248 (1989)

[VH97] M. Verlaan, A.W. Heemink, Tidal flow forecasting using reduced rank square root filters. Stoch. Environ. Res. Risk Assess. **11**(5), 349–368 (1997)

[VH05] I. Vajk, J. Hetthéssy, Subspace identification methods: review and re-interpretation, in *Proceedings of the 5th International Conference on Control and Automation, ICCA'05* (2005), pp. 113–118

[VHMVS07] S. Van Huffel, I. Markovsky, R.J. Vaccaro, T. Söderström, Total least squares and errors-in-variables modeling. Signal Process. **87**(10), 2281–2282 (2007)

[VHV89] S. Van Huffel, J. Vandewalle, Analysis and properties of the generalized total least squares problem $Ax \approx b$ when some or all columns in A are subject to error. SIAM J. Matrix Anal. Appl. **10**, 294–315 (1989)

[Vib95] M. Viberg, Subspace-based methods for the identification of linear time-invariant systems. Automatica **31**(12), 1835–1851 (1995)

[VKZ06] D. Vries, K.J. Keesman, H. Zwart, Explicit linear regressive model structures for estimation, prediction and experimental design in compartmental diffusive systems, in *Proceedings of the 14th IFAC Symposium on System Identification*, Newcastle, Australia (2006), pp. 404–409

[vOdM95] P. van Overschee, B. de Moor, Choice of state-space basis in combined deterministic-stochastic subspace identification. Automatica **31**(12), 1877–1883 (1995)

[Vri08] D. Vries, Estimation and prediction of convection-diffusion-reaction systems from point measurements. Ph.D. thesis, Systems & Control, Wageningen University (2008)

[vRLS73] D.L. van Rooy, M.S. Lynn, C.H. Snyder, The use of the modified Choleski decomposition in divergence and classification calculations, in *LARS Symposia, Paper 22* (1973)

[vS94] J.H. van Schuppen, Stochastic realization of a Gaussian stochastic control system. J. Acta Appl. Math. **35**(1–2), 193–212 (1994)

[VS04] J.H. Van Schuppen, System theory for system identification. J. Econom. **118**(1–2), 313–339 (2004) [vSK91] G. van Straten, K.J. Keesman, Uncertainty propagation and speculation in projective forecasts of environmental change—a lake eutrophication example. J. Forecast. **10**(2–10), 163–190 (1991)

[VV02] V. Verdult, M. Verhaegen, Subspace identification of multivariable linear parameter-varying systems. Automatica **38**(5), 805–814 (2002)

[VV07] M. Verhaegen, V. Verdult, *Filtering and System Identification: A Least Squares Approach* (Cambridge University Press, Cambridge, 2007)

[Wal82] E. Walter, *Identifiability of State Space Models*. Lecture Notes in Biomathematics, vol. 46. (Springer, Berlin, 1982)

[Wal03] E. Walter, Bound-based Identification, in *Encyclopedia of Life Science Systems Article 6.43.11.2*, ed. by H. Unbehauen. UNESCO EOLSS (2003)

[WC97] L. Wang, W.R. Cluett, Frequency-sampling filters: an improved model structure for step-response identification. Automatica **33**(5), 939–944 (1997)

[Wel77] P.E. Wellstead, Reference signals for closed-loop identification. Int. J. Control **26**(6), 945–962 (1977)

[Wel81] P.E. Wellstead, Non-parametric methods of system identification. Automatica **17**, 55–69 (1981)

[WG04] E. Wernholt, S. Gunnarsson, On the use of a multivariable frequency response estimation method for closed loop identification, in *Proceedings of the IEEE Conference on Decision and Control*, vol. 1 (2004), pp. 827–832

[Whi70] R.C. White, Fast digital computer method for recursive estimation of the mean. IEEE Trans. Comput. **19**(9), 847–848 (1970)

[Wig93] T. Wigren, Recursive prediction error identification using the nonlinear Wiener model. Automatica **29**(4), 1011–1025 (1993)

[Wil86a] J.C. Willems, From time series to linear system. Part I. Finite dimensional linear time invariant systems. Automatica **22**, 561–580 (1986)

[Wil86b] J.C. Willems, From time series to linear system. Part II. Exact modelling. Automatica **22**, 675–694 (1986)

[Wil87] J.C. Willems, From time series to linear system. Part III. Approximate modelling. Automatica **23**, 87–115 (1987)

[WP87] E. Walter, L. Pronzato, Optimal experiment design for nonlinear models subject to large prior uncertainties. Am. J. Physiol., Regul. Integr. Comp. Physiol. **253**(3), 22–23 (1987)

[WP90] E. Walter, L. Pronzato, Qualitative and quantitative experiment design for phenomenological models—a survey. Automatica **26**(2), 195–213 (1990)

[WP97] E. Walter, L. Pronzato, *Identification of Parametric Models from Experimental Data*. Communications and Control Engineering (Springer, Berlin, 1997)

[WZG01] Q.-G. Wang, Y. Zhang, X. Guo, Robust closed-loop identification with application to auto-tuning. J. Process Control **11**(5), 519–530 (2001)

[WZL09] J. Wang, Q. Zhang, L. Ljung, Revisiting Hammerstein system identification through the two-stage algorithm for bilinear parameter estimation. Automatica **45**(11), 2627–2633 (2009)

[YB94] P.C. Young, K.J. Beven, Data-based mechanistic modelling and the rainfall-flow non-linearity. Environmetrics **5**(3), 335–363 (1994)

[YG06] P.C. Young, H. Garnier, Identification and estimation of continuous-time, data-based mechanistic (dbm) models for environmental systems. Environ. Model. Softw. **21**(8), 1055–1072 (2006)

[You84] P.C. Young, *Recursive Estimation and Time-series Analysis: An Introduction*. Communications and Control Engineering (Springer, Berlin, 1984)

[You98] P. Young, Data-based mechanistic modelling of environmental, ecological, economic and engineering systems. Environ. Model. Softw. **13**(2), 105–122 (1998)

[YST97] Z.-J. Yang, S. Sagara, T. Tsuji, System impulse response identification using a multiresolution neural network. Automatica **33**(7), 1345–1350 (1997)

[Zar79] M.B. Zarrop, *Optimal Experiment Design for Dynamic System Identification* (Springer, Berlin, 1979)

[Zar81] M.B. Zarrop, Sequential generation of d-optimal input designs for linear dynamic systems. J. Optim. Theory Appl. **35**(2), 277–291 (1981)

[Zhu05] Q.M. Zhu, An implicit least squares algorithm for nonlinear rational model parameter estimation. Appl. Math. Model. **29**(7), 673–689 (2005)

[ZT00] Y. Zhou, J.K. Tugnait, Closed-loop linear model validation and order estimation using polyspectral analysis. IEEE Trans. Signal Process. **48**(7), 1965–1974 (2000)

[Zwa04] H.J. Zwart, Transfer functions for infinite-dimensional systems. Syst. Control Lett. **52**(3–4), 247–255 (2004)

[ZWR91] A. Zakhor, R. Weisskoff, R. Rzedzian, Optimal sampling and reconstruction of MRI signals resulting from sinusoidal gradients. IEEE Trans. Signal Process. **39**(9), 2056–2065 (1991)

Index

K.J. Keesman, *System Identification*,
Advanced Textbooks in Control and Signal Processing,
DOI 10.1007/978-0-85729-522-4,

Printed by Books on Demand, Germany